Lecture Notes in Artifici 43

Subseries of Lecture Notes in Computer Science
Edited by J. Siekmann

Lecture Notes in Computer Science

Edited by G. Goos and J. Hartmanis

S. Doshita K. Furukawa K. P. Jantke
T. Nishida (Eds.)

Algorithmic Learning Theory

Third Workshop, ALT '92
Tokyo, Japan, October 20-22, 1992
Proceedings

Springer-Verlag
Berlin Heidelberg New York
London Paris Tokyo
Hong Kong Barcelona
Budapest

Series Editor

Jörg Siekmann
University of Saarland
German Research Center for Artificial Intelligence (DFKI)
Stuhlsatzenhausweg 3
D-66123 Saarbrücken, Germany

Volume Editors

Shuji Doshita
Department of Information Science, Kyoto University
Sakyo-ku, Kyoto 606-01, Japan

Koichi Furukawa
Faculty of Environmental Information, Keio University
5322 Endo, Fujisawa-shi, Kanagawa, 252 Japan

Klaus P. Jantke
Fachbereich Informatik, Mathematik und Naturwissenschaften
Hochschule für Technik, Wirtschaft und Kultur Leipzig
Postfach 66, D-04251 Leipzig, Germany

Toyaki Nishida
Graduate School of Information Science
Advanced Institute of Science and Technology, Nara
8916-5, Takayaka-cho, Ikoma-shi, Nara 630-01, Japan

CR Subject Classification (1991): I.2.6, I.2.3, F.1.1

ISBN 3-540-57369-0 Springer-Verlag Berlin Heidelberg New York
ISBN 0-387-57369-0 Springer-Verlag New York Berlin Heidelberg

Printed in Germany

Typesetting: Camera ready by author
Printing and binding: Druckhaus Beltz, Hemsbach/Bergstr.
45/3140-543210 - Printed on acid-free paper

PREFACE

This volume contains the papers that were presented at the Third Workshop on Algorithmic Learning Theory (ALT'92), which was held at the CSK Information Education Center in Tokyo from October 20 to 22, 1992. In addition to 3 invited papers, this volume contains 19 papers accepted for presentation at the workshop.

The contributions in the proceedings were selected from 29 extended abstracts submitted in response to the call for papers, at the final selection meeting of the program committee held in Tokyo on June 19, 1992. The volume contains three invited papers: "Discovery Learning in Intelligent Tutoring Systems" (by S. Otsuki), "From Inductive Inference to Algorithmic Learning Theory" (by R. Wiehagen), and "A Stochastic Approach to Genetic Information Processing" (by A. Konagaya).

By now the importance of machine learning to the success of the next generation of AI systems has been widely recognized and accepted. At the same time, decades of theoretical research in inductive inference and its complexity-theoretic analogue have led to the emergence of algorithmic respectively computational learning theory. ALT is the Japanese series of international workshops focusing on these learning-theoretical issues. The ALT workshops have been held annually since 1990, and are organized and sponsored by the Japanese Society for Artificial Intelligence (JSAI). The main objective of these workshops is to provide an open forum for discussions and exchanges of ideas between researchers from various backgrounds in this emerging, interdisciplinary field of learning theory.

There are two concurrent series, AII initiated in 1986 in Europe, and COLT started in 1988 in the USA. It is our intention to integrate the international community of scientists interested in algorithmic respectively computational learning theory. A first step towards such an integration is to exchange all available information. We are grateful to Springer-Verlag for providing an opportunity to present the proceedings of ALT'92 to a wider international community. A further step may be to integrate even learning theory conferences for a higher concentration of scientific discussions and more efficient transfer of ideas between disciplines.

The editors are deeply grateful to all the program committee members and referees who took part in the evaluation and the selection of submitted papers. In particular, we wish to thank M. Numao, T. Shinohara, and Y. Takada for their excellent work. The program committee thanks all three invited lecturers for having accepted the invitation.

We thank all those who made this workshop possible, especially K. Miura, T. Nishino, and A. Sakurai. Finally, we also wish to express our gratitude to CSK for the assistance and support with local arrangements.

Tokyo, September 1993

S. Doshita
K. Furukawa
K.P. Jantke
T. Nishida

CONFERENCE CHAIR

S. Doshita (Kyoto Univ.)

PROGRAM COMMITTEE

K. Furukawa (Keio Univ.; Chairman)

N. Abe (NEC)
H. Aso (ETL)
M. Hagiya (Univ. of Tokyo)
Hideki Imai (Yokohama Nat. Univ.)
Y. Inagaki (Nagoya Univ.)
H. Ishizaka (Fujitsu Labs.)
A. Maruoka (Tohoku Univ.)
H. Motoda (Hitachi)
M. Numao (Tokyo Inst. of Tech.)
T. Shinohara (Kyushu Inst. of Tech.)
Y. Takada (Fujitsu Labs.)
O. Watanabe (Tokyo Inst. of Tech.)
S. Arikawa (Kyushu Univ.)
J. Arima (ICOT)
M. Haraguchi (Tokyo Inst. of Tech.)
Hiroshi Imai (Univ. of Tokyo)
M. Ishikawa (Kyushu Inst. of Tech.)
S. Kuhara (Kyushu Univ.)
S. Miyano (Kyushu Univ.)
S. Nishio (Osaka Univ.)
M. Sato (Tohoku Univ.)
Y. Shirai (Osaka Univ.)
E. Tomita (Univ. of Elec.-Commun.)
T. Yokomori (Univ. of Elec.-Commun.)

LOCAL ARRANGEMENTS COMMITTEE

T. Nishida (Kyoto Univ.; Chairman)

H. Isozaki (NTT)
T. Kurokawa (IBM Japan)
K. Miura (Gunma Univ.)
T. Nishino (JAIST)
A. Sakurai (Hitachi)
H. Takayama (Ritsumeikan Univ.)
T. Kawamura (ICOT)
K. Matsumoto (Toshiba)
T. Miyashita (NEC)
H. Ono (Hiroshima Univ.)
R. Sugimura (Matsushita Elec.)
Y. Tanaka (Hokkaido Univ.)

List of Referees for ALT'92

N. Abe
S. Arikawa
J. Arima
H. Arimura
H. Asoh
M. Hagiya
M. Harao
Y. Hirai
Hiroshi Imai
M. Ishikawa
H. Ishizaka
H. Isozaki
T. Kawabata
H. Kawahara
K. Kobayashi
S. Kuhara
T. Kurita
A. Maruoka
H. Matsubara
M. Matsuoka
T. Miyahara
S. Miyano
H. Motoda
Y. Mukouchi
T. Nishino
M. Numao
Y. Sakakibara
S. Sakurai
A. Shinohara
Y. Shirai
Y. Takada
J. Takeuchi
E. Takimoto
S. Tangkitvanich
N. Tanida
E. Tomita
O. Watanabe
S. Yamada
A. Yamamoto
K. Yamanishi
T. Yokomori

Sponsor:
Japanese Society for Artificial Intelligence (JSAI)

Cooperative Institutions :
Information Processing Society of Japan (IPSJ)
Institute for New Generation Computer Technology (ICOT)
Institute of Electronics, Information and Communication Engineers of Japan (IEICE)
Japan Neural Network Society (JNNS)
Japan Society for Software Science and Technology (JSSST)
Japanese Cognitive Science Society (JCSS)
Society of Instrument and Control Engineers of Japan (SICE)

Table of Contents

Invited Papers

Technical Papers

Learning via Query

Neural Networks

Inductive Inference

Analogical Reasoning

Approximate Learning

INVITED PAPERS

Discovery Learning in Intelligent Tutoring Systems

Setsuko Otsuki
Department of Artificial Intelligence,
Kyushu Institute of Technology
680-4, Kawazu, Iizuka 820, Japan

Abstract

A brief history of Intelligent Tutoring Systems and their necessary educational functions which have already been realized and not yet been realized are presented separately, then problems to be solved within the framework of ITS and problems that transcend the framework of ITS are discussed. Lastly, it is indicated that the problems will be solved by an amalgamation of an open-end system like a micro world and a discovery system with direct manipulation into ITS and that the central problem to realize the amalgamation is a discovery learning by a machine itself.

1 Introduction

The first generation of CAI was made public in the 1950's and presently it is called traditional CAI. It is widely known that the traditional CAI has often been criticized as a mere electronic page-turning machine[1], because the traditional CAI has no ability to answer student's questions, and its only means of evaluating student's responses and selecting the next presentation is a pre-defined selection tree, whose forks are designed in great detail to correspond with all the pre-supposed students' good and wrong answers to an imbedded problems. It was supposed that all of these mooted points stemmed from inability of the traditional CAI to solve given problems.

The first development of the second generation of CAI, (alias Intelligent tutoring system, ITS in abbr.), was done by J. R. Carbonell in 1970[2] in order to remove the inability of problem solving by introducing an inferential function based on a knowledge representation, in place of the procedural representation of the traditional CAI. One of the most remarkable contributions of the work was the proposition of necessity to develop the following four elementary techniques.

1) Inferential technique for problem solving.

2) Student modeling by diagnosing error origins.

3) Representation method for teaching expertise.

4) Mixed initiative knowledge communication.

For more than ten ensuing years, his proposition had influenced ITS researches continuously. In fact, most papers published during these ten years were researches on either one of the four elementary techniques or on systems composed of the four elementary techniques[3]. These publications are roughly divided into two parts; basic researches in the 1970's[4] and its development in the first half of the 1980's[5].

1) The 1970's.

- Aims: To remove the defects of traditional CAI.
- Methods: Problem solving by computers, Student modeling, Teaching expertise, Mixed initiative dialogue.
- Elementary techniques: Knowledge representation method, Overlay student modeling, Buggy student modeling, Coaching, Natural language dialogue , Qualitative reasoning.

2) The first half of the 1980's

- Aims: Development of the elementary techniques
- Method: Identification of error origin, Cognitive modeling of human understanding
- Elementary techniques: Introducing structures into knowledge representation, Hypothetical inference, MIS, Adaptive guiding, Qualitative reasoning, Usage of student model, Truth maintenance.

In the middle period of the 1980's, as the whole aspect of ITS became gradually clear, demands for a breakthrough in the ITS methodology increased from the viewpoints of both tractability and educational cognitive science[6),7]. Issues on this problem will be treated in section 4. The problem has been strongly influenced by GUI techniques, and a new generation of CAI has been gradually revealed since the latter half in the 1980's[7].

3) The latter half of the 1980's

- Aims: Discovery environment by student initiative, Multiformity in CAI.
- Method: Micro world, Open-end, High individualization
- Elementary techniques: GUI, Multimedia, Hypermedia, Direct manipulation, Navigation, Natural language understanding.

In the latter half of the 1980's, although multifarious CAIs have emerged for practical use, AI techniques developed in ITS have not been applied. Hence it is impossible for these CAIs to give a student deep understanding. Recently, It seems to me that amalgamation of the third generation techniques into ITS becomes more and more important, but concrete techniques have not been established yet .

In section 2 what has been done and not yet been done in ITS are described and mooted points will be discussed in section 3. In section 4 the significance of so called discovery system is described from the stand point of deep understandings and in section 5 necessity of amalgamation of two methodology to realize the discovery system is described.

2 What has been done and not yet been don in ITS

2.1 What has been don in ITS

As described above, ITS has been equipped with various powerful architectures for mixed initiative and highly individualized knowledge communication. The following list shows what ITS has achieved.

1) ITS is able to solve problems by using suitably composed knowledge base to the domain.
2) A subset of sentences in natural language which corresponds to knowledge representation is able to be understood and generated.
3) As a result of 1) and 2), ITS is able to explain students a problem solving process, and able to answer students' questions.
4) ITS is able to identify the correct knowledge used in student's problem solving processes.
5) ITS is equipped with identification methods of student's error origins, though they are not complete.
6) As a result of 4) and 5) ITS is able to construct student models which correspond to each individual student's knowledge.
7) By using a student model, ITS is able to predict the student's problem solving process. Hence it is able to judge the suitability of the problem to be presented to the student.
8) Thus ITS is able to provide the most suitable learning material to each individual student or able to select the most adaptable teaching strategy to each student's comprehension level.

2.2 What has been left over in ITS.

To put the other way around, the above list suggests that ITS includes the following unsolved problems, which will be divided into two portions; problems within the framework of ITS and problems which transcend the framework of ITS.

(1) Problems within the framework of ITS

1) Completeness of the error origin identification

A varicty of the identification methods has been developed, but non of them satisfies generality, tractability and completeness at the same time.

2) Completeness of student models

An important prerequisite of a student model is the complete representation of problem solving knowledge which the student has presently acquired in the domain. But the incompleteness of error origin identification described in 1) prevents ITS from constructing a complete student model.

(2) Problems that transcend the framework of ITS

1) Techniques which have been developed in ITS are confined to apply to a limited form of tutoring, that is, they are confined to support student's learning effectively by error identification techniques in the problem solving. It is difficult for ITS to step out of the frame of the problem solving.

2) As a result of 1), It is difficult for a student to acquire a fundamental comprehension about a concept, an axiom, a theorem or a procedure that the student has first encountered.

3) In general, it is impossible to grasp the student's state entirely within the problem solving frame. Hence, even if a student answers correctly it is impossible for ITS to distinguish whether the answer is a result of student's fundamental comprehension or a result of mere memorization of only solving procedure.

4) As a result, ITS has not been equipped with an effective architecture of supporting a student to acquire a new knowledge by discovery, which the cognitive and educational science esteem the most important form of human learning because the discovery learning provides students a base to find how to behave when they encounter inexperienced situations.

The above issues have played a big role of bringing a variety of the 3rd generation CAI. Although at present their state of architecture still remains within the applications of GUI or hypermedia , it is well recognized that both the architecture and cognitive meanings of the 3rd generation CAI have a complementary importance to ITS. The problem will be discussed in section 4 and 5.

3 Problems within the framework of ITS: error diagnosis and error origin diagnosis

3.1 Importance of error origin diagnosis

Supposing that a system derives all the correct answers of a problem, a student's answer to the problem must fall one of the following two cases; the answer coincides with one of the derived answers or not. If the student's answer doesn't coincide, we can decide that it is an erroneous answer. However, if the knowledge base used for the problem solving is not complete, that is, if there is no guarantee that the system derives all the correct answers of the problem, the disagreement of the answers between the system and the student does not mean that the student answered incorrectly.

If the system has special knowledge that derives incorrect answers and is distinguished from ordinary domain knowledge, it is able to derive the correct answers by using ordinary domain knowledge only and incorrect answers by using both special and ordinary domain knowledge. Hence, the system is able to identify the student's incorrect knowledge (we call the incorrect rule a "Buggy rule") used in the problem solving process, together with the correctly used knowledge so long as the student's answer coincides with the derived answer regardless of their correctness. In general, buggy rules have a character of strong domain dependence, hence at the authoring, time consuming works of collecting and analyzing the buggy rules are required for each domain. On the other hand, if the system has knowledge about student's error origins which derives buggy rules, the knowledge is not only able to identify the student's buggy rule, but also able to be applied to broad domains, because the error origins are divided into two parts; a domain independent part and a domain dependent part, hence they have characteristic of more domain independence than the buggy rule has. The problem of domain independence becomes important for authoring new systems.

Besides the characteristic of domain independence, identifying the error origin and storing it in the student model makes ITS possible to predict a student's error in the use of even an inexperienced rule .

3.2 Means to identify students' errors

(1) Methods to generate buggy rules

The following four methods identify a student's correct or incorrect rule according to their correct or incorrect answers, respectively.

1) Error identification by buggy rule.

At the authoring ITS, collecting as many kinds of error instances as possible in the real student 's answers, classifying them according to the error origins and representing the results as buggy rules are the main task of the method.

As mentioned above, the defect of the method is the characteristic of strong domain dependence.

2) Error identification by perturbation method

There have been many research works published concerning student's errors and their causes in the fields of both cognitive science and pedagogy[8),9)]. Analysis of these works has revealed the causes of student's erroneous behavior. For instance, students often neglect or confuse the constraints when they apply already acquired rules to a different situation, students often drop a part of subgoals when they substitute subgoals for a goal, thus they encounter an impasse, moreover in order to dissolve the impasse students often behave inconsistently to their previous answers, etc. The facts suggest that by formulating operators by using the results from these analysis of the real errors and applying them to correct rules, buggy rules are able to be constructed.

The perturbation method[10)] reconstructs student's errors by applying the cognitively probable operators to the correct rule. The problem of how widely the perturbation method can cover the student's error depends on the breadth and details of the analysis in the real instances. The defect of the method is the problem of operator combinations, which will be discussed later.

3) Unification of perturbation method with buggy method

If buggy rules which are generated by perturbation method are stored in a buggy catalogue pool, as shown in Fig.1, they can be used directly for identifying student's errors so that a computing time for generating buggy rules may be saved. Buggy rules collected by human experts are used as the initial state of the buggy catalogue pool.

4) Error identification by truth maintenance

This method is a different approach from the above three methods which recognize the

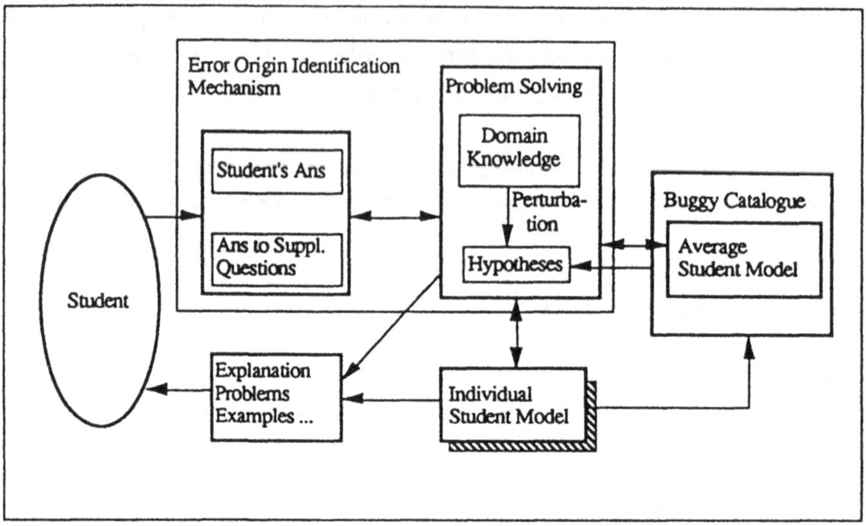

Fig.1 Student Modeling Mechnism

existence of a student's error first, then infer a student's erroneous rule by identifying the error origin. On the contrary, the fourth method infers a rule used by student first from a series of student's answers regardless of the correctness or incorrectness of the student's answer, then a truth maintenance system determines if the inferred rule is consistent with the domain rules or not.

Generating a student's rule from a series of student's answers means to make some kind of inductive inference. At present, the typical method uses MIS[11] (Model Inference System) for constructing student's rules and then ATMS[12] (Assumption Based Truth Maintenance System) for detecting the error candidate[13].

The defect of the method is susceptibility to influence from input noise which often occurs according to the change of student's comprehension and from pre-supposed terminology concerning the student's errors. In order to apply the method to the real system, tractability may cause another difficulty, which will be discussed later.

(2) Methods which do not construct the error rules

1) overlay method

So long as a student's answer coincides with one of the answers derived by the system, rules used for the derivation are considered to be acquired correctly by the student. Thus the method gathers the correctly acquired rules only. The defect point is that the method has no awareness of student's error.

2) Error identification by a classification tree

This method is the same as the medical diagnosing tree. Errors are analyzed by experts of education and an unique phenomenon to each error is selected to organize a fork in a diagnosing tree. This method depends strongly on the domain.

3.3 Criteria of error identification methods

The following four items are often used for assessment of error identification methods.

1) Generality (or independence of the domain)

Generality is an criterion of facilities in authoring ITSs.

2) Covering rate

Covering rate indicates the extent of the error range to which the system can identify student's errors. It is quite difficult to work out the precise rate because the whole kinds of human errors can not be detected even if the domain is limited. The covering rate is often used in comparison between different systems within a definite set of student's errors.

3) Tractability

One of the classification methods of error identification is whether real instances of human errors are taken into consideration or not. If instances are used, it is necessary to collect and analyze the instances, which needs almost the same time as or even more time than that of describing the domain knowledge, and yet it is impossible to guarantee the complete covering rate. In the case of perturbation method an overhead time of operator combinations is added to the execution time. To avoid the overhead a teaching sequence is introduced according to the relationship between a goal and subgoals of the rule, and a problem concerning a goal is not presented to the student until all of its subgoals have been correctly comprehended by the student. This method prevents a student from learning a new goal by using subgoals which have not yet been comprehended.

On the other hand although real instances of human errors are not taken into consideration, a student's rule is able to be inferred by using MIS if all the predicate names and arities which will appear in the student's erroneous answers are suitably determined beforehand. This strong verbal dependence causes to decrease the covering rate. Besides, it is necessary for MIS to employ a method like ATMS for detecting inconsistency of the derived rule to the domain knowledge, which requires a further computation and consumes considerable computing time because of the complexity of the time-space structure and the label computation.

4) Accuracy

Whether the obtained error origin hits the true student's error origin is also an important problem. This problem should be examined in each system separately.

4 Open-end educational systems

4.1 Micro world for supporting discovery learning

An open-end educational system like a micro world or a hyper media system with direct manipulation and visualization of internal states has a possibility of realizing a third generation CAI because of the following new features.

1) A student is allowed to observe invisible phenomena like force, gravity, velocity, sound waves, etc. schematically by manipulating objects directly.
2) As a result, it is possible for a student reorganize his/her mental model about the functions and the behavior of the objects by repeating a trial and error cycle of hypothesizing, experimenting and verifying. Thus, the system has the possibility of supporting discovery learning.
3) Visualization function which allows students observe a state and its transitions makes students understand the internal structure of the world, if suitable explanations are added by qualitative simulation.

Thus, the open-end system arouses a student interest in the objective world, establishes student's independence and supports student's discovery learning. In spite of these excellent features, the system has not yet obviated the following defects.

1) It is impossible for open-end system to answer student's questions concerning his/her trial and error operation.
2) There is no means to support a student when he/she impasses or loses the way in the experiment because the system has no means to infer the student's intention in the trial and error operation.
3) Highly individualized coaching is impossible because the system has no means to model a student.

The above discussion, together with the discussion in 2.2, shows the complementary characteristics of the open-end system and ITS.

4.2 Discussion on human learning from the epistemological view point

A learning process through which the human is conscious of the existence of an unknown rule, understands it and applies it by him/herself to a new problem is described by the following three steps.

The first step is an inductive knowledge acquisition from an instance based on trial and error. In this step, the obtained rule may be strongly influenced by the model of the world. It is the most important thing in the first step that the student should has already acquired all the premise knowledge of the target rule.

In the second step plural instances are employed to formulate a rule by generalizing rules obtained in the first step. The over specifica-

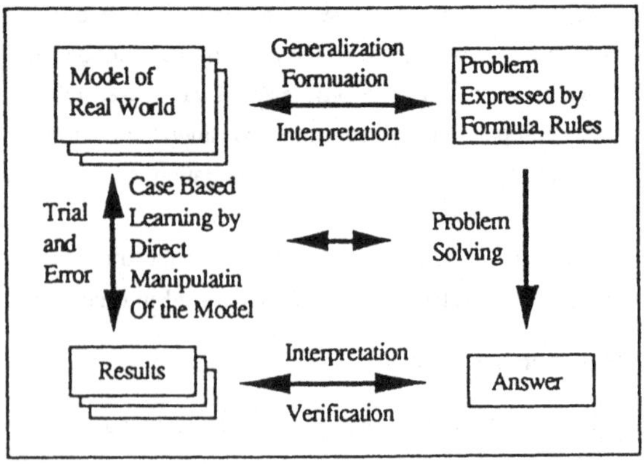

Fig. 2 A model of human learning

tion that may occur in the previous step should be reformulated in the second step.

The third step fixes the rule acquired in the previous step by applying the rule deductively to a variety of new problems so that the rule may be assimilated to his/her background knowledge. Thus it becomes unnecessary for students to think back to the premises or fundamentals of the rule every time they solve a problem.

Fig.2 explains the process schematically. The squares represent concepts and the arrows represent mental or concrete operations. Hence, experiments in a laboratory or natural phenomena in the real word are replaced by models of the upper-left squares in this figure.

An open-end educational system like a micro world or a hyper media system with direct manipulation and visualization of the internal states corresponds to the first step learning, while ITS corresponds to the thirds step, and a new method should be introduced to amalgamate these two methodology through the second step.

5 Discussion and conclusions

In order to solve problems described above, researches on amalgamating ITS to the open-end system are now proceeding[14),15),16)]. Our approach to solve the problem bases on the following ideas.

1) In order to grasp the student's intention in the open-end system, it is necessary for the system to have an ability of discovery learning through internal experiments of the micro world by using back ground knowledge base. The internal experimentation[17)], shown in Fig. 3, is one of the solutions.

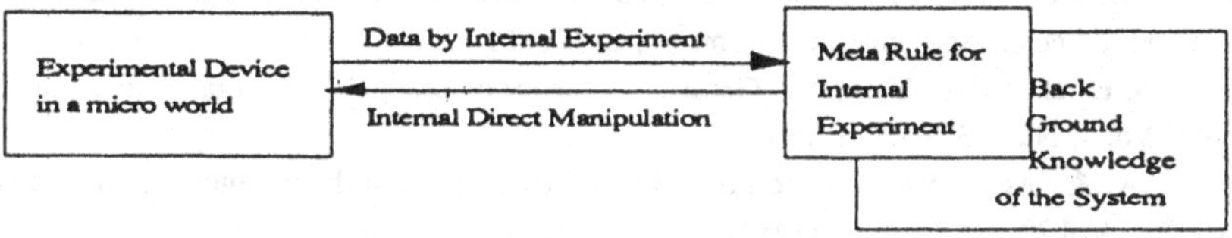

Fig. 3 Discovery learning through internal experiment

2) Premises of a target rule of an instance in the discovery environment, must be well-formed in the student's back ground knowledge. Suitability is able to be determined by a student model and well-formed domain knowledge. In this case, the well-formed relation is derived from the consistency between goal and sub goal relation.

3) System's support for student's discovery learning consists of the following two steps. Firstly, the knowledge region used for trial and error is gradually decreased from whole back ground knowledge to the premises of the target rule, and secondly, through meta-procedures for discovery learning which includes internal experiments the system acquires knowledge to support a student to organize a correct mental model.

4) ITS is used for deductive learning by using the already acquired knowledge, where if counter

examples are needed the instances used in the inductive learning of the first step is the most effective. The reason is that they are the common examples between the student and the system and that the student's comprehension level of generalization of the rule has been recorded in the student model.

5) In ITS only, if a student answers correctly it is impossible to determine whether the student has the fundamental comprehension or used the merely remembered procedures only. In amalgamated system the difference is obtained if a student model keeps records of processes for acquiring rules.

In the long run, all the problems discussed here resolve into the common problem of AI, machine learning in the big knowledge base.

References

[1] Oettinger, A.G.: Run. Computer Run: The Mythology of educational Innovation, Harvard University Press, Cambridge, MA., P.180, (1969).

[2] Carbonell, J.R.: "AI in CAI": An artificial intelligence approach to Computer-Assisted Instruction, IEEE trans. Man-Machine Systems 11(4), December, 190/202, (1970).

[3] Kearsley, G.: Intelligent CAI, in Encyclopedia of Artificial Intelligence, John Wiley & Sons, Inc., (1987).

[4] Sleeman, D. & Brown, J. S.: Intelligent Tutoring Systems, Academic Press, (1982).

[5] Wenger, E.: Artificial Intelligence and Tutoring Systems, Morgan Kaufmann Publishers, Inc., (1987).

[6] Self, J.: Bypassing the Intractable Problem of Student Modeling, Proc. of ITS'88, 18/24, (1988).

[7] Invited talks in Proc. of the 4th international conference on AI and Education, IOS, (1989).

[8] Matz, M.: Toward a Process Model for High School Algebra, in Intelligent Tutoring Systems (D.H. Sleeman, et. al. eds.), Academic Press, London, 25/50, (1982).

[9] Brown ,J.S. et. al.: Repair Theory: A Generative Theory of Bugs in Procedural Skills, Cognitive Science, Vol.4, No.4, 379/426, (1980).

[10] Takeuchi, A. & Otsuki, S.: Formation of Learner Model by Perturbation method and teaching knowledge, Vol.28, No.1, 54/63, (1987).

[11] E. Shapiro: Algorithmic Program Debugging, MIT Press, (1982).

[12] J. Doyle, A Truth Maintenance System, Artificial Intelligence, Vol.12, 231/272, (1979).

[13] Mizoguti, R. & Ikeda, M.: A Generic Framework for ITS And Its Evaluation. International Conference on ARCE, Advanced Research on Computers in Education, (eds. Lewis, R. & Otsuki, S.) Elsvier Science Publishers, pp.63-72, (1991).

[14] J. Self: Supporting The Disembedding of Learning, Proc. of East-West Conference on Emerging Computer Technologies in Education, April, (1992).

[15] W. J. Clancey: Guidon-Manage revisited: a socio-technical approach, in Intelligent Tutoring Systems, Lecture Notes in Computer Science 608, Springer-Verlag, (1992).

[16] Takeuchi, A., Shingae, T. & Otsuki, S.: Intelligent CAI Supporting both Inductive Learning and Deductive Learning, Japanese Society for AI, Sig-HICG'92.1,(1992).

[17] Shingae, T., Takeuchi, A. & Otsuki, S.: A Study on Inner Experiments and Formulation in a Micro World, Japanese Society for AI, Sig-IES'92.2, (1992).

FROM INDUCTIVE INFERENCE TO ALGORITHMIC LEARNING THEORY

Rolf Wiehagen
Department of Computer Science
Humboldt University
P.O.Box 1297 O-1086 Berlin

Abstract

We present two phenomena which were discovered in pure recursion-theoretic inductive inference, namely inconsistent learning (learning strategies producing apparently "senseless" hypotheses can solve problems unsolvable by "reasonable" learning strategies) and learning from good examples ("much less" information can lead to much more learning power). Recently, it has been shown that these phenomena also hold in the world of polynomial-time algorithmic learning. Thus inductive inference can be understood and used as a source of potent ideas guiding both research and applications in algorithmic learning theory.

1 Introduction

In her talk at STOC'92 Dana Angluin [3] described the relationship between inductive inference and algorithmic/computational learning theory as follows:
"Inductive inference is to computational learning theory roughly as computability theory is to complexity and analysis of algorithms. Inductive inference and computability theory are historically and logically prior to and part of their polynomially-obsessed younger counterparts, share a body of techniques from recursion theory, and are a source of potent ideas and analogies in their respective fields."
We want to present two phenomena which were discovered in pure recursion-theoretic inductive inference but which turn out to hold also in polynomial-time algorithmic learning theory. Hence inductive inference is really a field of "potent ideas" which can be fruitful for algorithmic learning theory. And we are convinced that these two phenomena will not be the only ones of this kind.

The first phenomenon.

The main problem of inductive inference we will deal with here consists in synthesizing of "global descriptions" for the objects to be learnt from examples.
Thus, one goal is the following. Let f be any computable function from $\mathbb{N}$ into $\mathbb{N}$. Given more and more examples $f(0), f(1), \ldots, f(n), \ldots$ a learning strategy is required to produce a sequence of hypotheses $h_0, h_1, \ldots, h_n, \ldots$ the limit of which is a correct global description of the function f. Since at any stage n of this learning process the strategy knows only the examples $f(0), f(1), \ldots, f(n)$, it seems reasonable to construct the hypothesis h_n such that for any $x \leq n$, the "hypothesis function" g described by h_n is defined and computes the value $f(x)$. Such a hypotheses h_n is called *consistent*. In other words, a hypothesis is consistent if all the information about the unknown object seen so far is completely and

correctly encoded into this hypothesis. Otherwise, a hypothesis is called to be *inconsistent*. Consequently, if the hypothesis h_n above is inconsistent, then there must be an $x \leq n$ such that $g(x) \neq f(x)$ (note that there are two reasons for that inequality: $g(x)$ may be undefined or $g(x)$ is defined but it differs from $f(x)$). Hence if a hypothesis is inconsistent then it is not only wrong at all but it is even wrong on an argument where the learning strategy already knows the correct value. At a first glance it seems that we can exclude strategies producing inconsistent hypotheses from our consideration at all. It appears that consistent strategies, i.e., strategies producing always consistent hypotheses, are the only reasonable kind of learning devices.
However, due to Barzdin [6], there are classes of recursive functions that can be learnt in the limit, but only by inconsistent strategies. Of course, the question arises immediately where the sense could be to output an inconsistent hypothesis. Informally, this "sense" consists in that there is no possibility to detect that a hypothesis is inconsistent. More exactly, the consistency/inconsistency of a hypothesis is undecidable, in general. Hence, in general, there is no algorithm working as a subroutine of the learning strategy which checks any hypothesis whether it is consistent or not, rejecting any inconsistent hypothesis and thereby giving the strategy one more chance to search for a consistent one.

Since the basic reason for the inconsistency phenomenon developped by Barzdin for recursive functions is just the undecidability of consistency, consequently, one more question arises, namely whether this phenomenon "survives" or whether we can really confine ourselves to consistent strategies, only, if we consider learning problems in domains where consistency is *decidable*. Moreover, what about this phenomenon, if we confine us to learning strategies working in polynomial time, i.e. "realistic" strategies?
We will show that also in these situations the inconsistency phenomenon does survive:
There are "natural" learning problems in domains where consistency is decidable which can be solved by polynomial-time strategies but provably by inconsistently working ones, only.
Observe that the reason for the inconsistency phenomenon is quite analogous in both cases, namely the algorithmic means do not suffice to produce consistent hypotheses always. In the world of arbitrary computable strategies the undecidability of consistency prevents them from finding consistent hypotheses, in general. Now, in the world of polynomial-time strategies the existence of problems which are hard to solve if $P \neq NP$ results in the strategies' inability to construct consistent hypotheses in polynomial time.
As far as we know the result above is the first one proving the existence of learning problems solvable in polynomial time only by inconsistent strategies. In our opinion this result strongly suggests to take inconsistent learning strategies seriously into consideration.

The second phenomenon.

In [11] the approach of inductive inference from good examples has been introduced and studied. Informally, good examples

- are considerably less examples than all examples, but they are "important",
- can effectively be computed from the object to be learnt,
- are sufficient for learning "rich" classes of objects.

In inductive inference only *finitely many* examples are allowed as good examples. In spite of the loss of "almost all" examples it turns out that, from good examples, considerably more function classes can be learnt than from all examples.
Again the question arises whether this surprising effect also holds for other object classes and for learning strategies running in polynomial time.
We give an affirmative answer to this question by stating finite learnability of

- the class of all pattern languages,
- a class of "almost all" finite automata

in polynomial time from good examples. Note that here good examples are again finitely many ones but their number is even bounded by a polynomial in the "size" (length of the pattern, number of states of the finite automaton, respectively) of the object to be learnt.

The paper is organized as follows. Section 2 gives notation, definitions and some basic results. Section 3 deals with the inconsistency phenomenon. In Section 4 we present learning from good examples. In Section 5 we discuss the results obtained.

2 Preliminaries

$\mathbb{N} = \{0, 1, 2, \ldots\}$ denotes the set of natural numbers. The set of all finite sequences of natural numbers is denoted by $\mathbb{N}^*$. For a finite set A, let card(A) denote the number of elements of A.
The classes of all partial recursive, recursive functions of one, two arguments are denoted by P, P^2, R, R^2, respectively.
A function $\psi \in P^2$ is called a numbering. We write ψ_i instead of $\lambda x \psi(i, x)$. Let $P_\psi = \{\psi_i \mid i \in \mathbb{N}\}$ denote the class of all partial recursive functions enumerated by ψ. A numbering $\phi \in P^2$ is called a Gödel numbering (cf. [26]) iff $P_\phi = P$ and, for any numbering $\psi \in P^2$, there is $c \in R$ such that, for any $i \in \mathbb{N}$, $\psi_i = \phi_{c(i)}$.
For a function $f \in R$ and $n \in \mathbb{N}$, let $f^n = cod(f(0), f(1), \ldots, f(n))$ where cod denotes an effective and bijective mapping from $\mathbb{N}^*$ onto $\mathbb{N}$.
We now define three basic types of inductive inference of recursive functions.

Definition *Let $U \subseteq R$ and let $\psi \in P^2$ be any numbering.*
U is called finitely *learnable with respect to ψ iff*
there is a strategy $S \in P$ such that, for any function $f \in U$, there is $n \in \mathbb{N}$ such that
(1) for all $x < n$, $S(f^x) = ?$,
(2) $\psi_{S(f^n)} = f$.

Here ? is a special symbol the output of which can be interpreted as saying by the learning strategy S "I don't know yet." It is required that the first "real" hypothesis be a correct ψ-program for the function f.
Let
$\text{FIN}_\psi = \{U \mid U$ is finitely learnable with respect to $\psi\}$,
$\text{FIN} = \bigcup_{\psi \in P^2} \text{FIN}_\psi$.
Finite learnability was introduced in [14]. The reader is also referred to [10], [16].

Definition *Let $U \subseteq R$ and let $\psi \in P^2$ be any numbering.*
U is called learnable in the limit *with respect to* ψ *iff*
there is a strategy $S \in P$ such that, for any function $f \in U$, there is $i \in \mathbb{N}$ such that
(1) $\psi_i = f$,
(2) $S(f^n) = i$ *for almost all* $n \in \mathbb{N}$.

Thus the sequence of hypotheses produced by the strategy S on the function f converges to a correct ψ-program of f. We note that no restriction is made that we should be able to algorithmically determine whether the sequence of hypotheses has already stabilized. It is easy to see that such a restriction would lead to the concept of finite learning.
Define
$\mathrm{LIM}_\psi = \{U \mid U$ is learnable in the limit with respect to $\psi\}$,
$\mathrm{LIM} = \bigcup_{\psi \in P^2} \mathrm{LIM}_\psi$.
As to some basic papers concerning learnability in the limit the reader is referred to [7], [8], [14].

Definition *Let $U \subseteq R$ and let $\psi \in P^2$ be any numbering.*
U is called behaviorally correct *learnable with respect to* ψ *iff*
there is a strategy $S \in P$ such that, for any function $f \in U$ and for almost all $n \in \mathbb{N}$,
$\psi_{S(f^n)} = f$.

Thus on the function f the strategy S produces a sequence of ψ-programs almost all of which compute f.
Define
$\mathrm{BC}_\psi = \{U \mid U$ is behaviorally correct learnable with respect to $\psi\}$,
$\mathrm{BC} = \bigcup_{\psi \in P^2} \mathrm{BC}_\psi$.
Behaviorally correct learnability was studied among others in [5], [8], [9], [25].
The following theorem proved in [5], [8], [25] gives an insight into the possibilities of finite, limit and behaviorally correct inference. Here pR denotes the set of all subsets of R.

Theorem 0 $\mathrm{FIN} \subset \mathrm{LIM} \subset \mathrm{BC} \subset pR$.

For survey papers on inductive inference the reader is referred to [4], [19], [22], [28].

3 Learning inconsistently

We first define consistent learning of recursive functions.

Definition *Let $U \subseteq R$ and let $\psi \in P^2$ be any numbering.*
U is called consistently *learnable in the limit with respect to* ψ *iff*
there is a strategy $S \in P$ such that
(1) U is learnable in the limit with respect to ψ by the strategy S,
(2) for any function $f \in U$, any $n \in \mathbb{N}$ and any $x \leq n$, $\psi_{S(f^n)}(x) = f(x)$.

Intuitively, a consistent strategy does correctly reflect all the data it has already seen. A hypothesis $S(f^n)$ such that, for any $x \leq n$, $\psi_{S(f^n)}(x) = f(x)$ is called consistent; otherwise it is called inconsistent.

Let
$\text{CONS}_\psi = \{U \mid U$ is consistently learnable in the limit with respect to $\psi\}$,
$\text{CONS} = \bigcup_{\psi \in P^2} \text{CONS}_\psi$.
The inconsistency phenomenon was developped by Barzdin [6].

Theorem 1 CONS $\subset$ LIM.

Thus there are classes of recursive functions which are identifiable in the limit, but only by inconsistent strategies. Note that also [7] observes the inconsistency phenomenon by stating that "many good machines do not have the overkill property" where "overkill" is synonymous to "consistency".
In general, a strategy attempting to learn functions consistently in the limit has to avoid two difficulties. One consists in changing the actual hypothesis the consistency of which seems "doubtful" to a new one which is definitely consistent at least at this moment. But changing "too often" results in the danger to converge not at all. Conversely, it is also dangerous to stay at a hypothesis the consistency of which is "doubtful", since this may result in converging to a really incorrect hypothesis. These difficulties are unavoidable, since in general the consistency of hypotheses is undecidable. This is the reason for the inconsistency phenomenon in inductive inference of recursive functions.
Actually, it turns out that a class $U \subseteq R$ is consistently learnable in the limit iff there is a suitable "space of hypotheses" with respect to which the problem of deciding the consistency of hypotheses is solvable at least relatively to the class U.
More exactly, let $U \subseteq R$ and let $\psi \in P^2$ be any numbering. We say that *U-consistency with respect to ψ is decidable* iff there is *cons* $\in P^2$ such that, for any $f \in U$, any $n \in \mathbb{N}$ and any $i \in \mathbb{N}$,
1) $cons(f^n, i)$ is defined,
2) $cons(f^n, i) = 1$ iff $\psi_i(x) = f(x)$ for any $x \leq n$.
Then in [31] the following result is contained.

Theorem 2 $U \in$ CONS *iff there is* $\psi \in P^2$ *such that*
(1) $U \subseteq P_\psi$,
(2) U-consistency with respect to ψ is decidable.

Note that, for infinitely many classes $U \in$ CONS, the halting problem with respect to the numbering ψ from Theorem 2 is *undecidable* (i.e., there is no $h \in R^2$ such that, for any $i, x \in \mathbb{N}$, $h(i,x) = 1$ iff $\psi_i(x)$ is defined). More exactly, let NUM $= \{U \mid U \subseteq R \;\&\; \exists\psi \in R^2\, [U \subseteq P_\psi]\}$ denote the family of all recursively enumerable classes of recursive funtions. Then, NUM $\subset$ CONS is proved in [30]. Now, it turns out that in order to learn an arbitrary class $U \notin$ NUM there are only "bad" spaces of hypotheses, namely, for any numbering $\psi \in P^2$ such that $U \subseteq P_\psi$ (being a necessary condition for learning U with respect to ψ), the halting problem with respect to ψ is undecidable (cf. Lemma 1 below, proved in [31]). Note that it is justified to call this a "bad" property of a space of hypotheses, since the decidability of the halting problem with respect to a numbering $\psi \in P^2$ easily implies the consistent learnability in the limit of *all* recursive functions from P_ψ with respect to ψ.

Lemma 1 *Let* $U \subseteq R$, $U \notin$ NUM.
Then, for any numbering $\psi \in P^2$ such that $U \subseteq P_\psi$, the halting problem with respect to ψ is undecidable.

Hence Theorem 2 points out a possibility to compensate the undecidability of the halting problem with respect to ψ by the decidability of U-consistency.
The reader is referred to [12], [17] and [32] where consistent identification was further investigated.

The results above could give the impression that the inconsistency phenomenon holds only in a "world" where problems such as the halting problem and the consistency problem are undecidable, in general. Therefore we now study the analogous problem in more "realistic" worlds where the consistency problem is always decidable. Moreover, to be "consequently realistic" we require from the learning strategies computability in polynomial time.
We will show that also in such a "realistic world" the inconsistency phenomenon does hold by pointing out a learning problem which is

- solvable consistently,
- unsolvable consistently in polynomial time, if $P \neq NP$,
- solvable inconsistently in polynomial time.

(For any information concerning P and NP the reader is referred to [13].)
The objects the learning of which we now will deal with are the pattern languages, cf. [2].
Other papers concerning learning of pattern languages are [18], [20], [27].
Let A be a finite, nonempty alphabet containing at least two elements. Let $X = \{x_i \mid i \in \mathbb{N}\}$ be a finite set of distinct variables such that $A \cap X = \emptyset$.
Then let Pat $= (A \cup X)^+$, the set of all nonempty words from $A \cup X$, denote the set of all *patterns*. We say that $w \in A^+$ is derivable from $p \in$ Pat iff p can be transformed into w by replacing any variable of p with a word from A^+. Thus $aabbb$ is derivable from ax_1x_2b, while $aabba$ is not.
Let $L(p)$ denote the set of all $w \in A^+$ derivable from $p \in$ Pat. Then $L(p)$ is called the *pattern language generated by p*.
Let PAT denote the set of all pattern languages.
In order to learn a pattern language L the learning strategy receives any labelled sequence of all words from A^+ where the label tells the strategy whether the corresponding word belongs to L or not.
Therefore we call a mapping I from $\mathbb{N}$ into $A^+ \times \{+,-\}$ an *informant* of L
iff 1) for any $w \in A^+$, there is $n \in \mathbb{N}$ and $\lambda \in \{+,-\}$ such that $I(n) = (w, \lambda)$,
2) for any $w \in A^+$, $n \in \mathbb{N}$, $\lambda \in \{+,-\}$, if $I(n) = (w, \lambda)$, then $w \in L$ iff $\lambda = +$.
Let $Info(L)$ denote the set of all informants of L.
For $I \in Info(L)$ and $n \in \mathbb{N}$, let $I^n = cod\,(I(0), \ldots, I(n))$ where cod denotes an effective and bijective mapping from the set of all finite sequences from $A^+ \times \{+,-\}$ onto $\mathbb{N}$.
Let $I_n = \{w \mid w \in A^+ \,\&\, \exists i \leq n\, \exists \lambda \in \{+,-\}\, [I(i) = (w, \lambda)]\}$,
$I_n^+ = I_n \cap L$, $I_n^- = I_n \setminus I_n^+$.

Definition PAT *is called* learnable in the limit from informant *(abbreviated:* PAT $\in$ INFO-LIM*) iff*
there is an effective strategy S *from* $\mathbb{N}$ *into* Pat *such that, for any* $L \in$ PAT *and any* $I \in Info(L)$,
1) for any $n \in \mathbb{N}$, $S(I^n)$ *is defined,*
2) there is $p \in$ Pat *such that* $L(p) = L$ *and, for almost all* $n \in \mathbb{N}$, $S(I^n) = p$.

Definition PAT *is called* consistently *learnable in the limit from informant (abbreviated:* PAT $\in$ INFO-CONS*) iff*
there is a strategy S such that
1) PAT $\in$ INFO-LIM *by S,*
2) for any $L \in$ PAT, $I \in Info(L)$ and $n \in \mathbb{N}$, $I_n^+ \subseteq L(S(I^n))$ and $I_n^- \cap L(S(I^n)) = \emptyset$.

Hence a strategy consistently learning pattern languages has always to produce "biconsistent" hypotheses generating *all* words from the unknown language seen so far and simultaneously generating *no* word from the language's complement seen so far. Of course, the consistency of a hypothesis is decidable, since the memership problem for pattern languages is decidable (i.e., given $w \in A^+$ and $p \in$ Pat, to decide whether or not $w \in L(p)$). On the other hand, it can be shown that the membership problem for pattern languages is *NP*-complete, cf. [2]. This suggests that any strategy learning PAT by solving the membership problem as a subroutine cannot run in polynomial time. But below we even will see that under $P \neq NP$ *no strategy at all* can learn PAT *consistently* in polynomial time.

Definition PAT *is called (consistently) learnable in the limit from informant* in polynomial time *(abbreviated:* PAT $\in$ Poly-INFO-LIM, PAT $\in$ Poly-INFO-CONS, *respectively) iff there is a strategy S and a polynomial pol such that*
1) PAT $\in$ INFO-LIM *(*PAT $\in$ INFO-CONS*) by S,*
2) for any $L \in$ PAT, $I \in Info(L)$ and $n \in \mathbb{N}$, time to compute $S(I^n) \leq pol(length(I^n))$.

Then in [31] the following theorem has been proved.

Theorem 3

(1) PAT $\in$ INFO-CONS.
(2) PAT $\notin$ Poly-INFO-CONS, *if $P \neq NP$.*
(3) PAT $\in$ Poly-INFO-LIM.

As far as we know this is the first learning problem at all which these properties have been proved for, especially, the consistent unsolvability in polynomial time under $P \neq NP$.
In our opinion, this result gives strong evidence of taking fast *inconsistent* learning strategies seriously into consideration.
Note that the reason for this result is quite analogous to the situation in inductive inference of recursive functions. While there, in general, the algorithmic means available do not suffice to produce consistent hypotheses *at all*, now the algorithmic means at hand are unable to construct consistent hypotheses *in polynomial time*, in general.

4 Learning from good examples

The idea of learning from good examples is to use *finite* sets of "well choosen" examples instead of the infinite sets of *all* examples to learn the unknown functions.

Definition *Let $U \subseteq R$ and let $\psi \in P^2$ be any numbering.*
U is called finitely learnable from good examples *with respect to ψ iff*

there is a numbering $ex \in P^2$, *a strategy* $S \in P$, *and a function* $z \in P$ *such that* $U \subseteq P_\psi$ *and, for any* $i \in \mathbb{N}$ *with* $\psi_i \in U$,
(1) ex_i *is a finite subfunction of* ψ_i, $z(i)$ *is defined, and* $z(i) = \text{card}(ex_i)$,
(2) for any finite subfunction ϵ *of* ψ_i, $\psi_{S(ex_i \cup \epsilon)} = \psi_i$.

Let us neglect the ϵ for a moment, i.e., take the special case $\epsilon = \emptyset$. Then it follows from condition (2) above that, for any function ψ_i from the class U, the strategy S "finitely" produces a correct ψ-program of ψ_i (which may be different from i) solely from ex_i—the finite set of good examples.
Furthermore it follows from condition (1) that, for any i such that $\psi_i \in U$, ex_i is effectively computable from i.
The ϵ we need in order to avoid "unfair coding tricks" such as $ex_i = \{(i, \psi_i(i))\}$ which would lead to trivial learning of the whole class R. On the other hand, in "real life" it seems to be seldom to get such a pure set ex_i of good examples. Often one gets additional correct, but non-necessary information (just the ϵ) and then one has to deal with the union of all the information, yielding another interpretation of the ϵ in the definition above.
A possible scenario of learning from good examples is the relationship between teacher and pupil. As a rule the teacher will not tell the pupil only the correct and final answer—s/he will not present him all s/he knows about the phenomenon to be learnt—say $\psi_i(0), \psi_i(1), \psi_i(2), \ldots$. Actually, s/he will offer some typical information, just "good examples", in order to enable the pupil to learn the unknown phenomenon by processing the good examples.
We will use the following abbreviations:
GEX-FIN$_\psi$ = $\{U \mid U$ is finitely learnable from good examples with respect to $\psi\}$,
GEX-FIN = $\bigcup_{\psi \in P^2}$ GEX-FIN$_\psi$.
Our next result from [11] shows that finite learning from good examples is more powerful than finite learning from all examples by "two orders of magnitude" corresponding to the inclusion FIN $\subset$ LIM $\subset$ BC from Theorem 0.

Theorem 4 GEX-FIN = BC.

Note that in [11] it has also been proved that $R \in$ GEX-LIM where GEX-LIM is defined in an analogous manner as GEX-FIN. Hence the loss of "almost all" examples again results in increasing the learning power considerably.
Consequently, it seems reasonable to ask whether this effect can also be achieved for other classes of objects to be learnt in polynomial time. The results below give a first affirmative answer to this question for pattern languages and for finite automata.

Definition *The class* PAT *of all pattern languages is said to be finitely learnable in polynomial time from good examples iff*
there is a computable mapping ex *from* Pat *into the set* $\text{fin}(A^+)$ *of all finite subsets of* A^+, *a computable strategy* S *from* $\text{fin}(A^+)$ *into* Pat *and a polynomial* pol *such that, for any* $p \in$ Pat,
1) $ex(p) \subseteq L(p)$ *and* $\text{card}\,(ex(p)) \leq pol\,(length(p))$,
2) for any finite W *with* $ex(p) \subseteq W \subseteq L(p)$,
$L\,(S(W)) = L(p)$ *and time to compute* $S(W) \leq pol\,(\sum_{w \in W} length(w))$.

Then in [21] the following result is proved.

Theorem 5 PAT *is finitely learnable in polynomial time from good examples.*

A similar result can be proved for finite automata. Therefore let DFA denote the set of all deterministic finite automata over some input alphabet X and output alphabet Y. Any automaton $A \in$ DFA is assumed to possess an initial state; hence A computes some function f_A from X^* to Y^*. Let Beh-DFA $= \{f_A \mid A \in \text{DFA}\}$ denote the set of all "behaviors" of automata from DFA. Let fin(Beh-DFA) denote the set of all finite subfunctions from the functions of Beh-DFA. Finally, for $A \in$ DFA, let Z_A denote the set of states of A.

Definition *A set* $\mathcal{A} \subseteq$ DFA *is said to be* finitely learnable in polynomial time from good examples *iff*
there is a computable mapping ex from DFA *into* fin(Beh-DFA), *a computable strategy* S *from* fin(Beh-DFA) *into* DFA, *and a polynomial pol such that, for any* $A \in \mathcal{A}$,
1) $ex(A) \subseteq f_A$ *and* $\text{card}\,(ex(A)) \leq pol\,(\text{card}(Z_A))$,
2) for any $g \in$ fin(Beh-DFA) *such that* $ex(A) \subseteq g \subseteq f_A$,
$f_{S(g)} = f_A$ *and time to compute* $S(g) \leq pol$(length of description of g).

Definition*A set* $\mathcal{A} \subseteq$ DFA *is called a* set of measure 1 *iff* $\lim_{k\to\infty} \text{card}(\mathcal{A}_k)/\text{card}(\text{DFA}_k) = 1$ *where for* $\mathcal{A}' \subseteq$ DFA *and* $k \in \mathbb{N}$, $\mathcal{A}'_k = \{A \mid A \in \mathcal{A}' \;\&\; \text{card}(Z_A) = k\}$.

Theorem 6 *There is* $\mathcal{A} \subseteq$ DFA *such that*
1) $\mathcal{A}$ *is a set of measure 1,*
2) $\mathcal{A}$ *is finitely learnable in polynomial time from good examples.*

An analogous result can be proved for finite automata of Moore type. For other work investigating the complexity of learning finite automata the reader is referred to [1], [15], [23], [24], [29].

On the other hand, good examples do not exist always. Therefore let $\mathcal{B}_n$ denote the set of all Boolean functions of n variables. Then, obviously, for any $f \in \mathcal{B}_n$, the set of *all* examples contains 2^n elements. If we require from a set of *good* examples that its cardinality has to be bounded by a polynomial in n, then it is easy to see that good examples for learning the whole class $\mathcal{B}_n$ cannot exist. Of course, this negative result does not exclude the possibility that for subclasses of $\mathcal{B}_n$ good examples (of polynomial size) do exist.

5 Conclusions

Firstly, we investigated the problem of consistent/inconsistent learning. In spite of the remarkable power of consistent learning it turns out that this power is not universal. There are learning problems which can be solved only by inconsistent strategies, i.e., strategies incorrectly reflecting the behavior of the unknown object to be learnt at intermediate steps of the learning process while the correct behavior of the object is already known.

This phenomenon was developped in a "very theoretical" domain, namely in inductive inference of recursive functions. The seemingly "senseless" work of inconsistent strategies can be explained here by the undecidability of consistency.

Now it turns out that the inconsistency phenomenon is also valid in more "realistic" situations, namely in domains where consistency is always decidable and the learning strategies

have to work in polynomial time. The reason is quite analogous to that of the world of arbitrary recursive functions. Now under $P \neq NP$, the NP-hardness of problems can prevent learning strategies from producing consistent hypotheses in polynomial time.
We even conjecture that an analogous effect can be shown for incremental learning of *finite* classes of *finite* objects such as Boolean functions.
Furthermore, we conjecture that there are learning problems even solvable in polynomial time consistently, but solvable "much faster" by inconsistent strategies.
In any case, we regard the results obtained as giving strong evidence to take fast *inconsistent* strategies seriously into account.
This also suggests directions of further research such as

- finding fast inconsistent learning techniques,
- deriving possibilities to identify that a given learning problem has no fast consistent solution.

Secondly, we dealt with learning from good examples. Again this paradigm was introduced and studied in pure recursion-theoretic inductive inference. The results obtained there show a considerable increase in learning power when only a "few", but "important" examples are available. Then we presented results that an analogous effect can be achieved for learning pattern languages and finite automata in polynomial time from good examples. Furthermore, we pointed out that good examples do not exist always. Hence directions of further research consist in

- finding criteria that, given some class of objects to be learnt, good examples do exist,
- if good examples do exist, then clarifying why they are good and to what kind of learning strategies this may lead (note that the strategy from Theorem 4 is an *enumeratively* working one, i.e., it searches the space of hypothesis more or less exhaustively, whereas the strategies from Theorem 5 and Theorem 6 construct their hypotheses "directly" from the good examples given).

Thus we hope to have presented two "good examples" that ideas derived in pure recursion-theoretic inductive inference can be helpful in polynomial-time algorithmic learning of various objects.
We conjecture that the possibilities of inductive inference serving as a source of ideas, phenomena, techniques... for algorithmic learning at all are far from being exhausted.

References

[1] Angluin, D., On the complexity of minimum inference of regular sets. *Information and Control* 39 (1978) 337–350.

[2] Angluin, D., Finding patterns common to a set of strings. *Journal of Computer and System Sciences* 21 (1980) 46–62.

[3] Angluin, D., Computational learning theory: Survey and selected bibliography. *Proc. ACM Symposium on Theory of Computing*, ACM Press, 351–368, 1992.

[4] Angluin, D. and Smith, C. H., Inductive inference: Theory and methods. *Computing Surveys* 15 (1983), 237–269.

[5] Barzdin, J., Two theorems on the limiting synthesis of functions. In [28], vol.1 (1974), 82–88 (in Russian).

[6] Barzdin, J., Inductive inference of automata, functions and programs. *Proc. Int. Congress of Mathematicians*, 455–460, 1974.

[7] Blum, L. and Blum, M., Toward a mathematical theory of inductive inference. *Information and Control* 28 (1975) 122–155.

[8] Case, J. and Smith, C., Comparison of identification criteria for machine inductive inference. *Theoretical Computer Science* 25 (1983) 193–220.

[9] Daley, R., On the error correcting power of pluralism in inductive inference. *Theoretical Computer Science* 24 (1983) 95–104.

[10] Freivalds, R., Finite identification of general recursive functions by probabilistic strategies. *Proc. Conf. Foundations of Computation Theory*, Akademie-Verlag, 138–145, 1979.

[11] Freivalds, R., Kinber, E. B. and Wiehagen, R., On the power of inductive inference from good examples. *Theoretical Computer Science* (to appear).

[12] Fulk, M., Saving the phenomena: Requirements that inductive inference machines not contradict known data. *Information and Computation* 79 (1988) 193–209.

[13] Garey, M. R. and Johnson, D. S., *Computers and Intractability*, Freeman and Company, 1979.

[14] Gold, E. M., Language identification in the limit. *Information and Control* 10 (1967) 447–474.

[15] Gold, E. M., Complexity of automaton identification from given data. *Information and Control* 37 (1978) 302–320.

[16] Jain, S. and Sharma, A., Finite learning by a team. *Proc. Third Annual Workshop on Computational Learning Theory*, Morgan Kaufmann, 163–177, 1990.

[17] Jantke, K. P. and Beick, H.-R., Combining postulates of naturalness in inductive inference. *Journal of Information Processing and Cybernetics (EIK)* 17 (1981) 465–484.

[18] Kearns, M. and Pitt, L., A polynomial-time algorithm for learning k-variable pattern languages from examples. *Proc Second Annual on Computational Learning Theory*, Morgan Kaufmann, 57–70, 1989.

[19] Klette, R. and Wiehagen, R., Research in the theory of inductive inference by GDR mathematicians—a survey. *Information Sciences* 22 (1980) 149–169.

[20] Ko, Ker-I, Marron, A. and Tzeng, W.-G., Learning string patterns and tree patterns from examples. *Proc. Seventh Int. Conf. on Machine Learning*, Morgan Kaufmann, 384–391, 1990.

[21] Lange, S. and Wiehagen, R., Polynomial-time inference of arbitrary pattern languages. *New Generation Computing* 8 (1991) 361–370.

[22] Osherson, D., Stob, M. and Weinstein, S., *Systems that learn.* MIT Press, 1986.

[23] Pitt, L., Inductive inference, DFA's, and computational complexity. *Proc. Int. Workshop on Analogical and Inductive Inference*, Lecture Notes in Artificial Intelligence 397 (1989) 18–44.

[24] Pitt, L. and Warmuth, M. K., The minimum consistent DFA problem cannot be approximated within any polynomial. *Tech. Report UIUCDCS-R-89-1499*, University of Illinois at Urbana-Champaign, Febr. 1989.

[25] Podnieks, K. M., Comparing various concepts of function prediction, part I, in [28], vol.1 (1974) 68–81 (in Russian).

[26] Rogers, H. Jr., *Theory of recursive functions and effective computability*, McGraw-Hill, 1967.

[27] Shinohara, T., Polynomial-time inference of extended regular pattern languages. *Proc. RIMS Symp. on Software Science and Engineering*, Lecture Notes in Computer Science 147 (1983) 115–127.

[28] *Theory of Algorithms and Programs*, vol.1, 2, 3. Barzdin, J., Ed., Latvian State University, Riga, 1974, 1975, 1977 (in Russian).

[29] Trakhtenbrot, B. A. and Barzdin, J., *Finite Automata: Behavior and synthesis.* North-Holland, 1973.

[30] Wiehagen, R., Limes-Erkennung rekursiver Funktionen durch spezielle Strategien. *Journal of Information Processing and Cybernetics (EIK)* 12 (1976) 93–99.

[31] Wiehagen, R. and Zeugmann, T., Too much information can be too much for learning efficiently. *Proc. Int. Workshop on Analogical and Inductive Inference*, Lecture Notes in Artificial Intelligence, Oct. 1992.

[32] Zeugmann, T., A-posteriori characterizations in inductive inference of recursive functions. *Journal of Information Processing and Cybernetics (EIK)* 19 (1983) 559–594.

A Stochastic Approach to Genetic Information Processing

Akihiko Konagaya

C&C Systems Research Labs., NEC Corp.,
4-1-1 Miyazaki, Miyamaeku, Kawasaki, Kanagawa 216, Japan
e-mail konagaya@csl.cl.nec.co.jp

Abstract

This paper stresses the importance of stochastic machine learning theory for analyzing genetic information such as protein sequences. It is commonly recognized that machine learning theory would play an essential role to extract important information from the enormous amounts of raw genetic information generated by biologists. However, it is also true that more flexible and robust learning methodologies are required to deal with divergence occurring on the genetic information.
For this purpose, we adopt stochastic knowledge representations and stochastic learning algorithms and show their effectiveness with a stochastic motif extraction system. The system aims to extract stable common patterns conserved in some protein category. In the system, common patterns (stochastic motifs) are represented by stochastic decision predicates, and a genetic algorithm with Rissanen's minimum description length principle is used to select "good stochastic motifs" from the viewpoint of increasing prediction performance.

1 Introduction

Rapid improvement of molecular biology technology has succeeded to generate enormous genetic information, such as nucleic sequences and protein sequences. To analyze the genetic information both computer technology and molecular biology technology are required. This leads to the appearance of new scientific domain named genetic information processing or bio informatics.

The goal of genetic information processing is to extract valuable information from genetic information. To achieve this, various computer-based systems have been developed; homology search systems to retrieve similar genetic sequences from a sequence database, secondary or tertiary protein structure inference systems for unknown genetic sequences, motif extraction systems to find common patterns conserved in protein categories, molecular orbitary or molecular dynamics to simulate the behavior of proteins, etc. To enhance these systems, much attention has been focused on artificial intelligence technology, especially for machine learning theory as well as database, image processing and numeric processing technologies. However, it should be noted that very few machine learning theory have succeeded so far to extract biologically meaningful information.

Let us consider the reasons by focusing on motif extraction from protein sequences. The purpose of motif extraction is to find common patterns in a protein category. Such patterns are important since they are conserved in the evolution process for some reason; in fact, there is a good correspondence between conserved patterns and protein active sites and/or special protein structures such as Zinc-Finger and Luesin-Zipper[1]. From the viewpoint of artificial intelligence, motif extraction can be considered as a kind of inductive learning process which finds rules from given sample sequences. However, extracting valuable motifs is not trivial because (1) almost all motifs have exceptions, (2) overfitting may occur when searching for the best fitting rules for sample sequences, and (3) combinatorial explosion may occur when searching for all motif candidates.

To overcome these difficulties, we adopt a stochastic knowledge representation and a stochastic learning algorithm. That is, we propose a "stochastic motif" that represents stochastic mapping from protein sequences to protein functions or protein structures. A stochastic motif may contain exceptions but is more stable and reliable for discriminating unknown sequences or predicting protein functions or structures. To represent the stochastic motif, we also proposed a stochastic decision predicate, a collection of Horn clauses with a probability to represent reliability of each clause. One of the difficulties of extracting stochastic motifs from protein sequences is overfitting to the given sample sequences. To avoid this, we adopt Rissanen's Minimum Description Length (MDL) principle. We can easily show that the best fitting stochastic motif is unstable in the sense that it depends on the sample sequences. The MDL principle solves this problem by balancing between the complexity of a motif and its classification errors. It gives a strategy of selecting an optimal stochastic motif on the basis of the sum of the bit lengths required to encode a stochastic motif and its logarithmic likelihood to the sample protein sequences.

To avoid the combinatorial explosion in the motif extraction, we use "genetic algorithms", which are a kind of probabilistic search algorithm based on the natural evolution process. The virtue of genetic algorithms is that they offer an efficient generate-and-test search by means of simple genetic operators that simulate "crossover", "mutation" and "selection". Our experimental results demonstrate that a genetic algorithm extracts stable stochastic motifs if the MDL principle is adopted for the design of the selection operator or fitness function.

The organization of the rest of this paper is as follows. Section 2 gives a background of stochastic motif extraction. Section 3 gives a representation for stochastic motifs, which we call *Stochastic Decision Predicates.* Section 4 gives a strategy for selecting a good stochastic motif using the MDL principle. Section 5 gives an algorithm for finding optimal stochastic motifs. Section 6 gives an overview of our stochastic motif extraction system. Section 7 presents experimental results on extracting stochastic motifs based on our proposed methodology. Finally, in section 8 we discuss current difficulties and future works. This work has been done as a part of the fifth generation computer systems project for the evaluation of the parallel inference machines.

2 Stochastic Motif

Divergence is one of the characteristics of nature. In practice, it seems difficult to find exact rules in biology. One of such example is a discrimination rule between birds and mammals. Birds can be characterized by simple rules such as having a beak or a bill and wings and laying eggs, and mammals can be characterized by having four legs and childbirth. But, a strange animal *platypus* has a bill and four legs, and lays eggs!

The same thing happens in motif extraction. We can easily find simple common patterns conserved in most sequences in some protein category. However, such simple common patterns almost always have exceptions. The exceptions can be eliminated if we introduce more complex patterns. However, this is not safe because the result may be sample dependent and less effective for the prediction of protein functions and protein structures in unknown sequences. To overcome this difficulty, we pursue stable motifs instead of precise motifs, and propose a "stochastic motif" which inherently includes exceptions, are more stable, and more naturally represent protein functions.

Let us show the example of a stochastic motif using cytochrome c, a protein which plays an important role in the respiratory chain. Figure 1 shows some known cytochrome c sequences for various species. Each character in the sequence corresponds to an amino acid. In most cytochrome c sequences, we can find the common pattern "$CXXCH$" where "C", "X", "H'

Species	Sequence of Cytochrome c
Human	..FIMKCSQCHTVEK..
Mouse	..FVQKCAQCHTVEK..
Chicken	..FVQKCSQCHTVEK..
Snake	..FSMKCGTCHTVEE..
Prawn	..FVQRCAQCHSAQA..
Yeast	..FKTRCLQCHTVEK..
Hemp	..FKTKCAECHTVGR..
Tetrahymena	..FDSQCSACHAIEG..
Rhodopila	..FHTICILCHTDIK..
Microbium	..VFKQCKICHQVGP..
Pseudomonas	..VFKQCMTCHRADK..

Figure 1: A part of cytochrome c sequences

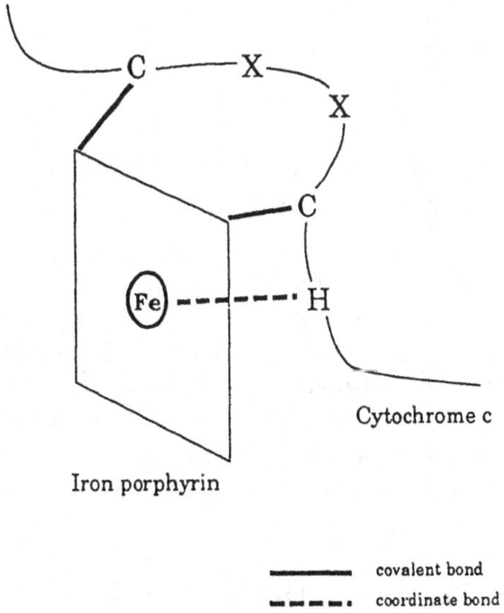

Figure 2: A heme c binding in a cytochrome c

stand for a cysteine, any amino acid, and a histidine, respectively, and the second "X" does not necessary to coincide with the first "X". In fact, the pattern "$CXXCH$" is biologically meaningful because it corresponds to a protein function; the two cysteines and the histidine binds to a heme which cytochrome c holds in the center (Figure 2).

As with other motifs, the pattern "$CXXCH$" also has exceptions. It does not exist in the cytochrome c of Euglena, and the pattern "$CXXCH$" exists in an adrenodoxin of a pig which is a different category from the cytochrome c. A stochastic motif can show the reliability of a pattern by calculating the ratio of the number of target protein sequences containing the pattern and the number of sequences containing the pattern, as follows. "If the pattern $\cdots$ "$CXXCH$" $\cdots$ is included in the sequence, then the sequence is cytochrome c with probability 130/227 and otherwise it belongs to other protein categories with probability 8072/8076." Note that the matched sequences for the first clause are eliminated from the total number of sequences to calculate frequency for the second clause.

3 Stochastic Decision Predicates

There are many ways to represent stochastic motifs. As a first step for a stochastic representation of motifs, we devised the stochastic decision predicate, a natural extension of a decision list with probabilities. The stochastic decision predicate consists of linearly ordered Horn clauses with probability parameters as follows.

```
motif(S,cytochrome_c) with 130/227.
      :- contain(S,''CXXCH'').
motif(S,others) with 8072/8076.
```

The general form is the following.

$$
\begin{array}{lll}
motif(S, C_1) & (\text{with } p_1) & :- Q_1^{(1)} \wedge \cdots \wedge Q_{k_1}^{(1)}. \\
motif(S, C_2) & (\text{with } p_2) & :- Q_1^{(2)} \wedge \cdots \wedge Q_{k_2}^{(2)}. \\
\cdots\cdots\cdots & & \\
\cdots\cdots\cdots & & \\
motif(S, C_{m-1}) & (\text{with } p_{m-1}) & :- Q_1^{(m-1)} \wedge \cdots \wedge Q_{k_{m-1}}^{(m-1)}. \\
motif(S, C_m) & (\text{with } p_m) & :- Q_1^{(m)} \wedge \cdots \wedge Q_{k_m}^{(m)}.
\end{array}
$$

Here we call each "$motif(S, C_i)$ (with p_i) :- $Q_1^{(i)} \wedge \cdots \wedge Q_{k_i}^{(i)}$." a *stochastic clause*. The stochastic clause can be read as S is categorized into C_i with probability p_i if $Q_1^{(i)}, \cdots, Q_{k_i}^{(i)}$ are all **true**. We assume sequential interpretation of the stochastic clauses in this paper. That is, $motif(S, C_i)$ is selected after $motif(S, C_1), \cdots, motif(S, C_{i-1})$ are examined. The body goals $Q_1^{(i)} \wedge \cdots \wedge Q_{k_i}^{(i)}$ $(i = 1, \cdots, m)$ represent a condition to discriminate a category C_i when S is given. Each goal $Q_j^{(i)}$ consists of the disjunction of goals $R_{1_j}^{(i)}; \cdots; R_{h_j}^{(i)}$ where $R_{h_j}^{(i)}$ represents some predicate that discriminates a category C_i, such as $contain(S, \sigma)$ which is *true* when S contains a pattern σ.

3.1 Semantics of Stochastic Decision Predicate

The semantics of stochastic decision predicates are given from the viewpoint of computational learning theory of stochastic rules[3]. A stochastic decision predicate represents a probabilistic mapping from protein sequences to categories. The probabilistic mapping can be regarded as a conditional probability distribution over the categories when a sequence is given, by introducing a probability structure on the sequence-category pairs. See the paper [4] for the formal approach to learning stochastic motifs.

4 The MDL Principle in Stochastic Motif Extraction

We adopt the MDL principle to avoid overfitting when extracting stochastic motifs. For example, as we have shown in section 2, the pattern "$CXXCH$" has exceptions in the cytochrome c. It is possible to avoid these exceptions by adding more conjunctions and disjunctions of patterns such as "$AAQCH$" and "$PGTKM$". However, care must be taken so that the obtained result does not become sample dependent, that is, overfit to the sample sequences. Therefore, we adopt the MDL principle to extract simple but stable stochastic motifs which may contain exceptions rather than precise motifs without exceptions.

The MDL principle originally comes from coding theory in communication. The basic idea is to optimize the number of bits when sending an information by finding a rule and its exceptions in the information. The MDL principle selects a rule such that minimizes the total bit length of the rule and the exceptions.

For example, suppose there is a binary string "101101100". Sending the string requires 9 bits if we do not use any rule. Less bits are sufficient if we compress the string as three repeats of "10*" and exceptions "110" for the third bit of each repeat instead of * in the rule. Total bits becomes 7.75 where the rule and the exception require 4.75 and 3 bits, respectively. We may find a more complex rule to reduce the number of exceptions, but such a rule might require a longer bit length to be encoded. Therefore, it is important to balance the complexity of rule and the number of exceptions to reduce the total bit length: this is the MDL principle.

In our methodology, we apply the MDL principle for extracting stochastic motifs as the way proposed by Yamanishi for learning stochastic rules: Yamanishi's MDL learning algorithm[3]. In Yamanishi's algorithm, the MDL principle selects a stochastic rule that balances the complexity of the stochastic rule and its likelihood of matching the sample data. We follow his algorithm with slight modification which mainly comes from the difference of stochastic rule representation: stochastic decision lists and stochastic decision predicates, and some practical reasons for applying the MDL learning algorithm to the motif extraction.

Our methodology selects a stochastic motif that balances the complexity of representation and likelihood of matching the sample sequences. The complexity of a stochastic motif representation is measured by the description lengths to encode the probability parameters and Horn clauses of a stochastic decision predicate. The likelihood of a stochastic motif is measured by the description length of likelihood, that is, by the logarithmic likelihood of categories when the sequences are given to the stochastic motif. The appendix describes the details of calculating the description lengths for a stochastic motif.

5 Genetic Algorithms

To overcome the combinatorial explosion in the motif extraction, we adopt a genetic algorithm, a stochastic search algorithm based on the natural evolution process[6]. Genetic algorithms simulate the survival of the fittest in a population of individuals which represent points in a search space. The individuals are often represented by binary strings. A function, often called a fitness function, gives values to the binary strings. The aim of a genetic algorithm is to find a global optimum of the fitness function when given an initial population of individuals by applying genetic operators in each generation. The genetic operators consist of the following operators: crossover, mutation and selection.

Crossover

The crossover operator produces two descendants by exchanging part of two individuals. This operator aims to make a better individual by replacing a part of an individual with a better part of another individual. For example, crossover of the strings "000110" and "110111" at the

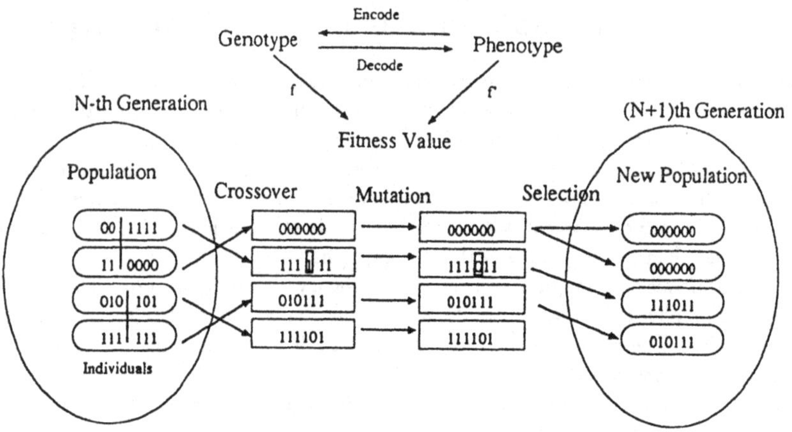

Figure 3: Mechanism of Simple Genetic Algorithms

third position produces the strings "000111" and "110110". The candidates of the crossover operation and the crossover position are randomly chosen.

Mutation

The mutation operator changes certain bit(s) in an individual. For example, the string "000110" becomes "001110" if mutation occurs at the third bit. The operation aims to escape from search spaces from which individuals cannot escape by means of only crossover operators.

Selection

The selection operator chooses good individuals in a population according to their fitness values and the given selection strategy. This operator aims to increase better individuals in the population while maintaining certain diversity. It simulates the survival of the fittest principle. The operator first calculates the relative fitness of all individuals. Then, several lesser individuals are discarded and the same number of better individuals are duplicated according to their relative fitness values. Note that the selection is probabilistic, not deterministic. So, better individuals have a higher chance of remaining or being duplicated but this is not guaranteed.

The performance of genetic algorithms is largely dependent on the design of the fitness function. The most interesting characteristic of our genetic algorithm is in its use of the MDL principle to calculate the fitness value of a stochastic motif. The description length gives the appropriate relative fitness values in the population, although smaller is better in this case.

6 The Stochastic Motif Extraction System

This section gives our overview of the stochastic motif extraction system. The target hypothesis space is the domain of stochastic decision predicates. The search strategy is the MDL principle. The search algorithm is an asynchronous parallel genetic algorithm which consists of the set of subpopulations in which individuals migrate asynchronously. In each subpopulation, individuals represent stochastic motifs in the target hypothesis space, and fitness function calculates the corresponding description lengths of the stochastic motifs represented by the stochastic decision predicates.

The search time depends considerably on the size of the hypothesis space. A large hypothesis space makes it difficult for us to find the optimal stochastic decision predicate in a reasonable time. Therefore, as the first step of motif extraction, we restricted the stochastic predicates to the following forms.

```
motif(S,proteinClass) with p1
      :- contain(S,pattern1) and
         contain(S,pattern2) ...
motif(S,others) with p2.
```

That is, we use a predicate *motif* which discriminates the target protein category *proteinClass* from other proteins (*others*) in the database. The discrimination conditions are represented by the conjunction of a predicate *contain*. As the pattern candidates in the *contain* predicate, we adopt 128 patterns that occur frequently in the target proteins.

The mapping from a stochastic decision predicate to a binary string is the following. Each bit corresponds to one of the 128 patterns. A bit 1 represents the occurrence of the pattern in a discrimination condition, and a bit 0 represents that the pattern does not occur in the discrimination condition. For example, suppose we use 3-bit length binary strings whose first, second, third bits correspond to the pattern "$CXXCH$", "$PXLXG$", "$GXKM$", respectively. Then, the binary string "100" represents the following stochastic decision predicate.

```
motif(S,proteinClass) with p1
      :- contain(S,"CXXCH").
motif(S,others) with p2.
```

The binary sting "011" represents the following stochastic decision predicate.

```
motif(S,proteinClass) with p1
      :- contain(S,"PXLXG") & contain(S,"GXKM").
motif(S,others) with p2.
```

According to this mapping, 128 bits binary strings can express 2^{128} kinds of stochastic decision predicates. As for the genetic operators, we adopt one-point crossover, one-point mutation and roulette wheel selection. Other runtime parameters are the following: the adjustment parameter is 1.0, the number of subpopulations is 63, subpopulation size is 16, the crossover rate is 1.0, the mutation rate is 0.01 and the migration rate is 0.5, that is, one individual per two generations in average.

7 Evaluation

Using the stochastic motif extraction system, we have already extracted 166 stochastic motifs from the protein categories that have more than 10 entries in the Protein Identification Resources (PIR32.0) with currently 9633 entries[1]. Table 1 shows a portion of the results.

In table 1, the line with percent (%) shows the name of protein category, super family number and the number of sequences in the category. The following line shows the common patterns extracted by the system, description lengths and distributions discriminated by the patterns. The column *DL* is the total description length of the extracted stochastic motif. The column *CL*, *PL* and *LL* are the description lengths of Horn clauses, a probability parameter and a logarithmic likelihood to the sample sequences, respectively.

Cytochrome c is a heme-binding protein that carries an electron in respiratory chain. *Cytochrome p450* is a mono-oxygenase containing a proto-heme. *Pepsin* is an acid protease secreted from the stomach. *Trypsin* is a protease secreted from a pancreas. *Globin* is an apo protein that constructs a hemoglobin when binding with a heme molecule. *Immunoglobulin C region*

[1] Annotated and classified entries by homology in pirl.dat.

Table 1: A Portion of Stochastic Motifs obtained from Protein Sequences

Patterns	DL	CL	PL	LL	N_1^+/N_1	N_2^+/N_2
% cytochrome c (1.0, 140) CXXCH	309.544	18.288	10.564	280.693	137/244	9386/9389
% cytochrome P450 (21.0, 33) FXXGXR & GXRXC & RXCXG	127.788	55.523	9.018	63.247	28/28	9600/9605
% pepsin (476.0, 19) FXXXFD & VPXXXC	80.802	38.575	8.700	33.526	17/18	9613/9615
% trypsin (458.0, 40) GWG & CXXDXG	124.490	34.253	9.435	80.802	37/50	9580/9583
% globin (902.0, 456) PXTXXXF & HGXXV	767.740	38.383	10.964	718.392	395/434	9138/9199
% immunoglobulin C region (892.0, 74) VXXFXP & CXVXH	357.216	37.575	9.895	309.746	53/95	9517/9538
% immunoglobulin V region (886.0, 268) DXXXYXC	692.147	20.095	10.871	661.181	237/379	9223/9254

is a constant region of immunoglobulin C. *Immunoglobulin V region* is a variable region of immunoglobulin C.

There are a lot of controversial issues in the biological significance of the obtained results from the view point of genetic information processing. However, the following observations would be more controversial those who are interested in machine learning.

7.1 Comparison of the MDL principle and the Maximum likelihood method

One of our concerns in the stochastic motif extraction is how the MDL principle works in genetic algorithms. To show this, prediction errors are compared to the maximum likelihood (ML) method using the cross validation technique ([7] p.75-76). In the ML method, good individuals are selected using only the description length of likelihood (LL) without consideration for the complexity of a stochastic decision predicate ($CL + PL$).

Using the cross validation technique, the prediction errors can be counted as follows. Firstly, let S_i be a disjoint subgroup of protein sequences S for certain N where $S = \cup_{i=1}^{N} S_i$. Let S_i' be a sample set which removes the i th subgroup from the original protein sequences ($S_i' = S - S_i$). Then, let M_i be a stochastic motif extracted from the sample set S_i', and count the number of prediction errors E_i^+ and E_i^- using the subgroup S_i as a test set, where E_i^+ shows the number of protein sequences that belong to the target protein category but is not *true* for the first clause of the stochastic motif M_i. E_i^- shows the number of protein sequences that do not belong to the target protein category but is *true* for the first clause of the stochatic motif M_i.

Table 2 shows the prediction errors for cytochrome c by cross validation method when divided into 10 subgroups. The results show that the stochastic motifs obtained using a genetic algorithm with the MDL principle are more stable than the ones obtained using a genetic algorithm with the ML method. As seen in table 2, the stochastic motifs obtained by the genetic algorithm with the ML method is sample dependent. It shows strong discrimination performance for the sample protein sequences ($\sum_{i=1}^{10} E_i^-$), but shows weak predictive performance for the test sequences ($\sum_{i=1}^{10} E_i^+$).

Contrary to our expectations, this result comes from the difference of convergence speed between GA with MDL and GA with ML as shown in figure 4. The upper, middle and lower lines represents the average description lengths of the worst, the average and the best individuals so far in each generation. It arises not from the overfitting caused by the ML method since the

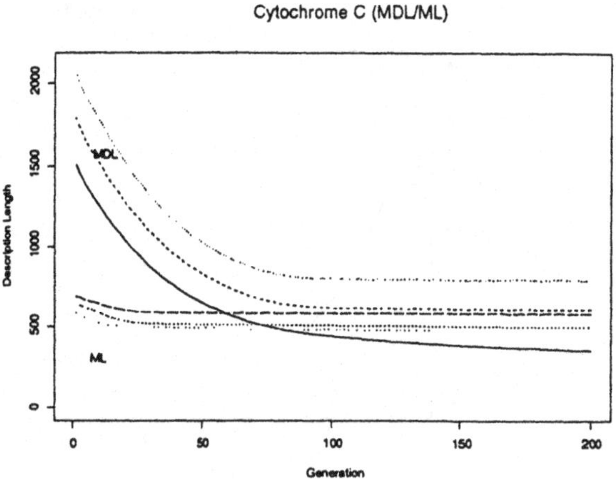

Figure 4: Average description lengths of the best stochastic motif encountered in each generation

Table 2: Prediction errors for Cytochrome C by Cross Validation Method

	MDL	ML
$\sum_{i=1}^{10} E_i^+$	3	57
$\sum_{i=1}^{10} E_i^-$	96	0
Total	99	57

optimal stochastic motif for cytochrome c is "$CXXCH$" in both the MDL principle and the ML method.

The difference of the convergence speed comes from the bias caused by the MDL principle. As shown in figure 5, the number of patterns in the best stochastic motif encountered continuously decrease in case of the MDL principle while it is almost constant in case of the ML method. This is natural since the description length of Horn clauses basically corresponds to the number of patterns. In other words, the MDL principle gives a bias for GA to select individuals with fewer patterns.

One might think it would be possible to reduce the search space if the best stochastic motif can be found in stochastic motifs with fewer patterns. This is true so far as we have examined. The largest stochastic motif has four patterns (Histon H1) and most have two or three patterns. However, it should be noted that we might underestimate the effect of model length (clause length (CL) in this case) and over-simplification may caused by the MDL principle. To show the intrinsic differences between the MDL principle and the ML method, further investigation is required.

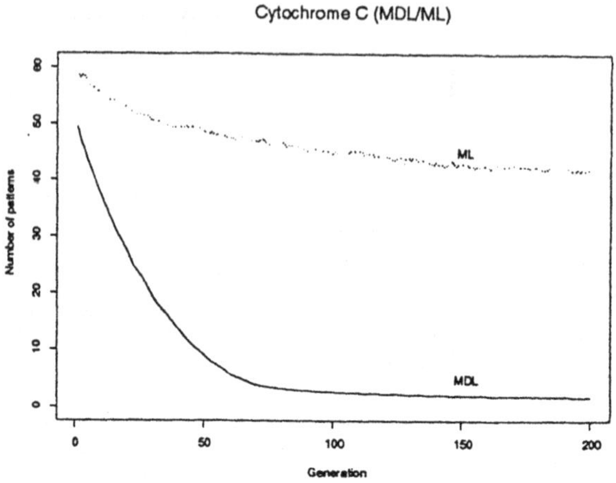

Figure 5: Average number of patterns of the best stochastic motif encountered in each generation

8 Discussion

The following works remain to deal with actual protein sequences on the basis of our methodology.

- The extension of stochastic decision predicate form: In our experience, the number of categories for discrimination is limited two, that is, the target category and the others. A stochastic decision predicate over two categories can be constructed by concatenating the obtained stochastic clauses for each protein category and recalculating the probabilistic parameter, although it causes another combinatorial problem: the order of protein categories. Another interesting extension is providing other predicates, such as a distance between patterns. However, one should be careful that such predicates are really useful for the approximation of protein functions.

- Disjunction of patterns: In the current implementation, no form is provided for the disjunction of patterns on the mapping from stochastic decision predicates to binary strings on the genetic algorithm. For example, the pattern "$CXXCH \vee AAQCH$" may be more appropriate since it eliminates three exceptions caused by Euglinae. Finding the pattern "$AAQCH$" is possible if we apply our algorithm to the protein sequences which eliminate the sequences that match "$CXXCH$". However, it should be noted that the pattern "$AAQCH$" is not so reliable since there are only three instances in the protein data base.

- More complex patterns: It is true that the patterns we used in our experiments are too simple to reflect protein functions. For example, it is a well known fact that in the heme-c binding motif "$CXXCH$", no histidine, cysteine, proline nor tryptophan occur in "XX" and that small amino acids tends to occur there. To represent such information, more complex stochastic motifs are required. Our early experience shows that hidden markov models (HMM) seem to be appropriate for this purpose.

Reducing hypothesis space: Since the MDL principle has a bias against selecting complex patterns, it is possible to eliminate complex patterns, for example, more than five patterns from the hypothesis space. However, we might overbias to the description length of Horn clauses. If this is true, we have to change the adjustment parameter, and also have to search a larger hypothesis space which may include complex patterns, with more than five patterns. In that case, genetic algorithms would be more powerful tools than conventional search algorithms.

9 Conclusion

The importance of stochastic approach for genetic information processing is described using a motif extraction system as an example. Our proposed methodology is characterized by the stochastic representation of motifs using stochastic decision predicates, the MDL principle to avoid overfitting and fast search algorithms using genetic algorithms. Our experimental results show that the methodology actually produces a computationally and biologically meaningful motif for cytochrome c, whose good predictive performance has been statistically proven by the cross validation method. We believe the methodology can also be applied to various kinds of discrimination problems in genetic information processing.

Acknowledgment The author wishes to express his sincere gratitude to Dr. K. Nitta of ICOT and to Dr. N. Koike of NEC Corporation for their encouragement and support. The author thanks Dr. K. Yamanishi and Mr. H. Kondo for their technical advice for the MDL principle and genetic algorithms. The author also thanks Mr. K. Yamagishi, Mr. S. Oyanagi, Mrs. A. Ikeda, Mrs. Y. Kobayashi and Miss K. Hikita for their great contribution to the analysis of real protein sequences.

References

[1] Aitken, Alastair, (1990). *Identification of Protein Consensus Sequences*, Ellis Horwood Series in Biochemistry and Biotechnology.

[2] Rissanen, J.(1978). Modeling by shortest data description. *Automatica, 14*, 465-471.

[3] Yamanishi, K.(1990). A learning criterion for stochastic rules. *Proceedings of the 3-rd Annual Workshop on Computational Learning Theory*, (pp. 67-81), Rochester, NY: Morgan Kaufmann. Its full version is to appear in Jr. on Machine Learning.

[4] Yamanishi, K. & Konagaya, A.(1991). Leaning Stochastic Motifs from Genetic Sequences. *in Proc. of the Eighth International Workshop of Machine Learning.*

[5] Rissanen, J.(1983). A universal prior for integers and estimation by minimum description length. *Annals of Statistics, 11*, 416-431.

[6] Goldberg,D.E., (1989). *Genetic Algorithms in Search, Optimization, and Machine Learning*, Addison-Wesley Publishing Company, Inc.

[7] Breiman, L., Friedman, J.H., Olshen, R.A., & Stone, C.J.(1984). Classification and regression trees. *Wadsworth Statistics/Probability Series.*

Appendix: How to Calculate Description Lengths of Stochastic Motifs

The description lengths are calculated as follows. Note that "log" denotes logarithm with base 2 in the following calculation. Let LL be description length of likelihood of categories C^N when the sequences S^N are given to the stochastic motif represented by a probability parameter θ and Horn clauses M. Let E_j be the set of sequences which are false for the $1,\cdots,j-1$th clauses and are true for the jth clause. Let N_j be the number of sequences in E_j and let N_j^+ be the number of sequences which are in E_j and belong to C_j, the category of the j−th clause. Then the likelihood of C^N when given S^N with respect to a stochastic decision predicate with a probability parameter θ and Horn clauses M, which we denote $P(C^N \mid S^N : \theta \prec M)$, is calculated as follows:

$$P(C^N \mid S^N : \theta \prec M) = \prod_{j=1}^{m} p_j^{N_j^+}(1-p_j)^{N_j - N_j^+}.$$

The description length LL is given by $-\log P(C^N \mid S^N : \hat{\theta} \prec M)$, which can be calculated, as follows:

$$LL = \sum_{i=1}^{m} N_i\{H(\tilde{p}_i) + D_{KL}(\tilde{p}_i \parallel \hat{p}_i)\} \tag{1}$$

where $\tilde{p}_i = N_i^+/N_i$ and $\hat{p}_i$ is an estimate of the true parameter p_i^*, which is set to be $\frac{N_i^+ + 1}{N_i + 2}$(the Bayes estimator). In addition, $H(\tilde{p}_i)$ and $D_{KL}(\tilde{p}_i \parallel \hat{p}_i)$ are entropy function and Kullback-Leibler divergence defined as follows: $H(\tilde{p}_i) = -\tilde{p}_i \log \tilde{p}_i - (1-\tilde{p}_i)\log(1-\tilde{p}_i)$, $D_{KL}(\tilde{p}_i \parallel \hat{p}_i) = \tilde{p}_i \log \frac{\tilde{p}_i}{\hat{p}_i} + (1-\tilde{p}_i)\log\frac{1-\tilde{p}_i}{1-\hat{p}_i}$

Let PL be the description length of the parameter $\hat{\theta} = (\hat{p}_1, \cdots \hat{p}_m)$ for a fixed Horn clauses M. Since the accuracy (variance) of the maximum likelihood estimator is $O(1/\sqrt{N})$, the description length PL is given by:

$$PL = \sum_{i=1}^{m} \frac{\log N_i}{2} \tag{2}$$

Let CL be the description length of the Horn clauses M.

In the motif extraction system, CL is given by:

$$\begin{aligned} CL &= \sum_{i=1}^{m}[\ \log^*(\sum_{j=1}^{k_i} h_j) + (\sum_{j=1}^{k_i} h_j - 1) \\ &\quad + \sum_{j=1}^{k_i}\sum_{l=1}^{h_j}\{\log\binom{L_l^j(i)}{X_l^j(i)} \\ &\quad + (L_l^j(i) - X_l^j(i)) * \log(\mid \mathcal{A} \mid -1)\} + \log r\ ] \end{aligned} \tag{3}$$

where $L_l^j(i)$ and $X_l^j(i)$ are the number of amino acids and of variables, respectively, in the pattern in the l−th predicate in the j−th disjunction region of the i−th clause. On the right-hand of (3), the first term denotes the description length of the number of *contain* predicates in the i−th clause. For any $d > 0$, $\log^* d$ denotes $\log d + \log\log d + \cdots$ where the sum is taken over all positive terms (Rissanen's integer coding scheme [5]). The second term of (3) denotes the description length of the sequence $\vee, \wedge, \wedge, \cdots$ in the i−th clause. The third term denotes the description length of the positions of variables in the pattern σ appearing in the predicate '$contain(S,\sigma)$.' The fourth term denotes the description length required to describe amino acids (not variables) included in the pattern σ appearing in the predicate '$contain(S,\sigma)$'. The last term $\log r$ denotes the description length of the category C appearing in the predicate '$motif(S,C)$'.

By summing (1), (2), and (3), we have the following description length DL:

$$DL \stackrel{\text{def}}{=} LL + \lambda\{PL + CL\}$$

where λ is the adjustment parameter.

TECHNICAL PAPERS

Learning via Query

On Learning Systolic Languages

Takashi YOKOMORI

Department of Computer Science
and Information Mathematics
University of Electro-Communications
1-5-1 Choufugaoka, Chofu, Tokyo 182, JAPAN
E-mail:yokomori@cs.uec.ac.jp

Abstract

We study the learning problem of systolic languages from queries and counterexamples. A systolic language is specified by a systolic automaton which is a kind of network consisting of uniformly connected processors(finite automata).

In this article, we show that the class of binary systolic tree languages is learnable in polynomial time from the learning protocol what is called minimally adequate teacher.

Since the class of binary systolic tree languages properly contains the class of regular languages, the main result in this paper gives a generalization of the corresponding Angluin's result for regular languages.

1 Introduction

In the recent research activity of inductive inference, Angluin has introduced the model of learning called *minimally adequate teacher* (MAT), that is, the model of learning via membership queries and equivalence queries, and has shown that the class of regular languages is efficiently learnable using deterministic finite automata (DFAs)([2]). More specifically, she has presented an algorithm which, given any regular language, learns from MAT a minimum DFA accepting the target in time polynomial in the number of states of the minimum DFA and the maximum length of any counterexample provided by the teacher.

Following Angluin's work, several extended results about polynomial-time MAT learnability for subclasses of context-free grammars have been reported ([3, 8, 14,

15]); however, the polynomial-time learnability of the whole class of context-free grammars is still open.

On the other hand, there has been a stream of research history on another type of automata where cellular automata were the fisrt target of research at the begining and multi-dimensional cellular arrays have took the place afterward. Among variations of multi-dimensional arrays, *systolic automata* have been introduced to study the formal properties of a large scaled multi-processor systems or VLSI systems. In fact, a numerous number of literatures on systolic automata and their algorithms have been reported with many applications such as various types of special-purpose VLSI processors for sorting, pattern matching, matrix computation, image processing, and so forth. Thus, systolic automata are generally taken as one of the most promising models of computation for parallel processing([5]).

The present paper deals with systolic automata whose underlying structure of processors is given as a binary tree, called *binary systolic tree automata*(BSTAs; [7, 6]), and considers the learning problem for the class of languages accepted by BSTAs from MAT mentioned above.

We show that the class of binary systolic tree languages(BSTLs) accepted by BSTAs is learnable in polynomial time from MAT. That is, we present an algorithm which, given a BSTL L, learns a BSTA M accepting L in time polynomial in n and m, where n is the number of states of a minimum BSTA equivalent to M and m is the maximum length of any counterexample provided during the learning process. This provides a generalization of the corresponding result for regular languages.

2 Preliminaries

2.1 Definitions and Notation

We assume the reader to be familiar with the rudiments of formal language theory(see, e.g., [4] or [12]).

For a given finite alphabet Σ, the set of all strings with finite length (including *zero*) is denoted by Σ^*. (An *empty* string is denoted by λ.) $lg(w)$ denotes the length of a string w. Σ^+ denotes $\Sigma^* - \{\lambda\}$. A *language* L over Σ is a subset of Σ^*.

For any set S, $|S|$ denotes the cardinality of S.

A *nondeterministic binary systolic tree automaton*(NBSTA) is denoted by $M = (\Sigma, \Sigma_o, \Sigma_o', \sharp, f, g)$, where the Σs are finite alphabets such that $\Sigma_o' \subseteq \Sigma_o$, $\sharp$ is a symbol not in Σ, and f and g are mappings of Σ_o^2 and $\Sigma \cup \{\sharp\}$ into 2^{Σ_o} and Σ_o, respectively.

The domain of f is extended to the set : $\bigcup_{n\geq 1} \Sigma_o^{2^n} = A$ as follows: Consider a string $x \in A$, then there exists $n \geq 1$ such that $lg(x) = 2^n$. Let $x = y_1 y_2 \cdots y_{2^{n-1}}$, where each y is a string of length 2. We apply f to each y, yielding a set of new strings consisting of all x_1 such that

$$x_1 \in f(y_1)f(y_2)\cdots f(y_{2^{n-1}}).$$

If the length of x_1 is greater than 1(i.e., $lg(x) > 2$), then we repeat the same procedure. Continuing in this manner, we finally obtain a set of strings of length 1 Z, i.e., $Z \subseteq \Sigma_o$. We then write $F(x) = Z$. In this manner, we obtain an extension (a mapping) F defined on A.

A mapping g is a coding from $(\Sigma \cup \{\sharp\})$ to Σ_o. A string $w \in \Sigma^*$ of length more than 1 is accepted by M iff $F(g(w\sharp^i)) \cap \Sigma_o' \neq \emptyset$, where $i(\geq 0)$ is the unique integer such that $lg(w\sharp^i) = 2^n$ for some $n \geq 1$, and $lg(w) > i$. A string w of length 1(the empty string λ) is accepted by M iff $g(w)$(resp., $g(\sharp)$) $\in \Sigma_o'$. (Thus, Σ_o' designates the set of accepting states.)

In a usual manner, we can define a deterministic version of NBSTAs by modifying f in an obvious way. Then, *deterministic* BSTAs(DBSTAs) are obtained as special forms of NBSTAs.

The set of strings accepted by M is denoted by $L(M)$ and is termed the *language accepted by* M. A language L is a *binary systolic tree language*(BSTL) iff $L = L(M)$ for some DBSTA M.

By $|\Sigma_o|$ we define the size of a NBSTA M, denoted by *size*(M). M is *minimum* iff *size*(M) is minimum, that is, for any NBSTA M' that is equivalent to M, *size*$(M) \leq$ *size*(M') holds.

Since we are concerned with the learning problem of BSTLs, without loss of generality, we may restrict our consideration to only λ-free languages.

Proposition 1 ([6]) *For any language L, L is accepted by a NBSTA iff it is accepted by a DBSTA.* □

Thus, in what follows, we may assume that any BSTL L can be specified by an NBSTA M. We simple write BSTA for NBSTA, and explicitly write DBSTA for a deterministic BSTA.

Example 1. Consider a DBSTA $M = (\{a,b,c\},\{A,B,C,\},\{C\},\sharp,f,g)$, where f and g are defined by the tables below:

f	A	B	C	$\sharp$
A	A	C	C	$\sharp$
B	$\sharp$	B	$\sharp$	$\sharp$
C	$\sharp$	C	$\sharp$	$\sharp$
$\sharp$	$\sharp$	$\sharp$	$\sharp$	$\sharp$

x	a	b	c	$\sharp$
$g(x)$	A	B	C	$\sharp$

For example, a string $w = aaabbbbb$ is accepted by M as shown in (a) of Figure 1, while strings $abab$, $aab\sharp$ are not, as shown in (b) and (c). It is seen that M accepts only strings of length 2^i for some $i \geq 1$. The language $L(M)$ accepted by M is $\{a^m b^n | m, n \geq 1, m + n = 2^i \text{ for some } i \geq 1\}$. This language is shown not to be even context-free. □

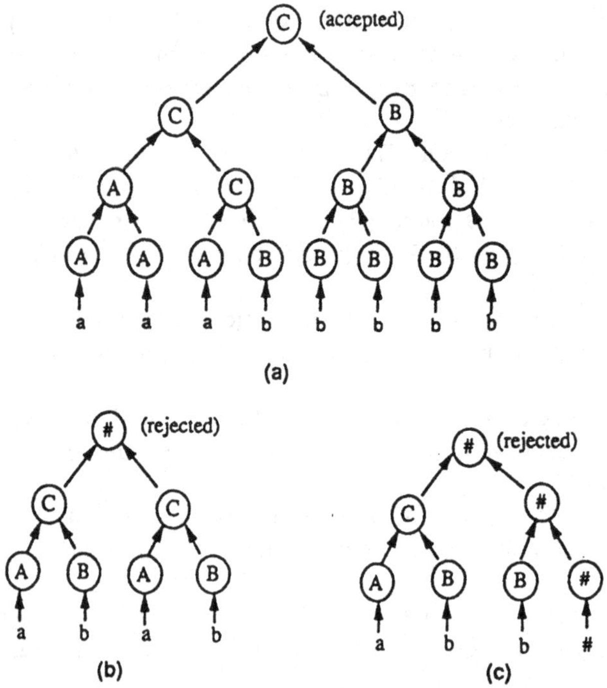

Figure 1. Computation Trees

Let $M = (\Sigma, \Sigma_o, \Sigma'_o, \sharp, f, g)$ be a BSTA. For $w \in \Sigma^*$, and $X \in \Sigma_o$, we denote by $\mathrm{Tr}(X, w)$ a *computation tree* which means a tree rooted by X representing the process of computing w by M(see Figure 1). When only X is emphasized, we denote it by $\mathrm{Tr}(X)$. Moreover, we simply write $\mathrm{Tr}(w)$ if X is clear or not necessary. For any tree T, $\mathrm{fr}(T)$ denotes a string over $\Sigma \cup \{\sharp\}$ at the frontier(leaves) of T.

In what follows, we often take f as a set of transition rules rather than a mapping, and denote $z \in f(x, y)$ by $(x, y) \xrightarrow[f]{} z$.

As a known result, we give the following.

Proposition 2 ([6]) *The class of BSTLs properly contains the class of regular languages. Further, the class is closed under Boolean operations, i.e., union, concatenation, and complement.* □

Further, it is easy to prove the following theorem.

Theorem 3 *For any BSTL L, there effectively exists a BSTA $M = (\Sigma, \Sigma_o, \Sigma'_o, \sharp, f, g)$ such that $L = L(M)$ and g is injective.* □

Hence, by fixing g as $g(x) = [x]$(for $\forall x \in \Sigma \cup \{\sharp\}$), in the sequel, we may denote a BSTA by $M = (\Sigma, \Sigma_o, \Sigma'_o, \sharp, f)$, where it is assumed that $g(\Sigma \cup \{\sharp\}) = \{[x] | x \in \Sigma \cup \{\sharp\}\} \subseteq \Sigma_o$.

2.2 MAT Learning

Let L_* be a target language to be learned over a fixed alphabet Σ. We assume the following types of queries in the learning process.

A *membership query* proposes a string $x \in \Sigma^*$ and asks whether $x \in L_*$ or not. The answer is either *yes* or *no*. An *equivalence query* proposes a BSTA M and asks whether $L_* = L(M)$ or not. The answer is *yes* or *no*, and in the latter case together with a counterexample w in the symmetric difference of L_* and $L(M)$. A counterexample w is *positive* if it is in $L_* - L(M)$, and *negative* otherwise.

The learning protocol consisting of membership queries and equivalence queries is called *minimally adequate teacher*(MAT). The purpose of the learning algorithm is to find a BSTA $M = (\Sigma, \Sigma_o, \Sigma'_o, \sharp, f)$ such that $L_* = L(M)$ with the help of the minimally adequate teacher.

3 Main Results

3.1 Learning BSTAs

[Convention] Throughout this section, we assume that for a target BSTL L_*, let $M_* = (\Sigma, \Sigma_{o*}, \Sigma'_{o*}, \sharp, f_*)$ be a BSTA such that $L_* = L(M_*)$. For a string $w \in \Sigma^*$, by the *hat version* $\hat{w}$ we denote the unique string $w\sharp^i$, where $i (\geq 0)$ is the smallest such that $lg(w\sharp^i) = 2^n$ for some $n \geq 0$. Further, let $\hat{\Sigma} = \Sigma \cup \{\sharp\}$.

(1) Introducing new states

Given a *positive* counterexample w of L_*, let $S(\hat{w})$ be a set of new states consisting of all leaves of the subtrees of the computation tree $\mathrm{Tr}(\hat{w})$ for $\hat{w}$ by M_*, that is,

$$S(\hat{w}) = \{ [x] \mid x = \mathrm{fr}(T) \text{ and } T \text{ is a subtree of } \mathrm{Tr}(\hat{w})\}.$$

The following lemma obviously holds.

Lemma 4 Let w be in $L_* = L(M_*)$. Then, for any state $X \in \Sigma_{o*}$ that appears in the computation tree $\mathrm{Tr}(\hat{w})$, there is an $[x]$ in $S(\hat{w})$ such that $\mathrm{fr}(\mathrm{Tr}(X)) = x$. □

(2) Constructing new candidate rules

Suppose that we have a conjectured BSTA $M = (\Sigma, \Sigma_o, \Sigma'_o, \sharp, f)$ at the current stage of learning process. From the set of new states $S(\hat{w})$ produced above and the current set of states Σ_o, we newly construct new candidate transition mapping f_{new} as follows:

$$([u],[v]) \xrightarrow[f_{new}]{} [z], \qquad \text{where } [u],[v],[z] \in \Sigma_o \cup S(\hat{w}), \text{ and at least one of the three is in } S(\hat{w}).$$

As seen below, a mapping f_{new} is added to the current mapping f to obtain new conjectured BSTA for the next stage of learning.

(3) Diagnosing transition mapping f

Let $M = (\Sigma, \Sigma_o, \Sigma'_o, \sharp, f)$ be a conjectured BSTA. For a string $w \in \Sigma^*$, let $\mathrm{Tr}([z], \hat{w})$ be a computation tree for w by M, where $r : ([u],[v]) \underset{f}{\rightarrow} [z]$ is a transition rule used at the top of $\mathrm{Tr}([z], \hat{w})$, and $x = \mathrm{fr}(\mathrm{Tr}([u]))$, $y = \mathrm{fr}(\mathrm{Tr}([v]))$. Then, a tree $\mathrm{Tr}([z], \hat{w})$ is *correct for L_** iff $z \notin L_*\sharp^*$ or $\hat{w}(= xy) \in L_*\sharp^*$. A comutation tree T (by M) is *incorrect for L_** iff it is not correct for L_*. Further, a rule r is *incorrect for L_** iff $\mathrm{Tr}([u])$ and $\mathrm{Tr}([v])$ are correct for L_*, but $\mathrm{Tr}([z])$ is incorrect for L_*. Finally, we define that for $\forall u \in \hat{\Sigma}$, $\mathrm{Tr}([u])$ is always correct for L_*.

Now, given a *negative* counterexample $w'(\in L(M) - L_*)$, the algorithm has to modify the conjectured BSTA so that it may not accept w' by removing wrong rules. In order to determine such wrong rules, the algorithm calls the diagnosing procedure whenever a negative counterexample is provided. Since w' is a negative counterexample, i.e., it is accepted by the current conjecture M, there is a computation tree $\mathrm{Tr}([z], \widehat{w'})$ for $\widehat{w'}$.

The diagnosing algorithm is a sepcial version of Shapiro's *contradiction backtracing algorithm* ([13]). This procedure takes as an input a computation tree $\mathrm{Tr}([z], \widehat{w'})$ and outputs a transition rule r in f of M which is incorrect for L_*, where $w' \in L(M) - L_*$.

Assume that $\mathrm{Tr}([u])$ and $\mathrm{Tr}([v])$ be the subtrees of $\mathrm{Tr}([z], \widehat{w'})$ such that $x = \mathrm{fr}(\mathrm{Tr}([u]))$, $y = \mathrm{fr}(\mathrm{Tr}([v]))$ and $\widehat{w'} = xy$, where $([u],[v]) \underset{f}{\rightarrow} [z]$ is a transition rule used at the top of $\mathrm{Tr}([z], \widehat{w'})$. Note that $\widehat{w'}$ is not in $L_*\sharp^*$ and that since w' is accepted by M, z is in $L_*\sharp^*$.

procedure diag($\mathrm{Tr}(w')$)
  If $\widehat{w'} = xy \in \hat{\Sigma}^2$,
  then output $r : ([u],[v]) \underset{f}{\rightarrow} [z]$ and halts;
  else make a disjunctive membership query
    if "$[u \notin L_*\sharp^*]$ or $[x \in L_*\sharp^*]$" holds true ? ;
    if the answer is "*no*"
    then call **diag**($\mathrm{Tr}(x)$) ;
    else make a disjunctive membership query
      if "$[v \notin L_*\sharp^*]$ or $[y \in L_*\sharp^*]$" holds true ? ;
      if the answer is "*no*"
      then call **diag**($\mathrm{Tr}(y)$) ;
      else output $r : ([u],[v]) \underset{f}{\rightarrow} [z]$ and halts

We are now ready to prove the correctness of the diagnosing algorithm.

Lemma 5 *Given a negative counterexample w' and a conjectured BSTA M, the diagnosing algorithm always halts and outputs a rule incorrect for L.*

Proof. It is clear that since the procedure **diag** is recursively called with a proper subtree of the computation tree $\mathrm{Tr}(\widehat{w'})$, the algorithm always halts and outputs some rule in f.

Let $r : ([u],[v]) \xrightarrow[f]{} [z]$ be a rule returned by the diagnosing algorithm, where $x = \mathrm{fr}(\mathrm{Tr}([u]))$, $y = \mathrm{fr}(\mathrm{Tr}([v]))$. It should be noted that from the property of the input for **diag**, a tree $\mathrm{Tr}([z])$ is always incorrect for L_*.

If x and y are in $\hat{\Sigma}$, then by definition, we have that $x = u$, $y = v$, and that $\mathrm{Tr}([u])$ and $\mathrm{Tr}([v])$ are correct for L_*. Thus, a rule r is incorrect for L_*.

Assume the other case. Then, since this is the first time the answers of membership queries are both "*yes*", at this moment we have that $([u \notin L_*\sharp^*]$ *or* $[x \in L_*\sharp^*])$ *and* $([v \notin L_*\sharp^*]$ *or* $[y \in L_*\sharp^*])$. In any of these four cases, it holds true that $\mathrm{Tr}([u])$ and $\mathrm{Tr}([v])$ are correct for L_*. Hence, a rule r is incorrect for L_*. □

(4) Learning algorithm *LESA*

The following is a learning algorithm *LESA*(*Learning Systolic Automata*) for BSTLs:

Input : a BSTL L_* over a fixed Σ
Output : a BSTA M such that $L_* = L(M)$
Procedure :
  initialize $M = (\Sigma, \emptyset, \emptyset, \sharp, f)$, where f is defined by null values;
  repeat
    make an equivalence query to the current $M = (\Sigma, \Sigma_o, \Sigma_o', \sharp, f)$;
    If the answer is "*yes*" **then** output M and halts
    else if the answer is a *positive* counterexample w
    then introduce the set of new states $S(\hat{w})$ from w;
      $\Sigma_o := \Sigma_o \cup S(\hat{w})$;
      $\Sigma_o' := \Sigma_o' \cup \{[\hat{w}]\}$;
      construct the set of new rules f_{new};
      $f := f \cup f_{new}$;
    else (the answer is a *negative* counterexample w')
      construct a computation tree $\mathrm{Tr}(\widehat{w'})$;
      call **diag**$(\mathrm{Tr}(\widehat{w'}))$ to find an incorrect rule r;
      $f := f - \{r\}$

3.2 The Correctness and Time Analysis of *LESA*

[Correctness]

In orde to prove the correctness of *LESA*, We first show that if a string w' which is not in L_* is accepted by the conjectured BSTA M, then there exists at least one incorrect rule for L_* in f, that is, if f has no incorrect rule for L_*, then M accepts no string w' such that $w' \notin L_*$, that is, it holds true that $L(M) \subseteq L_*$.

Before proceeding to the next lemma, the following fact should be noted. For $w \in \Sigma^*$, let $\mathrm{Tr}([z], \hat{w})$ be a computation tree by a conjecture M. Then, w is in $L(M)$ iff z is in $L_*\sharp^*$.

Lemma 6 *Let* $M = (\Sigma, \Sigma_o, \Sigma_o', \sharp, f)$ *be a conjectured BSTA and* L_* *be a target language. If all of the rules in* f *are correct for* L_*, *then it holds true that* $L(M) \subseteq L_*$.

Proof. Let w be a string such that $w \in L(M)$. By the induction on the height i of the computation tree $\mathrm{Tr}([z], \hat{w})$ for $\hat{w}(lg(\hat{w}) = 2^i)$, we show that w is also in L_*.

Suppose $i = 1$, that is, $\hat{w} = ab$. Then, from the manner of constructing M, it must hold that $([a],[b]) \underset{f}{\rightarrow} [z]$, where $z \in L_*\sharp^*$. If ab is not in $L_*\sharp^*$, a rule $([a],[b]) \underset{f}{\rightarrow} [z]$ is incorrect for L_*, which contradicts the assumption on f. Hence, $\hat{w}(= ab)$ is in $L_*\sharp^*$. Thus, w is in L_*.

Assume that the lemma holds for any w' such that $lg(\widehat{w'}) < lg(\hat{w})$. Further, assume that $\mathrm{Tr}([z], \hat{w})$ has its subtrees $\mathrm{Tr}([u], x)$ and $\mathrm{Tr}([v], y)$, where $\hat{w} = xy$.

We claim that trees $\mathrm{Tr}([u], x)$ and $\mathrm{Tr}([v], y)$ are correct for L_*. (Because suppose, otherwise, that $\mathrm{Tr}([u], x)$ is incorrect for L_*, that is, $u \in L_*\sharp^*$ and $x \notin L_*\sharp^*$. By the induction hypothesis, $x \in L(M)\sharp^*$ implies that $x \in L_*\sharp^*$, contradicting that $x \notin L_*\sharp^*$. Hence, it must hold that $x \notin L(M)\sharp^*$. However, this implies that $u \notin L_*\sharp^*$, again contradicting that $u \in L_*\sharp^*$. Hence, $\mathrm{Tr}([u], x)$ is correct for L_*. The same argument can apply to $\mathrm{Tr}([v], y)$. Thus, the claim holds true.)

Now, let us suppose that $\hat{w}(= xy)$ is not in $L_*\sharp^*$. Then, since z is in $L_*\sharp^*$, a tree $\mathrm{Tr}([z], \hat{w})$ is incorrect for L_*. Hence, we have a rule $([u],[v]) \underset{f}{\rightarrow} [z]$ incorrect for L_*, contradicting the assumption on f. Thus, it is proved that $\hat{w}$ is also in $L_*\sharp^*$. This completes the proof. □

Let $M_* = (\Sigma, \Sigma_{o*}, \Sigma'_{o*}, \sharp, f_*)$ be a minimum BSTA such that $L_* = L(M_*)$.

As a terminology, for states $[x] \in \Sigma_o$ and $X \in \Sigma_{o*}$, we say that $[x]$ is *well-corresponding to X* iff there exists a computation tree $\mathrm{Tr}(X, x)$ for x by M_*. A rule $r : ([x],[y]) \underset{f}{\rightarrow} [z]$ is *well-corresponding to* a rule $(X, Y) \underset{f_*}{\rightarrow} Z$ in f_* iff $[x]$, $[y]$ and $[z]$ are well-corresponding to X, Y and Z, respectively.

Lemma 7 *The number of positive examples needed to identify a correct BSTA is at most $|\Sigma_{o*}|$(the cardinality of the state set of M_*).*

Proof. Let $M = (\Sigma, \Sigma_o, \Sigma'_o, \sharp, f)$ be a conjectured BSTA from *LESA*.

For $w \in L_*$, let $S_*(\hat{w})$ be the set all the states in Σ_{o*} appearing in the computation tree $\mathrm{Tr}(\hat{w})$ for w by M_*. By the nature of *LESA*, if a positive counterexample w is given, then there exists at least one state p that is not contained in Σ_o but needed to accept w. By Lemma 4, $S(\hat{w})$ contains all states well-corresponding to the states in $S_*(\hat{w})$ of M_*. Further, for each new state p in $S(\hat{w})$, all the rules containing p are added to the rule set f, and only incorrect rules in f are removed by the diagnosing procedure.

To sum up, whenever a positive counterexample w is given, there exists at least one state p in $S(\hat{w}) - \Sigma_o$ that is necessary for accepting w and is well-corresponding to some state in Σ_{o*}. Thus, this justifies the claim and completes the proof. □

Note that this lemma implies that after receiving at most $|\Sigma_{o*}|$ number of positive counterexamples, Σ_o includes sufficient number of states to accept the target language L.

When at most $|\Sigma_{o*}|$ number of positive counterexamples are given, the set of transition rules f includes all the rules that are well-corresponding to those in f_*.

In other words, any string in L_* is accepted by the conjectured BSTA M. Thus, at that time it holds that $L_* \subseteq L(M)$, and hence no more positive counterexample is required and provided. By Lemma 5, each time a negative counterexample is given, one incorrect rule for L_* is determined and removed from f of the current conjecture M. Further, we know that only incorrect rule is removed at any stage. Therefore, the number of required negative counterexamples is at most the maximum number of rules of conjectured BSTA. By Lemma 6, if all incorrect rules are removed, the resulting conjectured BSTA M accepts no string which is not in L_*, that is, $L(M) \subseteq L_*$.

From these facts described above, we conclude that *LESA* always converges and outputs a correct BSTA. Thus, we obtain the main theorem;

Theorem 8 *For any BSTL L_*, the learning algorithm LESA eventually terminates and outputs a BSTA M accepting L_*.* □

[Time Analysis]

We note that for a positive counterexample w, the number of states newly introduced from w in *LESA* is obviously bounded by $O(lg(\widehat{w})^2)$, i.e., it holds that $|S(\widehat{w})| \leq O(lg(\widehat{w})^2)$.

Let m be the maximum length of any counterexample provided during the learning process. Further, let $n(= |\Sigma_{o*}|)$ be the number of states of a minimum BSTA accepting a target L_*.

Lemma 9 *The total number $|f_{total}|$ of all rules introduced in the learning algorithm is bounded by $O(n^3m^6)$.*

Proof. Let $|\Sigma_{o,max}|$ be the maximum number of states of any conjectured BSTA. Let $\{w_1, ..., w_t\}$ be the set of positive counterexamples provided by *LESA* in the entire process of learning. (Note that, by Lemma 7, t is at most $|\Sigma_{o*}|$.) Then, from the above observation, $|\Sigma_{o,max}| \leq \sum_{i=1}^{t} O(lg(\widehat{w_i})^2) \leq O(|\Sigma_{o*}|m^2)$ is obtained. (Note that $lg(\widehat{w_i}) < 2lg(w_i)$.) Further, from the manner of constructing f,

$$\begin{aligned} |f_{total}| &\leq |\Sigma_{o,max}|^3 \\ &\leq O((|\Sigma_{o*}|m^2))^3 \\ &= O(n^3m^6) \end{aligned}$$

is obtained. □

Now, suppose a negative counterexample w' is given. Then, the learning algorithm *LESA* constructs the computation tree $\mathrm{Tr}(\widehat{w'})$ by parsing $\widehat{w'}$ via the conjectured BSTA $M = (\Sigma, \Sigma_o, \Sigma'_o, \sharp, f)$ at that time. Since there exists an algorithm that constructs a computation tree for $\widehat{w'}$ in time proportional to $|f|\,\mathrm{lg}(\widehat{w'})^3$, each parsing procedure requires at most $O(|f_{total}|m^3)$ times.

It is obvious that the total time complexity of *LESA* is dominated by that of *parsing* negative counterexamples and to obtain the computation trees for them.

Each time a negative counterexample is provided, the number of transition rules f of the conjectured BSTA from *LESA* is reduced exactly by one. Further, by Lemma 9, the number of transition rules f of any conjecture is at most k, where $k = O(n^3m^6)$. Hence, the total time required for parsing is bounded by

$$km^3 + (k-1)m^3 + \cdots + m^3 = \frac{1}{2}k(k+1)m^3.$$

Thus, we have :

Theorem 10 *The total running time of LESA is bounded by a polynomial in m the maximum length of any counterexample provided during the learning process and n the number of states of a minimum BSTA for a target language.* □

4 Discussions

We have shown that the class of binary systolic tree languages(BSTLs) accepted by BSTAs is learnable in polynomial time from MAT. That is, we have presented an algorithm *LESA* which, given a BSTL L, learns a BSTA M accepting L in time polynomial in n and m, where n is the number of states of a minimum BSTA equivalent to M and m is the maximum length of any counterexample provided during the learning process. This provides a generalization of the corresponding result for regular languages.

The algorithm *LESA* can induce in a straightforward manner its modified version which is viewed as a learning algorithm on the "in the limit" basis. That is, we can think of *LESA* as an algorithm which learns a correct BSTA *in the limit* from membership queries and the *complete presentation*(i.e., positive and negative examples) of a target language. Further, it learns any BSTL L *consistently, conservatively*, and *responsively*, and may be implemented to run in time polynomial in n (the number of states of a minimum BSTA for L) and m (the maximum length of any example provided so far). Note that Angluin's algorithm for learning regular languages also has its variation of this type([2]).

There are not a few related works: Among others, firstly, the basic strategy of *LESA* comes from [1] in which the learnability of k-bounded context-free grammars is shown from equivalence queries and *nonterminal* membership queries. [8] proposes an algorithm for learning simple deterministic languages based on the similar type of learning strategy. This learning strategy is also used in [14] and [16] to show the MAT learnability for the subclass of context-free languages called context-deterministic languages and for the class of regualr languages, respectively. The class of context-deterministic languages contains all regular languages and is incomparable to any other subclass with the MAT learnability result. In [16], instead of DFAs, a proposed algorithm learns NFAs in polynomial time in the size of the minimum DFA for the target and the maximum length of any counterexample.

Table summarizes the known MAT learnability and related results for various classes of languages.

	Language class	Representations	Learning Protocol
Angluin (1987)	regular	DFA	MAT
Angluin (1987)	context-free	k-bounded CFG	MAT+α
Bermann & Roos(1987)	one-counter	DOCA	MAT
Takada (1988)	even-linear	(Reduction to DFA)	MAT
Ishizaka (1989)	simple deterministic	CFG	MAT
Sakakibara (1990)	context-free	CFG	MAT on structured strings
Sakakibara (1990)	a subclass of context-sensitive	EFS	MAT + predicate oracle
Nishino (1990)	a subclass of context-sensitive	LFG	MAT + predicate oracle
Shirakawa & Yokomori(1992)	c-deterministic	CFG	MAT
Yokomori (1992)	regular	NFA	MAT
this article	BSTL	BSTA	MAT

(not exhaustive)

It is an interesting problem to be investigated how far we can go along the line of a learning strategy proposed here in the MAT learning paradigm.

Further, exploring a subclass of BSTLs efficiently learnable in the limit from positive data is another interesting subject, and in fact, such a subclass called *reversible* BSTLs is introduced and shown to be efficiently learnable in the limit from positive data([17]).

Acknowledgements

This work is supported in part by Grants-in-Aid for Scientific Research No. 04229105 from the Ministry of Education, Science and Culture, Japan.

References

[1] D. Angluin. Learning k-bounded context-free grammars. Res. Rep. 557, Dept. of Comput. Sci., YALE Univ, 1987.

[2] D. Angluin. Learning regular sets from queries and counterexamples. *Information and Computation*, 75:87–106, 1987.

[3] P. Berman and R. Roos. Learning one-counter languages in polynomial time. In *28th IEEE Symp. on FOCS*, pages 61–67, 1987.

[4] M. A. Harrison. *Introduction to Formal Language Theory.* Addison-Wesley, Reading, MA, 1978.

[5] H.T.Kung, R.F.Sproull, and G.L.Steel Jr.(eds.). *VLSI Systems and Computations.* Computer Science Press, 1981.

[6] K. Culik II, J. Gruska, and A. Salomaa. Systolic automata for VLSI on balanced trees. *Acta Informatica*, 18:335–344, 1983.

[7] K. Culik II, A. Salomaa, and D. Wood. Systolic tree acceptors. *R.A.I.R.O. Theoretical Informatics*, 18:53–69, 1984.

[8] H. Ishizaka. Polynomial time learnability of simple deterministic languages. *Machine Learning*, 5:151–164, 1990.

[9] T. Nishino. Mathematical analysis of lexical-functional grammars—complexity, parsability, and learnability. In *Proceedings of Seoul International Conference on Natural Language Processing*, 1990.

[10] Y. Sakakibara. Learning context-free grammars from structural data in polynomial time. *Theoretical Computer Science*, 76:223–242, 1990.

[11] Y. Sakakibara. On learning smullyan's elementary formal systems: Towards an efficient learning for context-sensitive languages. *Advances in Software Science and Technology*, 2:79–101, 1990.

[12] A. Salomaa. *Formal Languages.* Academic Press, New York, NY, 1973.

[13] E. Shapiro. Inductive inference of theories from facts. Res. Rep. 192, Dept. of Comput. Sci., YALE Univ, 1981.

[14] H. Shirakawa and T. Yokomori. Polynomial-time MAT learning of c-deterministic context-free grammars. Res. Rep. 92-04, Dept. of Comput. and Inform. Math., Univ. of Electro-Communications, 1992.

[15] Y. Takada. Grammatical inference for even linear languages based on control sets. *Information Proccesing Letters*, 28:193–199, 1988.

[16] T. Yokomori. Learning non-determinisitc finite automata from queries and counterexamples. (*to appear*) In *Proceedings of International Workshop on Machine Intelligence '92,* Glasgow, August, 1992.

[17] T. Yokomori. On learning systolic languages. Res. Rep. 92-07, Dept. of Comput. and Inform. Math., Univ. of Electro-Communications, 1992.

A Note on the Query Complexity of Learning DFA *

(Extended Abstract)

José L. Balcázar *Josep Díaz* *Ricard Gavaldà*
Dept. Llenguatges i Sistemes Informàtics
Universitat Politècnica Catalunya
Pau Gargallo 5, 08028 Barcelona, Spain

Osamu Watanabe
Department of Computer Science
Tokyo Institute of Technology
Meguro-ku, Tokyo 152, Japan

Abstract. It is known [1] that the class of deterministic finite automata is polynomial time learnable by using membership and equivalence queries. We investigate — the query complexity — the number of membership and equivalence queries for learning deterministic finite automata. We first show two lower bounds in two different learning situations. Then we investigate the query complexity in general setting, and show some trade-off phenomenon between the number of membership and equivalence queries.

1. Introduction

Query learning was introduced by Angluin [1] and is currently one of the most important models in computational learning theory. It differs from other models, such as PAC-learning or inductive inference, in that the learning process, the *learner*, obtains information about the concept to learn by making *queries* to some *teacher*.

The problem we study in this paper is that of inferring deterministic finite automata. Pitt [5] surveys the status of this important problem in several learning models. For the case of query learning, Angluin proved an important positive result.

Proposition 1.1. [1] There exists a polynomial time (Mem,Equ)-learner for dfa.

*This research was partially supported by the ESPRIT II Basic Research Actions Program of the EC under contract No. 3075 (project ALCOM). The third author was supported in part by the National Science Foundation under grant CCR89-13584 while visiting University of California, Santa Barbara. The fourth author was supported in part by Takayanagi Foundation of Electronics and Science Technology. E-mail addresses: balqui@lsi.upc.es, diaz@lsi.upc.es, gavalda@lsi.upc.es, watanabe@cs.titech.ac.jp

Remark. A *(Mem,Equ)-learner* is a learner that makes two types of queries: membership queries (Mem) and equivalence queries (Equ). Also *dfa* is an abbreviation for "deterministic finite automaton/automata".

On the other hand, Angluin also showed that neither membership query nor equivalence query alone is good enough to learn dfa in polynomial time.

Proposition 1.2.
(1) [2] No polynomial time (Mem)-learner exists for dfa.
(2) [3] No polynomial time (Equ)-learner exists for dfa.

In this paper, extending the proof techniques for the above results, we investigate the number of membership and equivalence queries — the query complexity — for learning dfa.

We show lower bounds in two different learing situations. First we investigate the number of membership queries necessary for *randomized* learners to learn dfa by using only membership queries. It is shown that randomized computation cannot improve the deterministic lower bound [2]; that is, every randomized learner must ask exponentially many membership queries for learning dfa. Secondly, we consider the situation where learners can alternate between membership and equivalence queries only a constant number of times, and prove again an exponential lower bound on the number of queries.

Next we consider general case, i.e., the case where no restriction (except the number of queries) is assumed on the way of asking membership and equivalence queries. Here we prove some "trade-off" between the number of membership and equivalence queries.

It is easy to see that Angluin's dfa learning algorithm (witnessing Proposition 1.1) needs at most n equivalence queries for learning n state dfa. We first show that it is possible to reduce the number of equivalence of queries, say, to $n/f(n)$ while increasing that of membership queries by $2^{f(n)}$ factor. On the other hand, it is also proved that one cannot reduce equivalence queries to $n/f(n)$ queries without increasing membership queries approximately $2^{O(f(n))}$. Thus, for example, while we can construct a polynomial time dfa learning algorithm that asks $n/c\log n$ equivalence queries, it is impossible to reduce equivalence queries more than $O(\log n)$ factor without using exponential number of membership queries.

The results in this paper are all proved in the notion of "bounded learning" [8, 9]. Roughly speaking, the goal of a learning algorithm in bounded learning is to output an hypothesis that only needs to be correct up to a given length, given as an input parameter. This learning notion is somewhat different from the original notion studied in Angluin's papers, but we take this learning notion because we can avoid tedious and minor problems in the original notion. (See [9] for the justification of the bounded learning notion.) However, the concept classes that we use to prove our lower bounds are finite and have a fixed length, thus our results also hold under Angluin's learning notion.

In this abstract some proofs are omitted. See [4] for the complete proofs.

2. Preliminaries

In this paper we follow standard definitions and notations in formal language theory and computational complexity theory; in particular, those for finite automata are used without definition. The reader will find them in standard textbooks.

Let Σ denote $\{0,1\}$, and throughout this paper, we use Σ for our alphabet. The length of a string x is denoted by $|x|$. The cardinality of a finite set A is written as $\|A\|$. Symbols $A^{\leq m}$ and $A^{=m}$ are used to denote the sets $\{x \in A : |x| \leq m\}$ and $\{x \in A : |x| = m\}$ respectively. By "deterministic finite automaton" we mean a "complete" deterministic finite automaton over Σ^*. That is, we assume that a state transition function is total.

Notions and Notations for Query Learning

We briefly explain the notions and notations for discussing query learning formally. We basically follow the style established in [8, 9].

A learning problem is specified by using "representation class" [6]. A ***representation class*** is a triple (R, Φ, ρ), where $R \subseteq \Sigma^*$ is a *representation language*, $\Phi : R \to 2^{\Sigma^*}$ is a *semantic function* or *concept mapping*, and $\rho : R \to \mathbf{N}$ is a *size function*. For example, a representation class for dfa is formally defined as follows: $DFA = (R_{\mathrm{dfa}}, \Phi_{\mathrm{dfa}}, \rho_{\mathrm{dfa}})$, where R_{dfa} is the set of dfa encoded in Σ^*, and for any $r \in R_{\mathrm{dfa}}$, $\Phi_{\mathrm{dfa}}(r)$ and $\rho_{\mathrm{dfa}}(r)$ are respectively the set accepted by dfa r and the number of states in dfa r. Following common convention, we write $\Phi_{\mathrm{dfa}}(r)$ as $L(r)$ and $\rho_{\mathrm{dfa}}(r)$ as $|r|$.

Our computation model for learning is the "learning system". A *learning system* $\langle S, T\rangle$ is formed by a *learner* S and a *teacher* T that are organized as in Figure 1.

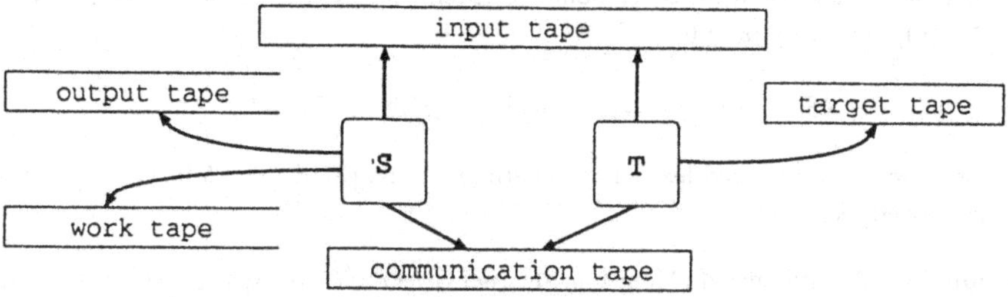

Figure 1: Learning System

The tapes except the communication tape and the target tape are used in a ordinary way. The communication tape is used for the communication between S and T. That is, a query from S and a answer from T are written on it. The target tape is used to keep a representation r of a target concept. The situation where r being on the target tape intuitively means that T knows the concept that is represented by r. A teacher T who knows r is written as "$T(r)$". Prior to the execution, some input ω and a target representation r are given respectively on the input tape and the target tape. Then the computation of $\langle S, T\rangle$ (which is written as $\langle S, T(r)\rangle(\omega)$) starts from S, executes S and T in turn, and finally halts at S. If S outputs y on its output tape and halts normally, then we say that $\langle S, T(r)\rangle(\omega)$ *outputs* y (and write $\langle S, T(r)\rangle(\omega) = y$).

For query types, we consider membership query (Mem) and equivalence query (Equ). For each membership query, a teacher T is supposed to answer "yes" or "no"; on the other hand, for each equivalence query, T is supposed to answer "yes" or some counterexample to the query. A learner is called, e.g., *(Mem,Equ)-learner* if it asks membership and equivalence queries. A tuple (Mem,Equ) is called a *query-answer* type.

Now we are ready to define our "learnability" notion. For simplifying our discussion, we explain and define notions by using (Mem,Equ) for a typical query-answer type. However, these notions are defined similarly for the other query-answer type.

In this paper, we consider only "bounded learning", which has been introduced in [8, 9] as one reasonable query learning notion. (Since bounded learning is assumed throughout of this paper, "bounded" is often omitted.) Intuitively, in bounded learning, for a given parameter $m \geq 0$, the goal of a learner is to obtain a representation that denotes a target set up to length m. The parameter m is called a *length bound.* More precisely, "bounded learning" is defined as follows. For any target representation r and, for a given equivalence query r', we say that $T(r)$ answers r' correctly up to length ℓ if T gives a counterexample if it exists in $\Sigma^{\leq \ell}$ and "yes" otherwise. A teacher T is called a *(consistent) bounded (Mem,Equ)-teacher* for *DFA* if for given target representation r and length bound m, $T(r)$ answers each membership query correctly w.r.t. r, and $T(r)$ answers each equivalence query correctly up to length m. By considering a bounded teacher, we can avoid the case where a learner is given unnecessarily long counterexamples and the case where a learner abuses teacher's power of searching through infinite number of strings.

Definition 2.1. A (Mem,Equ)-learner S *learns* $C = (R, \Phi, \rho)$ (or, C is *learnable* by S) *in the bounded learning sense* if for every bounded (Mem,Equ)-teacher T for C, every $r \in R$, every $n \geq \rho(r)$, and every $m \geq 0$,

$$\langle S, T(r)\rangle(n, m) = r' \text{ such that } \Phi(r')^{\leq m} = \Phi(r)^{\leq m}.$$

In this paper a randomized learner is sometimes investigated. The following definition is for randomized learners.

Definition 2.2. A randomized (Mem,Equ)-learner S *learns* $C = (R, \Phi, \rho)$ *with the success probability* $\geq \delta$ if for every bounded (Mem,Equ)-teacher T for C, every $r \in R$, every $n \geq \rho(r)$, and every $m \geq 0$,

$$\Pr\{\ \langle S, T(r)\rangle(n, m) = r' \text{ such that } \Phi(r')^{\leq m} = \Phi(r)^{\leq m}\ \} \ \geq\ \delta.$$

Now the definition of "polynomial time learnability" is straigthforward. A learner is *polynomial time* if for some polynomial p and for all input $\langle n, m\rangle$, it halts within $p(n+m)$ steps. A representation class C is *polynomial time (Mem,Equ)-learnable in the bounded learning sense* if C is learnable by some polynomial time (Mem,Equ)-learner.

Finally, we define "query complexity". Intuitively, "query complexity" is the number of queries asked by S in the worst case. More precisely, for any learner S for *DFA*, query complexity $\#query_S$ is defined as follows: Let $\mathcal{T}$ be the family of bounded teachers for *DFA* of S's query-answer type. For any $T \in \mathcal{T}$, any $r \in R_{\text{dfa}}$, and any $n, m \geq 0$, let $\#query_{\langle S,T(r)\rangle}(n, m)$ be the number of queries asked during the computation $\langle S, T(r)\rangle(n, m)$. Now for any $n, m \geq 0$,

$$\#query_S(n,m) \;=\; \max\{\ \#query_{(S,T(r))}(n,m)\ :\ T \in \mathcal{T}, r \in R_{\mathrm{dfa}}\ \}.$$

Membership query complexity $\#mem\text{-}query_S$ and equivalence query complexity $\#equ\text{-}query_S$ are defined similarly.

3. Two Lower Bound Results

Here we show two lower bounds on the query complexity in different learning situations. These two lower bound results can be regarded as, respectively, extension of each part of Proposition 1.2.

First we investigate the number of membership queries necessary by a randomized (Mem)-learner for *DFA*.

Theorem 3.1. For any δ, $0 < \delta \leq 1$, let S be any randomized (Mem)-learner S that learns *DFA* with success probability $\geq \delta$. Then for any $k > 0$, we have the following bound:

$$\#mem\text{-}query_S(k+2,k) \;\geq\; \delta 2^k - 1.$$

Remark. Proposition 1.2 (1) is a special case, i.e., $\delta = 1$, of this theorem.

Proof. Consider any δ, $0 < \delta \leq 1$. We show that any learner with the query complexity better than the above lower bound cannot learn *DFA* with success probability $\geq \delta$.

Let T_0 be a (Mem)-teacher for *DFA*. Let S be any (Mem)-learner, and suppose that for some $k > 0$, $\#mem\text{-}query_S(k+2,k) < \delta 2^k - 1$. We consider the problem of learning the empty set or singleton sets $\{w\}$, where $|w| = k$. Note that there is some dfa with $k+2$ states for the empty set. Also for every $w \in \Sigma^k$, there is some dfa with $k+2$ states that accepts $\{w\}$. (Recall that we are considering complete dfa; thus an "error state" is necessary.) The representation of a dfa for the empty set is denoted as $r_\emptyset$, and the one for $\{w\}$ is denoted as r_w. We show that for some $w \in \Sigma^k$, the probability that $\langle S, T_0(r_w)\rangle(k+2,k)$ outputs a correct answer (i.e., some dfa representation of $\{w\}$) is less than δ.

Now consider the execution $\langle S, T_0(r_\emptyset)\rangle$ on input $(k+2,k)$. For any string $w \in \Sigma^k$, we say that w is *δ-well treated* if

$$\Pr\left\{ \begin{array}{l} \text{either } \langle S, T_0(r_\emptyset)\rangle(k+2,k) \text{ queries } w, \\ \text{or } \langle S, T_0(r_\emptyset)\rangle(k+2,k) \text{ outputs some } r \text{ s.t. } L(r) = \{w\} \end{array} \right\} \;\geq\; \delta.$$

Notice that if a string w is not δ-well treated, then it is no hope to have $\Pr\{\ \langle S, T_0(r_w)\rangle(k+2,k)$ outputs a correct answer $\} \geq \delta$. On the other hand, the following claim states that the number of δ-well treated strings is not large enough. (The proof is omitted in this abstract.)

Claim. The number of δ-well treated strings is less than $\dfrac{\#mem\text{-}query_S(k+2,k)+1}{\delta}$.

Now from the assumption $\#mem\text{-}query_S(k+2,k) \leq \delta 2^k - 1$, it is clear that some $w \in \Sigma^k$ is not δ-well treated. Therefore, we have some w such that S cannot learn r_w with success probability $\geq \delta$. $\square$

Corollary 3.2. For any δ, $0 < \delta \leq 1$, no polynomial time randomized (Mem)-learner S exists that learns *DFA* with success probability $\geq \delta$.

Next we consider the case where the number of alternations between membership and equivalence queries is limited.

Theorem 3.3. Let S be any (Mem,Equ)-learner for *DFA* such that the size of dfa asked as S's equivalence query is bounded by some nondecreasing polynomial in input parameters. Furthermore, suppose that S alternates between equivalence and membership queries a bounded number of times a. Then for any $c > 0$ and any sufficiently large $k > 0$, we have the following bounds:

$$\#\textit{equ-query}_S(3k^3, 2(k+a)^2+1) \geq c^k \quad \text{or} \quad \#\textit{mem-query}_S(3k^3, 2(k+a)^2+1) \geq 2^k.$$

Remark. Proposition 1.2 (2) is proved as a special case of this theorem.

For the proof, we use the technique established in [3]. Thus, we first recall some definitions and facts from [3]. For any $k > 0$, and any i, $1 \leq i \leq n$, define $L(i,k)$ to be the set of strings of length $2k$ whose ith bit is equal to the $(k+i)$th bit. Consider any set $L(i_1,k)L(i_2,k)\cdots L(i_k,k)$, where $1 \leq i_1,...,i_k \leq k$. Then it is easy to show that the set is accepted by some dfa with $3k^2+2$ states. Let R_k denote the set of dfa representations r such that $|r| = 3k^2+2$ and $L(r) = L(i_1,k)L(i_2,k)\cdots L(i_k,k)$ for some $1 \leq i_1,...,i_k \leq k$. From the above discussion, any set of the form $L(i_1,k)L(i_2,k)\cdots L(i_k,k)$ has a dfa representation in R_k; thus, $\|R_k\| = k^k$.

The following lemma, which states the lower bound of the number of equivalence queries for learning $r \in R_k$, plays a key role for proving our theorem. (The proof is omitted in this abstract.)

Lemma 3.4. Let S be any (Equ)-learner for *DFA* such that for some nondecreasing polynomial p, S on input $\langle 3k^2+2, 2k^2\rangle$ never asks a dfa (as an equivalence query) with more than $p(k)$ states. Then there exists a bounded (Equ)-teacher T_1 for *DFA* such that for any $c > 0$ and any sufficiently large k, $\langle S, T_1(r)\rangle(3k^2+2, 2k^2)$ needs to ask at least c^k queries for some $r \in R_k$.

Proof of Theorem 3.3. Let S be any (Mem,Equ)-learner as assumed in the theorem. More specifically, for some constant $a \geq 0$, we assume that S alternates between equivalence and membership queries at most a times, and for some nondecreasing polynomial p, S on input $\langle k^3, 2(k+a)^2+1\rangle$ never asks a dfa with more than $p(k)$ states. We prove the theorem by induction on a.

(Induction Base: $a = 0$) In this case no alternation occurs. Thus, the learner is either a (Mem)-learner, and the lower bound essentially follows from Theorem 3.1 (where $\delta = 1$), or is an (Equ)-learner, and then the lower bound follows from Lemma 3.4.

(Induction Step: $a \geq 1$) For the induction hypothesis, we assume that the lower bound hold for $a-1$, and furthermore that the target sets used for proving the lower bound consist of strings of length $\leq 2(k+a-1)^2+1$.

Now suppose that the lower bound does not hold for a; that is, for some c, and for infinitely many k, we have $\#\textit{equ-query}_S(k^3, 2(k+a)^2+1) < c^k$ and $\#\textit{mem-query}_S(k^3, 2(k+a)^2+1) < 2^k$. From this we lead a contradiction. Let us first describe some class of target sets, for which a contradiction is derived, and then discuss the size of dfa accepting these sets.

We distinguish two cases depending on the type of the first query.

Case 1: membership queries are asked first.

By induction hypothesis, assume there is a set of representations $R_{a,k}$ that cannot be learnt with $a-1$ alternations starting with equivalence queries. Inductively, we also assume that every dfa in $R_{a,k}$ accepts only strings of lengths $\leq 2(k+a-1)^2+1$. Consider sets of representations for languages of the form $L = wL(r)$, where $r \in R_{a,k}$ and $|w| = 4(k+a+1)+2 \geq k$.

Simulate the initial membership query phase of S answering always "no". The number of queries is less than the total number of membership queries, so at the end of the phase we can select some w that has never appeared as prefix of a query. If S is still able to learn the representations for $wL(r)$ when $r \in R_{a,k}$, then a trivial modification of S learns $R_{a,k}$ with $a-1$ query alternations, which contradicts the induction hypothesis. Observe the additional fact that the strings in $wL(r)$ have lengths $\leq 2(k+a)^2+1$ from our choice of the length of w.

Case 2: equivalence queries are asked first.

Inductively, assume again that there is a set of representations $R_{a,k}$ that cannot be learnt with $a-1$ alternations starting now with membership queries. Assume as well that every dfa in $R_{a,k}$ accepts only strings of lengths $\leq 2(k+a-1)^2+1$. We now consider now representations for sets of the form $0L(r) \cup 1L(r')$, where $r \in R_{a,k}$ and r' belongs to certain R' as we explain next.

From the set R_k used in Lemma 3.4, we define R_{k+a} by substituting $k+a$ to k. Define $R' = R_{k+a}$. Now consider the first phase of equivalence queries of S. By the bound on the number of equivalence queries, we know that after this phase there remain at least two representations in R' that S cannot distinguish. Moreover, during the process, S has obtained only counterexamples of length $2(k+a)^2$ from some teacher; hence, if it is able now to learn the part $0L(r)$, then a trivial modification learns $R_{a,k}$ in $a-1$ alternations, contradicting the induction hypothesis.

For any $r \in R_k$ and $r' \in R'$, consider the length of the strings in $0L(r) \cup 1L(r')$. By induction hypothesis, $0L(r)$ contains only strings of length $\leq 2(k+a-1)^2+2$, which is less than $2(k+a)^2+1$. Also by construction, $1L(r')$ contains only strings of length $2(k+a)^2+1$ Hence, $0L(r) \cup 1L(r')$ consists of strings of length $\leq 2(k+a)^2+1$.

The above constructions of target sets themselves show that the strings involved are never longer than $2(k+a)^2+1$ as needed by the theorem. On the other hand, it is provable by induction that those target sets have dfa representation of size $\leq k^3$. Therefore, the above lower bound argument holds for S's computation on $\langle k^3, 2(k+a)^2+1\rangle$. □

Corollary 3.5. No polynomial time (Mem,Equ)-learner exists for *DFA* that alternates membership and equivalence queries a constant number of times.

4. Trade-off Between the Number of Membership and Equivalence Queries

In this section, we consider general case, and discuss some trade-off relation between the number of membership and equivalence queries.

First we generalize Angluin's learning algorithm to reduce equivalence queries while spending some more membership queries. More specifically, our generalized algorithm takes $\langle n, m, h\rangle$ as input and learns a target dfa in the bounded learning sense, by asking n/h equivalence queries and $2^h \cdot p_1(n+m)$ membership queries, where p_1 is some fixed polynomial. Furthermore, the algorithm runs in polynomial time w.r.t. the number of queries.

Theorem 4.1. There is a (Mem,Equ)-learner S_0 for *DFA* with the following complexity: for every $n, m, h > 0$,
(a) $\#\mathit{equ\text{-}query}_{S_0}(n, m, h) \leq \frac{n}{h}$,
(b) $\#\mathit{mem\text{-}query}_{S_0}(n, m, h) \leq 2^h \cdot p_1(n+m)$, and
(c) S_0 on input $\langle n, m, h\rangle$ halts within $p_2(\#\mathit{query}_{S_0}(n, m, h))$.
Where p_1 and p_2 are polynomials depending on S_0.

Proof. We first recall some facts about Angluin's algorithm. In the algorithm, an *observation table* plays an important role for constructing hypotheses. An observation table is a tuple (S, E, T), where S and E are finite and prefix-closed sets, and T maps $(S \cup S \cdot \Sigma) \times E$ to $\{0, 1\}$. At certain points, the algorithm builds a dfa $M = M(S, E, T)$ (i.e., a hypothesis) from the table, and presents M to the teacher as an equivalence query. If the answer is "yes", the algorithm halts. Otherwise, it uses a received counterexample to expand S, E, and T, in a way such that the next equivalence query must have at least one more state than the previous one. Furthermore, the algorithm has the following property: If the target set is accepted by a n state dfa, then when the constructed hypothesis has n states at some point, it must accept exactly the target language. From these properties, it is clear that Angluin's algorithm needs at most n equivalence queries.

We design S_0 so that it adds at least h states to M between each two consecutive equivalence queries. If we can still ensure that n bounds the number of states of M, then clearly S_0 needs at most n/h equivalence queries for obtaining the target dfa.

To do this, we define the notion of *observation table with h lookahead.* It is a tuple (S, E, T) as before, but table T maps $(S \cdot \Sigma^{\leq h}) \times E$ to $\{0, 1\}$. Note that Angluin's tables are tables with 1 lookahead. S_0 acts as Angluin's algorithm, but keeping an observation table with h lookahead. This requires filling $\|\Sigma^{\leq h}\|$ entries in T with membership queries each time that S increases.

Clearly, this modification does not affect the correctness of the learner; that is, as Angluin's learning algorithm. S_0 learns *DFA* correctly. Furthermore, maintaining this additional information roughly increases the number of necessary membership queries by $\|\Sigma^{\leq h}\| \cdot p_1(n+m)$ for some polynomial p_1. Thus, the entire membership query complexity satisfies the theorem with some polynomial p_1. It is also easy to show that S_0 halts in time polynomial in the total number of queries.

Now it remains to show that at least h states are added after each equivalence query since this implies the desired upper bound on the number of equivalence queries. To show this property, it is enough to prove the following stronger version of lemma in [1]. (The proof of the lemma is omitted in this abstract.) □

Lemma 4.2. (Adapted from Lemma 4 in [1]) Assume that (S, E, T) is a closed and consistent observation table with h lookahead. Suppose that dfa $M = M(S, E, T)$ has k states. If M' is any dfa consistent with T that has less than $k + h$ states, then M' is isomorphic to M.

From this theorem, it is straightforward to derive the following two upper bound results.

Corollary 4.3. Let $f(n)$ be any polynomial time computable function such that $f(n) < n$. There exist a (Mem,Equ)-learner S_f for *DFA*, and a polynomial q_f such that for every $n, m > 0$,

$$\#\textit{equ-query}_{S_f}(n,m) \leq \frac{n}{f(n)} \text{ and } \#\textit{mem-query}_{S_f}(n,m) \leq 2^{f(n)} \cdot q_f(n+m).$$

Corollary 4.4. For any $c > 0$, there exist a polynomial time (Mem,Equ)-learner S_c for *DFA* such that $\#\textit{equ-query}_{S_c}(n,m) \leq \dfrac{n}{c \log n}$ for every $n, m > 0$.

Concerning these upper bound results, a natural question is whether they can be improved. For example, we may ask whether the number of equivalence queries can be reduced by more than $O(\log n)$ factor. However, it is shown in the following that such reduction is not possible without increasing the number of membership queries more than polynomial. That is, there is a certain type of trade-off between the number of membership and equivalence queries.

For showing such a trade-off phenomenon, we first prove the following lower bound. (The proof is omitted in this abstract.)

Theorem 4.5. For any (Mem,Equ)-learner S for *DFA*, and for any $n, m > 0$, we have the following bounds:

$$\#\textit{equ-query}_S(n,m) \geq \left\lfloor \frac{n-2}{m} \right\rfloor \text{ or } \#\textit{mem-query}_S(n,m) \geq 2^m - \left\lfloor \frac{n-2}{m} \right\rfloor.$$

The following corollary contrasts with Corollary 4.3.

Corollary 4.6. Let $f(n)$ be any function such that $f(n) < n$ and $f(n)$ becomes arbitrarily large as n increases, and let S be any (Mem,Equ)-learner for *DFA*. Then for some constants $c_1, c_2 > 0$, and for infinitely many $n, m > 0$, we have

$$\#\textit{equ-query}_S(n,m) \geq \frac{n}{f(n)} \text{ or } \#\textit{mem-query}_S(n,m) \geq 2^{c_1 f(n)} - \frac{c_2 n}{f(n)}$$

Proof. Define nondecreasing sequence $m_1, m_2, \ldots$ so that $m_n \geq f(n)/2$ and $n/f(n) \leq \lfloor (n-2)/m_n \rfloor \leq 3n/f(n)$. Then the corollary follows from Theorem 4.5 for these pairs of n and m_n. (In this rough estimation, $c_1 = 1/2$ and $c_2 = 3$.) □

Thus, roughly speaking, the reduction of the equivalence query complexity by $1/f(n)$ factor always costs us about $2^{O(f(n))}$ membership queries. One interesting example is the following case, which shows the limitation of Corollary 4.4.

Corollary 4.7. Let $f(n)$ be any function such that $f(n) < n$ and $f(n)/\log n$ becomes arbitrarily large as n increases. Then there exists no polynomial time (Mem,Equ)-learner S for *DFA* with the query complexity $\#\textit{equ-query}_S(n,m) \leq \frac{n}{f(n)}$.

References

[1] Angluin, D., Learning regular sets from queries and counterexamples, *Information and Computation*, Vol. 75, 1987, pp.87-106.

[2] Angluin, D., Queries and concept learning, *Machine Learning*, Vol. 2, 1988, pp.319–342.

[3] Angluin, D., Negative results for equivalence queries, *Machine Learning*, Vol. 5, 1990, pp.121–150.

[4] Balcázar, J.L., Díaz, J., Gavaldà, R., and Watanabe, O., A note on the query complexity of learning DFA, Technical Report 92TR-0009, Dept. of Computer Science, Tokyo Institute of Technology, 1992.

[5] Pitt, L., Inductive inference, DFAs, and computational complexity, in *Proc. International Workshop on Analogical and Inductive Inference AII'89*, Lecture Notes in Artificial Intelligence 397, Springer-Verlag, 1989, pp.18–42.

[6] Pitt, L., and Warmuth, M., Reductions among prediction problems: on the difficulty of prediction automata, in *Proc. 3rd Structure in Complexity Theory Conference*, IEEE, 1988, pp.60–69.

[7] Valiant, L., A theory of the learnable, *Communications of the ACM*, Vol. 27, 1984, pp.1134–1142.

[8] Watanabe, O., A formal study of learning via queries, in *Proc. 17th International Colloquium on Automata, Languages and Programming*, Lecture Notes in Computer Science 443, Springer-Verlag, 1990, pp.137–152.

[9] Watanabe, O., A framework for polynomial time query learnability, Technical Report 92TR-0003, Dept. of Computer Science, Tokyo Institute of Technology, 1992.

POLYNOMIAL-TIME MAT LEARNING OF MULTILINEAR LOGIC PROGRAMS

Kimihito ITO Akihiro YAMAMOTO
Department of Electrical Engineering
Hokkaido University
Sapporo 060, JAPAN
Phone: Int. +81-11-716-2111 ext. 6473
Fax : Int. +81-11-707-9750
email : itok@huee.hokudai.ac.jp
yamamoto@huee.hokudai.ac.jp

Abstract : In this paper we give a MAT(Minimally Adequate Teacher) learning algorithm of multilinear logic programs. MAT learning is to infer the unknown model M_U that the teacher has, with membership queries and equivalence queries. In the class of multilinear programs, we show some programs which have not been proved to be MAT learnable previously. If a multilinear program P_U represents M_U that the teacher has, then the total running time of our learning algorithm is bounded by a polynomial in m, w and h, where m is the number of predicates in P_U, h is the number of non-linear clauses in P_U, and w is a parameter depending on counter-examples to equivalence queries. We also show multilinear programs with outputs are MAT learnable by extending the algorithm.

1. Introduction

In this paper we give a polynomial-time learning algorithm of multilinear programs in the framework of MAT learning. We also show multilinear programs with outputs are MAT learnable in polynomial time. Angluin[1] proposed the Minimally Adequate Teacher (MAT for short), who is assumed to answer correctly two types of queries from the Learner, membership queries and equivalence queries. A learning framework with MAT is called MAT learning. In [1], regular languages are shown to be MAT learnable. To learn logic programs is here to identify a program P for an initially unknown model M_U of a predicate R so that $M_U = M(P, R)$, where $M(P,R) = \{R(t_1, \cdots, t_n) ; R(t_1, \cdots, t_n)$ is in the least Herbrand model of $P\}$. Sakakibara[9] gave a polynomial-time MAT learning algorithm for linear monadic logic programs. Many researches on learning logic programs have been developed in the framework of identification in the limit, e.g.[7]

[11]. We prefer the MAT learning to develop our theory because we pay attention to the process of finding auxiliary predicates given in the algorithm in [9] and to the termination property of the algorithm.

However the learning algorithm in [9] is much restricted in the standpoint of learning logic programs. A linear monadic logic program is a finite set of clauses in which all atomic formulas are of the form $p(x)$ or $p(f(x_1, \cdots, x_k))$, where p is a monadic predicate symbol, f is a function symbol with k arguments, x is a variable, and $x_1, \cdots, x_k$ are mutually distinct variables. It is reasonable, from [6], to assume any linear monadic logic program as a set of clauses of the form $p(f(x_1, \cdots, x_k)) \leftarrow q_1(x_1), \cdots, q_k(x_n)$. The answer for linear monadic logic programs to any goal is either *yes* or *no*. Moreover we can not represent a model $M_{eq} = \{R(t, t)$; t is ground term$\}$ with any linear monadic program. Even if we introduce a function symbol *eq* with two arguments, and change M_{eq} into $M_{eq}' = \{R(eq(t, t))$; t is a ground term in which *eq* does not occur$\}$, there is no linear monadic program satisfying $M_{eq}' = M(P, R)$. If there were such a program P, then P should have a clause of the form $p(eq(x, y)) \leftarrow q(x), r(y)$ with $q(t)$ and $r(s)$ in $M(P)$ for some distinct terms t and s, and this contradicts the condition.

We solve the problems above by introducing a class of programs, called multilinear programs. In the class we can treat predicate symbols with n arguments. The answer for programs in the class can be substitutions for the variables in given goals. For example, a program $P = \{eq(a, a) \leftarrow,\ eq(b, b) \leftarrow,\ eq([], []) \leftarrow,\ eq([x_1|x_2], [y_1|y_2]) \leftarrow eq(x_1, y_1), eq(x_2, y_2)\}$ is in the class of multilinear program of dimension 2, if $\{a, b, [], [\_|\_]\}$ is the set of function symbols in the assumed language. This P satisfies $M_{eq} = M(P, R)$. The linear monadic programs treated in [9] are regarded as multilinear programs of dimension 1.

In the following section we define the fundamental concepts, and in Section 3 we define multilinear programs and their comprehensive form. In Section 4 we give the MAT learning algorithm of multilinear programs and analyze its time complexity. In Section 5 we discuss the multilinear programs with outputs, and in the last section we describe a problem arising in our theory and our future plan to make use of our results.

2. Preliminaries

In this paper we use the concepts and results on logic programing from Lloyd[5]. Let L be a first order language with a finite set Σ of function symbols and a finite set Π of predicate symbols. We treat each constant symbol as a function symbol with no argument. We assume there exists at least one constant symbol in L. We also assume a special predicate symbol R^n with its arity n is given for each $n \geqq 1$. $T(\Sigma)$ denotes the set of all ground terms in L. $LT(\Sigma, x)$ denotes the set of all terms in L in which a variable x occurs once. For every function or a predicate symbol f, we denote the number of arguments by $arity(f)$, the set $\{f \in \Sigma$; $arity(f) = n\}$ by Σ^n, and the set $\{p \in \Pi$; $arity(p) = n$,

$p \neq R^n\}$ by Π^n. The outer most function symbol in a term t is denoted by $init(t)$. For example, if $t=f(s_1,\cdots,s_k)$, then $init(t)=f$. We denote a tuple of terms $t_1,\cdots,t_k$ by $\langle t_1,\cdots,t_k\rangle$. A set $\{X_1:=t_1,\cdots,X_n:=t_n\}$ denotes a substitution that replaces X_i with t_i for $1\leqq i\leqq n$, where X_i are mutually distinct variable. As in the Prolog language, [], $[t \mid s]$, and $[t]$ represent the terms *nil*, *cons*(t, s), and *cons*(t, *nil*) respectively.

Let $A, B_1, \cdots, B_k$ ($k\geqq 0$) be atomic formulas. A clause of the form $A\leftarrow B_1, \cdots, B_k$ is a *definite clause*. A is called its *head* and $B_1, B_2, \cdots, B_k$ its *body*. A *logic program*, or a *program* for short, is a finite set of definite clauses. We denote all the predicates which occur in a program P by $\Pi(P)$. We denote the *least Herbrand model* for a program P by $M(P)$, and define $M(P, p)=\{p(t_1,\cdots,t_n)\,;\,p(t_1,\cdots,t_n)\in M(P)\}$ for $p\in\Pi(P)$. It is well known that if $A\leftarrow B_1, B_2, \cdots, B_k (k\geqq 0)$ is a ground instance of a clause in a program P and $B_1, B_2, \cdots, B_k\in M(P)$, then $A\in M(P)$, and conversely that if $A\in M(P)$, then there is a ground instance $A\leftarrow B_1, B_2, \cdots, B_k (k\geqq 0)$ of a clause in a program P such that $B_1, B_2, \cdots, B_k\in M(P)$. In this paper symbols and terms are sometimes superscripted like s^k. The cardinality of a set S is denoted by $|S|$.

Definition Let $t^1,\cdots,t^n$ be ground terms. The n-tuple $\langle t^1,\cdots,t^n\rangle$ is *linear* if

$$init(t^1)=\cdots=init(t^n).$$

Definition Let $t^1,\cdots,t^n$ be ground terms. The set $part(t^1,\cdots,t^n)$ is defined as follows:

(1) $\langle t^1,\cdots,t^n\rangle\in part(t^1,\cdots,t^n)$.

(2) If $\langle t^1,\cdots,t^n\rangle=\langle f(s^1{}_1,\cdots,s^1{}_k),\cdots,f(s^n{}_1,\cdots,s^n{}_k)\rangle$, then

$$\langle s^1{}_1,\cdots,s^n{}_1\rangle,\cdots,\langle s^1{}_k,\cdots,s^n{}_k\rangle\in part(t^1,\cdots,t^n).$$

(3) If $\langle s^1,\cdots,s^n\rangle\in part(t^1,\cdots,t^n)$ and $\langle u^1,\cdots,u^n\rangle\in part(s^1,\cdots,s^n)$, then

$$\langle u^1,\cdots,u^n\rangle\in part(t^1,\cdots,t^n).$$

(4) Only the elements that are proved to be in $part(t^1,\cdots,t^n)$ by using above rules (1)~(3) repeatedly are in $part(t^1,\cdots,t^n)$.

Definition Let S be a finite set of n-tuples of ground terms. We define $\Sigma(S)$ as follows:

$$\Sigma(S)=\{\langle f(t^1{}_1,\cdots,t^1{}_k),\cdots,f(t^n{}_1,\cdots,t^n{}_k)\rangle\,;\,f\in\Sigma^k,\langle t^1{}_1,\cdots,t^n{}_1\rangle,\cdots,\langle t^1{}_k,\cdots,t^n{}_k\rangle\in S, k\geqq 0\}.$$

In this paper, two clauses C and C' are regarded as identical iff C and C' can be the same by a variable renaming substitution and a permutation of atoms in their bodies. For a mapping Ψ from Π to Π, we extend it to a mapping from the set of all programs to itself as the following way:

(1) If $p(t^1,\cdots,t^n)$ is an atomic formula, then $\Psi(p(t^1,\cdots,t^n))=\Psi(p)(t^1,\cdots,t^n)$.

(2) If $A\leftarrow B_1,\cdots,B_n$ is a definite clause, then

$$\Psi(A\leftarrow B_1,\cdots,B_n)=\Psi(A)\leftarrow\Psi(B_1),\cdots,\Psi(B_n).$$

(3) If P is a logic program, then $\Psi(P)=\{\Psi(C)\,;\,C\in P\}$.

Definition P is *isomorphic* to P' if there exists a bijection Ψ from $\Pi(P)$ to $\Pi(P')$ such that $\Psi(P)=P'$.

3. Multilinear Programs

We define a class of multilinear programs of dimension n. The elements of Π^n are called *state predicates*, and R^n is called a *terminal predicate*.

Definition A *transitive clause of dimension n* is a definite clause of the form

$$p(f(x^1{}_1,\cdots,x^1{}_k),\cdots,f(x^n{}_1,\cdots,x^n{}_k))\leftarrow p_1(x^1{}_1,\cdots,x^n{}_1),\cdots,p_k(x^1{}_k,\cdots,x^n{}_k)$$

where $p,\ p_1,\ \cdots,\ p_k\in\Pi^n,\ k\geqq 0,\ f\in\Sigma^k$ and $x^i{}_j$ $(1\leqq i\leqq n, 1\leqq j\leqq k)$ are distinct variables.

Definition A *non-linear clause of dimension n* is a definite clause of the form

$$p(t^1,\cdots,t^n)\leftarrow$$

where $\langle t^1,\cdots,t^n\rangle$ is non-linear and $p\in\Pi^n$.

Definition A *terminal clause of dimension n* is a definite clause of the form

$$R^n(x^1,\cdots,x^n)\leftarrow p(x^1,\cdots,x^n)$$

where $x^1,\cdots,x^n$ are distinct variables and $p\in\Pi^n$.

Definition A logic program P is called a *multilinear program of dimension n* if

(1) $P=\phi$ or P is a non-empty set consisting of transitive clauses, non-linear clauses, and terminal clauses, and

(2) P has no two clauses $p(t^1,\cdots,t^n)\leftarrow A_1,\cdots,A_m$ and $q(t^1,\cdots,t^n)\leftarrow A_1,\cdots,A_m$ such that $p,q\in\Pi^n$, $p\neq q$, and $m\geqq 0$.

Example 1 Let $\Pi^2=\{eqc,\ del\}$, $\Sigma^0=\{a,\ b,\ nil\}$, $\Sigma^2=\{cons\}$, and $\Sigma=\Sigma^0\cup\Sigma^2$. The program P_{del_last} given below is a multilinear program of dimension 2.

$$P_F=\left\{\begin{array}{l} eqc(a,a)\leftarrow \\ eqc(b,b)\leftarrow \\ del([x_1|x_2],[y_1|y_2])\leftarrow eqc(x_1,y_1),\ del(x_2,y_2)\end{array}\right\},\qquad P_N=\left\{\begin{array}{l} del([a],[])\leftarrow \\ del([b],[])\leftarrow\end{array}\right\},$$

$$P_E=\{R^2(x,\ y)\leftarrow del(x,\ y)\},$$

$$P_{del\_last}=P_F\cup P_N\cup P_E.$$

Lemma 1 *Let P be a multilinear program of dimension n, $t^1,\cdots,t^n$ be ground terms, and p,q be in Π^n. If $p(t^1,\cdots,t^n)\in M(P)$ and $q(t^1,\cdots,t^n)\in M(P)$, then $p=q$.*

Next we introduce comprehensive multilinear programs.

Definition Let P be a multilinear program of dimension n. P is called *comprehensive* if for any $f \in \Sigma^k$ $(k \geqq 1)$ and for any $p_1, \cdots, p_k \in \Pi^n$, there exists a definite clause in P of the form $p(f(x^1{}_1,\cdots,x^1{}_k), \cdots, f(x^n{}_1,\cdots,x^n{}_k)) \leftarrow p_1(x^1{}_1,\cdots,x^n{}_1), \cdots, p_k(x^1{}_k,\cdots,x^n{}_k)$.

Lemma 2 *Let P be a comprehensive multilinear program of dimension n, and $\langle t^1,\cdots,t^n \rangle$ be any n-tuple of ground terms.*

(1) *If all elements in $part(t^1,\cdots, t^n)$ are linear, then there exists $p \in \Pi(P)$ such that $p(t^1,\cdots,t^n) \in M(P)$.*

(2) *If for each non-linear tuple $\langle s^1,\cdots, s^n \rangle$ in $part(t^1,\cdots, t^n)$, there is a non-linear clause $q(s^1,\cdots, s^n) \leftarrow$ in P, then $p(t^1,\cdots, t^n) \in M(P)$ for some $p \in \Pi(P)$.*

Definition Let P be a multilinear program of dimension n. The *comprehension of P*, denoted by $comh(P)$, is defined as follows:

$$P_{j1}=\left\{ \begin{array}{l} Fail(f(x^1{}_1,\cdots,x^1{}_k), \cdots, f(x^n{}_1,\cdots,x^n{}_k)) \\ \quad \leftarrow p_1(x^1{}_1,\cdots,x^n{}_1), \cdots, p_k(x^1{}_k,\cdots,x^n{}_k) \end{array} \middle| \begin{array}{l} f \in \Sigma^k, p_1, \cdots, p_k \in \Pi(P) \text{ s.t. no } p \in \Pi(P) \text{ satisfies} \\ \quad p(f(x^1{}_1,\cdots,x^1{}_k), \cdots, f(x^n{}_1,\cdots,x^n{}_k)) \qquad \in P \\ \qquad \leftarrow p_1(x^1{}_1,\cdots,x^n{}_1), \cdots, p_k(x^1{}_k,\cdots,x^n{}_k) \end{array} \right\},$$

$$P_{j2}=\left\{ \begin{array}{l} Fail(f(x^1{}_1,\cdots,x^1{}_k), \cdots, f(x^n{}_1,\cdots,x^n{}_k)) \\ \quad \leftarrow p_1(x^1{}_1,\cdots,x^n{}_1), \cdots, p_k(x^1{}_k,\cdots,x^n{}_k) \end{array} \middle| \begin{array}{l} p_i = Fail \text{ or } p_i \in \Pi(P) \text{ for } i=1,\cdots,k \\ \text{and at least one } p_i = Fail \end{array} \right\},$$

$comh(P) = P \cup P_{j1} \cup P_{j2}$, where $Fail$ is a new predicate symbol.

Example 2 Let P_{del_last} be the multilinear program defined in Example 1. Then

$comh(P_{del\_last}) = P_{del\_last} \cup P_{j1} \cup P_{j2}$, where

$$P_{j1}=\left\{ \begin{array}{l} Fail([x_1 \mid x_2], [y_1 \mid y_2]) \leftarrow eqc(x_1, y_1), eqc(x_2, y_2) \\ Fail([x_1 \mid x_2], [y_1 \mid y_2]) \leftarrow del(x_1, y_1), eqc(x_2, y_2) \\ Fail([x_1 \mid x_2], [y_1 \mid y_2]) \leftarrow del(x_1, y_1), del(x_2, y_2) \end{array} \right\},$$

$$P_{j2}=\left\{ \begin{array}{l} Fail([x_1 \mid x_2], [y_1 \mid y_2]) \leftarrow eqc(x_1, y_1), Fail(x_2, y_2) \\ Fail([x_1 \mid x_2], [y_1 \mid y_2]) \leftarrow del(x_1, y_1), Fail(x_2, y_2) \\ Fail([x_1 \mid x_2], [y_1 \mid y_2]) \leftarrow Fail(x_1, y_1), eqc(x_2, y_2) \\ Fail([x_1 \mid x_2], [y_1 \mid y_2]) \leftarrow Fail(x_1, y_1), del(x_2, y_2) \\ Fail([x_1 \mid x_2], [y_1 \mid y_2]) \leftarrow Fail(x_1, y_1), Fail(x_2, y_2) \end{array} \right\}.$$

Lemma 3 *For a multilinear program P of dimension n, $M(P, R^n) = M(comh(P), R^n)$.*

4 MAT Learning of Multilinear Logic Programs of Dimension n

Let M_U be a model such that $M_U = M(P_U, R^n)$ for some multilinear program P_U of dimension n. We show the learning algorithm which efficiently learns the initially unknown M_U from any *minimally adequate teacher* if Σ and R^n is given. As in [1], the teacher is assumed to answer correctly *membership queries* and *equivalence queries*. A *membership query* is an n-tuple $\langle t^1,\cdots, t^n \rangle$. The answer is *yes* or *no* depending on whether $R^n(t^1,\cdots, t^n)\in M_U$ or not. An *equivalence query* is a multilinear program P, which we call a *conjecture*. The answer is *yes* or *no* depending on whether $M(P, R^n)=M_U$. If it is *no*, then a *counter-example* is also provided which is an n-tuple of ground terms $\langle t^1,\cdots, t^n\rangle$ such that $R^n(t^1,\cdots, t^n)$ is in the symmetric difference of $M(P, R^n)$ and M_U.

4.1 Observation Tables

Definition Let S be a finite subset of $\{T(\Sigma)\}^n=T(\Sigma)\times \cdots \times T(\Sigma)$ (n times). S is called *closed wrt subtuple* if

$$part(t^1,\cdots, t^n)\subseteq S \quad \text{for any } \langle t^1,\cdots, t^n\rangle\in S.$$

Definition Let $X^1,\cdots, X^n$ be mutually distinct variables, E be a finite subset of $LT(\Sigma,X^1)\times \cdots \times LT(\Sigma, X^n)$ in which there exists a $\langle X_1,\cdots, X_n\rangle$. E is *closed wrt linear generalization to* S if $e\in E-\{\langle X_1,\cdots, X_n\rangle\}$ implies there exists an $e'\in E$ such that

$$e=e'\{X^1:=f(t^1_1,\cdots, t^1_{i-1}, X^1, t^1_i,\cdots, t^1_{k-1}),\cdots, X^n:=f(t^n_1,\cdots, t^n_{i-1}, X^n, t^n_i,\cdots, t^n_{k-1})\}$$

for some $f\in\Sigma^k$, $\langle t^1_1,\cdots, t^n_1\rangle,\cdots,\langle t^1_{k-1},\cdots, t^n_{k-1}\rangle\in S$, and $i\in N$.

Definition An *observation table* is a triplet $T=(S, E, MS)$, where S is a finite subset of $\{T(\Sigma)\}^n$ which is closed wrt subtuple, E is a finite subset of $LT(\Sigma, X^1)\times\cdots\times LT(\Sigma, X^n)$ which is closed wrt linear generalization to S, and MS is a mapping from $(S\cup\Sigma(S))\times E$ to $\{0, 1\}$.

An observation table can be visualized as a two dimension array with rows labelled by elements of $(S\cup\Sigma(S))$, columns labelled by elements of E, with the entry for row $\langle t^1,\cdots, t^n\rangle$ and column e equal to $MS(\langle t^1,\cdots, t^n\rangle, e)$. $MS(\langle t^1,\cdots, t^n\rangle, e)$ is defined as 1 if an $e\{X_1:=t^1, X_2:=t^2,\cdots, X_n:=t^n\}$ is in M_U, and as 0 otherwise. We denote, by $\mathrm{row}(t^1,\cdots, t^n)$, the function $r(e)=MS(\langle t^1,\cdots, t^n\rangle, e)$.

Definition An observation table T is *closed* if for each $\langle t^1,\cdots, t^n\rangle\in\Sigma(S)$ there exists an $\langle s^1,\cdots, s^n\rangle$ in S such that $\mathrm{row}(t^1,\cdots, t^n)=\mathrm{row}(s^1,\cdots, s^n)$.

Definition An observation table T is *consistent* provided that whenever $\langle t^1,\cdots, t^n\rangle$ and $\langle s^1,\cdots, s^n\rangle$ are elements of S such that $\mathrm{row}(t^1,\cdots, t^n)=\mathrm{row}(s^1,\cdots, s^n)$,

$$\begin{aligned}&\mathrm{row}(f(u^1_1,\cdots, u^1_{i-1}, s^1, u^1_i,\cdots, u^1_{k-1}),\cdots, f(u^n_1,\cdots, u^n_{i-1}, s^n, u^n_i,\cdots,u^n_{k-1}))\\ =\ &\mathrm{row}(f(u^1_1,\cdots, u^1_{i-1}, t^1, u^1_i,\cdots, u^1_{k-1}),\cdots, f(u^n_1,\cdots, u^n_{i-1}, t^n, u^n_i,\cdots,u^n_{k-1}))\end{aligned}$$

for all $f\in\Sigma^k$, $\langle u^1_1,\cdots, u^n_1\rangle,\cdots,\langle u^1_{k-1},\cdots, u^n_{k-1}\rangle\in S$, and $1\leq i\leq k$.

Definition An observation table T is called *minimum wrt non-linear tuple* if there exists an $\langle s^1,\cdots, s^n\rangle\in S$ such that

$$\langle t^1,\cdots,t^n\rangle\in part(s^1,\cdots, s^n) \text{ and } MS(\langle s^1,\cdots, s^n\rangle,\langle X^1,\cdots, X^n\rangle)=1$$

for each non-linear $\langle t^1,\cdots, t^n\rangle\in S$.

Definition An observation table T is called *canonical* if T is closed, consistent, and minimum wrt non-linear tuple.

Definition Let T be a canonical observation table. The *corresponding multilinear logic program* $Pr(T)$ is defined as follows :

$\Pi(Pr(T))=\{p_{\mathrm{row}(t^1,\cdots,t^n)} \mid \langle t^1,\cdots,t^n\rangle\in S\}\cup\{R\}$,

$$P_F=\left\{\begin{array}{l} p_{\mathrm{row}(f(t^1_1,\cdots,t^1_k),\cdots,f(t^n_1,\cdots,t^n_k))}(f(x^1_1,\cdots,x^1_k),\cdots,f(x^n_1,\cdots,x^n_k))\leftarrow \\ \qquad p_{\mathrm{row}(t^1_1,\cdots,t^n_1)}(x^1_1,\cdots,x^n_1),\cdots,p_{\mathrm{row}(t^1_k,\cdots,t^n_k)}(x^1_k,\cdots,x^n_k) \\ \mid f\in\Sigma^k, \langle t^1_1,\cdots,t^n_1\rangle,\cdots,\langle t^1_k,t^n_k\rangle\in S, x^i_j\ (1\leq i\leq n, 1\leq j\leq k) \text{ are mutually} \\ \qquad \text{distinct variables} \end{array}\right\},$$

$$P_N=\left\{ p_{\mathrm{row}(t^1,\cdots,t^n)}(t^1,\cdots,t^n)\leftarrow \ \middle|\ \langle t^1,\cdots,t^n\rangle\in S\cup\Sigma(S) \text{ and is non-linear} \right\},$$

$$P_E=\left\{ \begin{array}{l} R^n(x^1,\cdots,x^n) \\ \quad \leftarrow p_{\mathrm{row}(t^1,\cdots,t^n)}(x^1,\cdots,x^n) \end{array} \ \middle|\ \begin{array}{l} \langle t^1,\cdots,t^n\rangle\in S, MS(\langle t^1,\cdots,t^n\rangle,\langle X^1,\cdots,X^n\rangle)=1, \\ x^1,\cdots,x^n \text{ are mutually distinct variables} \end{array}\right\},$$

$Pr(T)=P_F\cup P_N\cup P_E$.

Note that $Pr(T)$ is comprehensive.

Lemma 4 *Let T be a canonical observation table. For any $\langle t^1,\cdots,t^n\rangle$ in $S\cup\Sigma(S)$ of T,*

$$p_{\mathrm{row}(t^1,\cdots,t^n)}(t^1,\cdots,t^n)\in M(Pr(T)).$$

Definition Let P be a multilinear program of dimension n and T be an observation table. P is *consistent with* T if for each $\langle t^1,\cdots,t^n\rangle$ in $S\cup\Sigma(S)$ of T and each e in E,

$$e\{X^1:=t^1,\cdots,X^n:=t^n\}\in M(P,R^n)\Leftrightarrow MS(\langle t^1,\cdots,t^n\rangle,e)=1.$$

Lemma 5 *Let T be a canonical observation table. Then $Pr(T)$ is consistent with T.*

Proof We prove the result by induction on *the depth of variables* in $e\in E$ of T defined as :

(1) The depth of variables in e is 0 if $e=\langle X^1,\cdots,X^n\rangle$.

(2) If the depth of variables in $e'\in E$ is d and $e=e'\{X^1:=f(t^1_1,\cdots,t^1_{i-1},X^1,t^1_i,\cdots,t^1_{k-1}),\cdots, X^n:=f(t^n_1,\cdots,t^n_{i-1},X^n,t^n_i,\cdots,t^n_{k-1})\}$, then the depth of variables in e is $d+1$.

Base step : Assume $e=\langle X^1,\cdots,X^n\rangle$. By Lemma 4, for any $\langle t^1,\cdots,t^n\rangle\in S\cup\Sigma(S)$, $p_{\mathrm{row}(t^1,\cdots,t^n)}(t^1,\cdots,t^n)$ is in $M(Pr(T))$. Since T is closed, for any $\langle t^1,\cdots,t^n\rangle\in S\cup\Sigma(S)$, there exist $\langle s^1,\cdots,s^n\rangle\in S$ such that $\mathrm{row}(t^1,\cdots,t^n)=\mathrm{row}(s^1,\cdots,s^n)$. Therefore,

$MS(\langle t^1,\cdots,t^n\rangle,\langle X^1,\cdots,X^n\rangle)=1\Leftrightarrow MS(\langle s^1,\cdots,s^n\rangle,\langle X^1,\cdots,X^n\rangle)=1$

$\Leftrightarrow$there exists a definite clause in $Pr(T)$ such that $R(x^1,\cdots,x^n)\leftarrow p_{\mathrm{row}(s^1,\cdots,s^n)}(x^1,\cdots,x^n)$

$\Leftrightarrow R(t^1,\cdots,t^n)\in M(Pr(T))$.

Induction step : Assume that the result holds for any $e\in E$ in which the depth of variables is at most h. Let e be an element of E in which the depth of variables is $h+1$.

Then, since E is closed wrt linear generalization to S, $e=e'\{X^1:=f(t^1_1,\cdots,t^1_{i-1},X^1,t^1_i,\cdots,t^1_{k-1}),\cdots,X^n:=f(t^n_1,\cdots,t^n_{i-1},X^n,t^n_i,\cdots,t^n_{k-1})\}$ for some $f\in\Sigma^k$, $\langle t^1_1,\cdots,t^n_1\rangle,\cdots,\langle t^1_{k-1},\cdots,t^n_{k-1}\rangle\in S$, $i\in N$, and $e'\in E$ in which depth of variables is h. Since T is closed, any $\langle t^1,\cdots,t^n\rangle\in S\cup\Sigma(S)$, there exists $\langle s^1,\cdots,s^n\rangle\in S$ such that $\text{row}(t^1,\cdots,t^n)=\text{row}(s^1,\cdots,s^n)$. Therefore,

$MS(\langle t^1,\cdots,t^n\rangle,e)=1 \Leftrightarrow MS(\langle s^1,\cdots,s^n\rangle,e)=1$

$\Leftrightarrow MS(\langle f(t^1_1,\cdots,t^1_{i-1},s^1,t^1_i,\cdots,t^1_{k-1}),\cdots,f(t^n_1,\cdots,t^n_{i-1},s^n,t^n_i,\cdots,t^n_{k-1})\rangle,e')=1$

$\Leftrightarrow R^n(e'\{X^1:=f(t^1_1,\cdots,t^1_{i-1},s^1,t^1_i,\cdots,t^1_{k-1}),\cdots,X^n:=f(t^n_1,\cdots,t^n_{i-1},s^n,t^n_i,\cdots,t^n_{k-1})\})\in M(Pr(T))$

by the induction hypothesis

$\Leftrightarrow R^n(e'\{X^1:=f(t^1_1,\cdots,t^1_{i-1},t^1,t^1_i,\cdots,t^1_{k-1}),\cdots X^n:=f(t^n_1,\cdots,t^n_{i-1},t^n,t^n_i,\cdots,t^n_{k-1})\})\in M(Pr(T))$□

Definition Let P be a multilinear program of dimension n. The number of non-linear clauses in P is denoted by $N(P)$.

Lemma 6 *Let T be a canonical observation table of dimension n, and P be any comprehensive multilinear program of dimension n. If P is consistent with T, $|\Pi(P)|\leq|\Pi(Pr(T))|$, and $N(P)\leq|\{\langle t^1,\cdots,t^n\rangle;\langle t^1,\cdots,t^n\rangle\in S$ and is non-linear$\}|$, then $Pr(T)$ is isomorphic to P.*

Proof We prove this lemma by exhibiting an isomorphism Ψ. Let p be a state predicate in $Pr(T)$. Firstly we show that for each $\langle t^1,\cdots,t^n\rangle\in S\cup\Sigma(S)$, if $p(t^1,\cdots,t^n)\in M(Pr(T))$, then there exists q such that $q(t^1,\cdots,t^n)\in M(P)$. If all elements of $part(t^1,\cdots,t^n)$ are linear, then by Lemma 2(1) clearly there exists such q. Assume that there exists a non-linear element $\langle s^1,\cdots,s^n\rangle$ in $part(t^1,\cdots,t^n)$. Since T is minimum wrt non-linear tuple, there exists $\langle u^1,\cdots,u^n\rangle$ such that $\langle s^1,\cdots,s^n\rangle\in part(u^1,\cdots,u^n)$ and $R(u^1,\cdots,u^n)\in M(Pr(T))$. Since P is consistent with T, $R^n(u^1,\cdots,u^n)\in M(P)$. Therefore, there is a non-linear clause in P such that $q(s^1,\cdots,s^n)\leftarrow$. By Lemma 2(2) there is a q such that $q(t^1,\cdots,t^n)\in M(P)$.

We write the q obtained above by $q_{\langle t^1,\cdots,t^n\rangle}$. We define, for each predicate $p_{row(t^1,\cdots,t^n)}$ in $Pr(T)$, $\Psi(p_{row(t^1,\cdots,t^n)})=q_{\langle t^1,\cdots,t^n\rangle}$. By Lemma 1, Ψ is a bijection from $\Pi(Pr(T))$ to $\Pi(P)$. In the way of constructing Ψ, we get $N(P)=|\{\langle t^1,\cdots,t^n\rangle;\langle t^1,\cdots,t^n\rangle\in S$ *and is non-linear*$\}|$. Since both $Pr(T)$ and P are comprehensive, and by the definition of $q_{\langle t^1,\cdots,t^n\rangle}$ $Pr(T)$ is isomorphic to P. □

4.2 MAT Learning Algorithm of Multilinear Programs and it's Time Analysis

Now we give the MAT learning algorithm of multilinear programs of dimension n is given as follows.

Algorithm 1 .

$S:=\phi$, $E:=\{\langle X^1,\cdots X^n\rangle\}$, $T=(S,E,MS)$;
$\{S\subseteq\{T(\Sigma)\}^n,\ E\subseteq LT(\Sigma,X^1)\times\cdots\times LT(\Sigma,X^n)\}$
Make a equivalence query proposing $Pr(T)$ as conjecture;
if the reply is *yes* then halt and output $Pr(T)$;

if the counter-example is $\langle t^1, \cdots, t^n \rangle$ then add all elements of $part(t^1, \cdots, t^n)$ to S;
construct the initial observation table T using membership queries;
repeat
  while T is not closed or not consistent
    if T is not consistent **then**
      find $\langle s^1, \cdots, s^n \rangle, \langle t^1, \cdots, t^n \rangle \in S, e \in E, \langle u^1{}_1, \cdots, u^n{}_1 \rangle, \cdots, \langle u^1{}_{k-1}, \cdots, u^n{}_{k-1} \rangle \in S, f \in \Sigma^k,$
      $i \in N$ such that
        $\mathrm{row}\langle s^1, \cdots, s^n \rangle = \mathrm{row}\langle t^1, \cdots, t^n \rangle$ and
        $MS(\langle f(u^1{}_1, \cdots, u^1{}_{i-1}, s^1, u^1{}_i, \cdots, u^1{}_{k-1}), \cdots, f(u^n{}_1, \cdots, u^n{}_{i-1}, s^n, u^n{}_i, \cdots, u^n{}_{k-1}) \rangle, e)$
        $\neq MS(\langle f(u^1{}_1, \cdots, u^1{}_{i-1}, t^1, u^1{}_i, \cdots, u^1{}_{k-1}), \cdots, f(u^n{}_1, \cdots, u^n{}_{i-1}, t^n, u^n{}_i, \cdots, u^n{}_{k-1}) \rangle, e)$;
      add $e\{X^1 := f(u^1{}_1, \cdots, u^1{}_{i-1}, X^1, u^1{}_i, \cdots, u^1{}_{k-1}), \cdots, X^n := f(u^n{}_1, \cdots, u^n{}_{i-1}, X^n, u^n{}_i, \cdots, u^n{}_{k-1})\}$
      to E;
      extend T by asking membership queries for missing elements ;
    endif
    if T is not closed **then**
      find $\langle t^1, \cdots, t^n \rangle \in \Sigma(S)$ such that for each $\langle s^1, \cdots, s^n \rangle \in S$,
        $\mathrm{row}\langle s^1, \cdots, s^n \rangle \neq \mathrm{row}\langle t^1, \cdots, t^n \rangle$;
      add $\langle t^1, \cdots, t^n \rangle$ to S;
      extend T by asking membership queries for missing elements ;
    endif
  endwhile
  Make an equivalence query proposing $Pr(T)$;
  If the reply is *no* and the counter-example is $\langle t^1, \cdots, t^n \rangle$ then
    add all elements of $part(t^1, \cdots, t^n)$ to S;
    extend T by asking membership queries for missing elements ;
    endif
until the reply is *yes* to the conjecture $Pr(T)$;
halt and output $Pr(T)$.

Next we analyze the time complexity of *Algorithm 1*. Put $w = \max\{|part(t^1, \cdots, t^n)|$; $\langle t^1, \cdots, t^n \rangle$ is the counter-example to an equivalence query$\}$. The time complexity of *Algorithm 1* depends on w. Let P_U be a multilinear program such that $M_U = M(P_U, R^n)$ for the model M_U the teacher has. Let $m = |\Pi(comh(P_U))|$, $h = N(comh(P_U))$, $l = |\Sigma|$ and $d = \max\{arity(f) ; f \in \Sigma\}$. The observation table in the algorithm is always minimum wrt non-linear tuple, because a non-linear $\langle t^1, \cdots, t^n \rangle$ is added to S only if some counter-example $\langle s^1, \cdots, s^n \rangle$ is given such that $R^n(s^1, \cdots, s^n) \in M_U$ and $\langle t^1, \cdots, t^n \rangle \in part(s^1, \cdots, s^n)$. Since P_U is always consistent with T, by Lemma 6, $|\Pi(Pr(T))|$ is at most m, and $N(Pr(T))$ is at most h. Whenever T is not closed or not consistent, $|\{\mathrm{row}(s^1, \cdots, s^k) ; \langle s^1, \cdots, s^k \rangle \in S\}|$, which is identical to $|\Pi(Pr(T))|$, increases at least one. Suppose the reply to the equivalence query for $Pr(T)$ is *no*. Let the next canonical observation table be T'. By Lemma 6, $|\Pi(Pr(T'))| \geqq |\Pi(Pr(T))|$, or $N(Pr(T')) \geqq N(Pr(T))$. Thus $|E|$ is at most m, $|S|$ is at most $m + w(m+h)$, and $|\Sigma(S)|$ is at most $l(m + w(m+h))^d$. Hence the size of the observation table is at most

$$((m + w(m+h)) + l(m + w(m+h))^d)m = O(\max(w^d m^{d+1}, w^d m^d h^d)).$$

The number of membership queries is at most the size of T. The number of equivalence queries and the number of making $Pr(T)$ is at most $m+h$. Judgment of whether T is closed and consistent can be done in time polynomial in size of T. From the definition of $comh(P)$, $N(P_U)=N(comh(P_U))$, and $|\Pi(comh(P_U))| \leqq |\Pi(P_U)|+1$. Hence we have the following theorem.

Theorem 1 *Let M_U be the model that the teacher has. Suppose there exists a multilinear program P_U such that $M_U=M(P_U, R)$. Then Algorithm 1 terminates with any membership queries and equivalence queries, and its output is P such that $M_U=M(P, R)$. Let $m=|\Pi(P_U)|$, $h=N(P_U)$, and $w=\max\{|part(t^1,\cdots, t^n)| ; \langle t^1,\cdots, t^n\rangle$ is the counter-example to an equivalence query$\}$. The total running time of Algorithm 1 is bounded by a polynomial in m, w and h.*

Example 3 Assume the teacher has the multilinear program in Example 1, and that Σ and R^2 are given to *Algorithm 1*. In Fig.1 we show the observation table when *Algorithm 1* has terminate. The output $Pr(T)$ is isomorphic to $comp(P_{del\_last})$ in Example 2, where $\Psi(p_{01}) = eqc$, $\Psi(p_{10})=del$, and $\Psi(p_{00})=Fail$.

5. Multilinear Programs with Outputs

By adding output argument to the predicate R^n, we extend the results in Section 4. Let Δ be a finite subset of $T(\Sigma)$.

Definition An *output clause of dimension n* is a definite clause of the form

$$R^{n+1}(x^1,\cdots, x^n, a) \leftarrow p(x^1,\cdots, x^n)$$

where $x^1,\cdots, x^n$ is distinct variables, $p\in\Pi^n$, $a\in\Delta$.

Definition A logic program P is called a *multilinear program of dimension n with outputs* if

(1) $P=\phi$ or P is non empty set consisting of transitive clauses, non-linear clauses, and output clauses, and

(2) P has no two clauses $p(t^1,\cdots, t^n)\leftarrow A_1,\cdots,A_n$ and $q(t^1,\cdots, t^n)\leftarrow A_1,\cdots,A_n$ such that $p, q\in\Pi^n$, $p\neq q$, and $m\geqq 0$.

Definition An *output query* is an n-tuple $\langle t^1,\cdots, t^n\rangle$. The answer is *output* or *no* depending on which $a\in\Delta$ is satisfy that $R^{n+1}(t^1,\cdots, t^n, a)\in M_U$. An *equivalence query* is a multilinear program with outputs P. The answer is *yes* or *no* depending on whether $M(P, R^{n+1})=M_U$. If it is *no*, then a *counter-example* is also provided, which is a tuple of ground terms $\langle t^1,\cdots, t^n\rangle$ such that $R^{n+1}(t^1,\cdots, t^n, a)$ is in the symmetric difference of $M(P_U, R^{n+1})$ and M_U.

By defining MS as a mapping from $S\cup\Sigma(S)\times E$ to $\{0\}\cup\Delta$, and by replacing membership queries with output queries, we get the MAT learning algorithm of

			E: X,Y	E: $[X\|[b]],[Y]$	
S	$[a,b]$	$[a]$	1	0	
S	a	a	0	1	$p_{01}(a,a)\leftarrow$
S	$[b]$	$[]$	1	0	$p_{10}([b],[])\leftarrow$
S	$[[a\|a],b]$	$[[a\|a]]$	0	0	⋮
S	$[a\|a]$	$[a\|a]$	0	0	
S	$[a]$	$[]$	1	0	$p_{10}([a],[])\leftarrow$
Σ(S)	b	b	0	1	$p_{01}(b,b)\leftarrow$
Σ(S)	$[a,a,b]$	$[a,a]$	1	0	$p_{10}([x\|y],[z\|w])\leftarrow$
Σ(S)	$[a,a]$	$[a]$	1	0	$p_{01}(x,z),\ p_{10}(y,w)$
	⋮	⋮	⋮	⋮	⋮

Fig. 1 An observation table

multilinear programs with outputs in the same way as in Section 4. The comprehension $comh(P)$ of a multilinear program P with outputs can be defined in the same way as in Section 3.

Theorem 2 *Let M_U be the model that the teacher has. Suppose there exist a multilinear program with outputs P_U such that $M_U=M(P_U, R^{n+1})$. Then there is a MAT learning algorithm which terminates with any membership queries and equivalence queries, and the output of which is P such that $M_U=M(P, R^{n+1})$. Let $m=|\Pi(P_U)|$, $h=N(P_U)$, and $w=\max\{|part(t^1,\cdots,t^n)|\ ;\ \langle t^1,\cdots,t^n\rangle$ is the counter-example to an equivalence query$\}$. The total running time of the algorithm is bounded by a polynomial in m, w and h.*

Example 4 Following P_{last} is a multilinear program of dimension 1 with outputs.

$$P_F=\left\{\begin{array}{lll} nil([])\leftarrow & atom_a(a)\leftarrow & atom_b(b)\leftarrow \\ l_a([x\,|\,y])\leftarrow atom_a(x),\ nil(y) & & l_b([x\,|\,y])\leftarrow atom_b(x),\ nil(y) \\ l_a([x\,|\,y])\leftarrow atom_a(x),\ l_a(y) & & l_b([x\,|\,y])\leftarrow atom_b(x),\ l_b(y) \\ l_a([x\,|\,y])\leftarrow atom_b(x),\ l_a(x) & & l_b([x\,|\,y])\leftarrow atom_a(x),\ l_b(y) \end{array}\right\},$$

$$P_O=\{\,R^2(x,a)\leftarrow l_a(x) \qquad R^2(x,b)\leftarrow l_b(x)\},$$

$$P_{last}=P_F\cup P_O.$$

6. Concluding Remarks

In this paper we show the multilinear program of dimension n is MAT learnable in polynomial time. The total running time of the learning algorithm depends on $N(P_U)$ and $|\Pi(P_U)|$, which are defined with a logic program P_U, not the model M_U. In the framework of model inference, these parameters should depend on M_U. In order to overcome this problem we need to show there exists a program P_U in $\{P\ ;\ M_U=M(P, R^n)\}$ which makes $N(P_U)$ and $|\Pi(P_U)|$ minimum.

To explain our future plan, we consider the inference of a program which outputs elements in a list in reverse order. By replacing R^2 in Example 1 by *del_last*, and R^2 in Example 4 by *last*, the reverse program can be given as follows:

$$P(reverse) = \left\{ \begin{array}{l} reverse([\,], [\,]) \leftarrow \\ reverse([a|x], y) \leftarrow reverse(x, z), last(y,a), del\_last(y,z) \\ reverse([b|x], y) \leftarrow reverse(x, z), last(y,b), del\_last(y,z) \end{array} \right\}$$

$$\cup P(last) \cup P(del\_last).$$

So, by using our results in this paper as subprocedures, we have a plan to construct a learning algorithm which can infer the reverse program above.

References

[1] Angluin, D. : Learning regular sets from queries and counter-examples, *Information and Computation*,*75* : 87-106, 1987.

[2] Brainerd, W.S. : The Minimalization of Tree Automata, *Information and Control*, 13: 483-491, 1968.

[3] Ishizaka, H. : Learning Simple Deterministic Languages, in *Proceedings of the Second Annual Workshop on Computional Learning Theory* :162-174, 1989.

[4] Ito, K. : *Logic Programs which are MAT Learnable*, Graduation Thesis, Faculty of Engineering, Hokkaido University, 1992.

[5] Lloyd, J. W. : *Foundations of Logic Programming*, Springer-Verlag,1984.

[6] Marque-Pucheu, G. : Rational Set of Trees and the Algebraic Semantics of Logic Programming, *Acta Informatica*, *20* : 249-260, 1983.

[7] Muggleton, S. : Inductive Logic Programming, in *Proceedings of the First International Workshop on Algorithmic Learning Theory* :42-62, 1990.

[8] Sakakibara, Y. : Learning Context-free Grammars from Structural Data in Polynomial Time, in *Proceedings of 1st Workshop on Computational Learning Theory* : 296-310, 1988.

[9] Sakakibara, Y. : Inductive Inference of Logic Programs Based on Algebraic Semantics, *New Generation Computing*,*7* : 365-380, 1990.

[10] Sakakibara, Y., Nishino, T. : EXACT Learning – Concept Learning from Queries (in Japanese), *Journal of Information Processing Society Japan*, 32 : 246-255, 1991.

[11] Shapiro, E. Y. : Inductive Inference of Theories from Facts, *Technical Report, 192*,Yale University Computer Science Dept. 1981.

Neural Networks

ITERATIVE WEIGHTED LEAST SQUARES ALGORITHMS FOR NEURAL NETWORKS CLASSIFIERS

Takio Kurita
Electrotechnical Laboratory
1-1-4 Umezono, Tsukuba, 305 Japan
E-mail: kurita@etl.go.jp

Abstract

This paper discusses learning algorithms of layered neural networks from the standpoint of maximum likelihood estimation. Fisher information is explicitly calculated for the network with only one neuron. It can be interpreted as a weighted covariance matrix of input vectors. A learning algorithm is presented on the basis of Fisher's scoring method. It is shown that the algorithm can be interpreted as iterations of weighted least square method. Then those results are extended to the layered network with one hidden layer. It is also shown that Fisher information is given as a weighted covariance matrix of inputs and outputs of hidden units for this network. Tow new algorithms are proposed by utilizing this information. It is experimentally shown that the algorithms converge with fewer iterations than usual BP algorithm. Especially UFS (unitwise Fisher's scoring) method reduces to the algorithm in which each unit estimates its own weights by a weighted least squares method.

1. Introduction

Feed-forward neural networks with error back-propagation learning algorithm (BP) [1, 2] have been successfully applied to many problems including pattern recognition, robotics and control, vision, image analysis, etc. Squared-error criterion is usually used as cost function and is minimized to determine the weights of the network by the steepest descent method. For classification of K classes, desired outputs are usually set to K-dimensional binary vectors in which one element is unity corresponding to the correct class and all others are zero. In this case, outputs of this network are interpreted as estimates of Bayesian *a posteriori* probabilities [3].

The most popular alternative cost function for pattern classifiers is the cross-entropy between actual outputs and desired outputs [3, 4, 5]. This is derived by the assumption that desired outputs are independent, binary, random variables, and that the actual network outputs represent the conditional probabilities that these binary, random variables are one. It can also be interpreted as minimizing the Kullback-Liebler probability distance measure, maximizing mutual information, or as maximum likelihood parameter estimation [4, 5, 6, 7]. This cost function has yielded similar error rates with squared-error in a phoneme classification experiments [8]. It is, however, demonstrated that number of iterations required to converge is

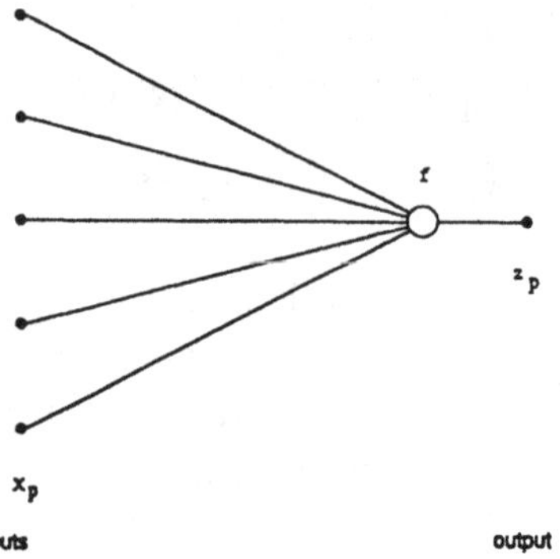

Figure 1. A neural network with only one unit.

fewer than squared-error criterion [9]. It is also shown that a cross-entropy cost function is minimized when outputs of the networks estimate Bayesian *a posteriori* probabilities [3]. This cost function was also used to determine the number of hidden units by information criteria [10].

This paper shows that iterative weighted least squares algorithms can be derived from the cross-entropy cost function by applying maximum likelihood estimation procedure. In section 2 we will study the network which consists of only one neuron from the view of maximum likelihood estimation in detail and show that Fisher information of the network is given as a weighted covariance matrix of inputs vectors [11]. A learning algorithm is presented on the basis of Fisher's scoring method, which is a kind of Newton-Raphson method and Fisher information is used instead of Hessian matrix. The algorithm can be interpreted as iterations of weighted least squares method [12]. Then we extend these results to layered neural networks with one hidden layer in section 3. Fisher information is also given as a weighted covariance matrix of inputs and outputs of hidden units. Tow new algorithms which utilize Fisher information are presented and are experimentally compared with BP algorithm.

2. The network with only one neuron

Here the networks which consists of only one neuron are studied from the view of maximum likelihood estimation. An example of the network with only one neuron is shown in Fig. 1.

2.1. Likelihood

Suppose that the input-output function of the neuron is logistic. Then the output z_p of the network is computed for an input $x_p = (x_{p1}, \ldots, x_{pI})^T$ as

$$z_p = \frac{exp(\eta_p)}{1 + exp(\eta_p)}, \tag{1}$$

where $\eta_p = \sum_{i=1}^{I} a_i x_{pi}$ and a_i is the weight from the i-th input.

Let the set of learning samples be $\{< x_p, t_p > | p = 1, \ldots, P\}$, where we assume that the teacher's signal t_p is given as binary (0 or 1). If we interpret the output z_p as an estimate of conditional probability given an input vector x_p, then the log-likelihood of the network for the set of learning samples is given by

$$l = \sum_{p=1}^{P} \{t_p \ln z_p + (1 - t_p) \ln(1 - z_p)\}. \tag{2}$$

Thus the maximum likelihood estimate of the weights for the set of learning samples is computed as the one that maximizes this log-likelihood.

This can also be interpreted as minimizing the cross-entropy between actual outputs and desired outputs, minimizing Kullback-Liebler probability distance measure, or as maximizing mutual information [4, 5, 6, 7].

2.2. Fisher information

In the maximum likelihood estimation, Fisher information plays an important rule. Here we calculate Fisher information of the network with only one neuron explicitly.

The first and the second order derivatives of the log-likelihood (2) are given as

$$\begin{aligned} \frac{\partial l}{\partial a_i} &= \sum_{p=1}^{P} \delta_p x_{pi}, \\ \frac{\partial^2 l}{\partial a_l \partial a_i} &= -\sum_{p=1}^{P} \omega_p x_{pl} x_{pi}, \end{aligned} \tag{3}$$

where $\delta_p = t_p - z_p$ and $\omega_p = z_p(1 - z_p)$. In matrix form, we have

$$\begin{aligned} \nabla l &= \sum_{p=1}^{P} \delta_p \boldsymbol{x}_p = X^T \delta, \\ \nabla^2 l &= -\sum_{p=1}^{P} \omega_p \boldsymbol{x}_p \boldsymbol{x}_p^T = -X^T W X \end{aligned} \tag{4}$$

where $X^T = [\boldsymbol{x}_1, \ldots, \boldsymbol{x}_P]$, $W = \mathrm{diag}(\omega_1, \ldots, \omega_P)$ and $\delta = (\delta_1, \ldots, \delta_P)^T$.

Fisher information for the weights $\boldsymbol{a} = (a_1, \ldots, a_I)^T$, namely minus the expected value of the Hessian matrix $\nabla^2 l$, is given by

$$F = -E(\nabla^2 l) = X^T W X. \tag{5}$$

This means that Fisher information for the weights of the network is a weighted covariance matrix of input vectors $\{\boldsymbol{x}_p\}$. The weight of each sample is given by $\omega_p = z_p(1 - z_p)$ that has the maximum $\frac{1}{4}$ at $z_p = \frac{1}{2}$ and the minimum 0 at $z_p = 0$ or $z_p = 1$. This means that inputs whose output of the network is uncertain (z_p is near $\frac{1}{2}$) contribute more to Fisher information than those whose output is certain (z_p is near 0 or 1).

2.3. Iterative weighted least square algorithm

To obtain the weights which maximize the log-likelihood, it is necessary to have an optimization algorithm. Fisher's scoring method is a kind of Newton-Raphson method and Fisher information is used instead of Hessian matrix. For the network with only one neuron Fisher information is the same as Hessian except the sign. Therefor Fisher's scoring method and Newton-Raphson method reduce to the same algorithm [12].

Let the current estimate of the weights be $\boldsymbol{a}$. Then the estimate is repeatedly updated by $\boldsymbol{a}^* = \boldsymbol{a} + \delta \boldsymbol{a}$, where the increment $\delta \boldsymbol{a}$ is determined by solving the linear equation

$$F \delta a = \nabla l. \tag{6}$$

Since Fa is given by $Fa = X^T W \eta$, the new estimate $\boldsymbol{a}^*$ can be obtained by solving the linear equation

$$X^T W X a^* = X^T W \eta + X^T \delta = X^T W(\eta + W^{-1}\delta), \tag{7}$$

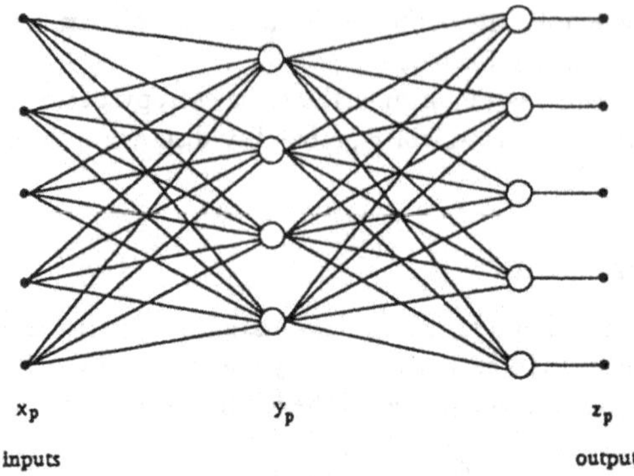

Figure 2. A layered neural network with one hidden layer.

where $\eta = (\eta_1, \ldots, \eta_P)^T$. This equation can be interpreted as the normal equation of a weighted least squares method to estimate $\eta + W^{-1}\delta$ from the input vectors X.

Thus the learning algorithm based on Fisher's scoring method for the network with only one neuron can be interpreted as iterations of a weighted least squares method.

From the equation (7) we can derive the following simple method to determine initial weights a_0 of the iterative weighted least squares algorithm. Suppose that all weights are initially zero, namely $a = o$. Then $W = \frac{1}{4}I$, $\eta = o$, and $\delta = t - \frac{1}{2}1$. Inserting these values into equation (7), we have the following equation

$$a_0 = 4(X^TX)^{-1}X^T(t - \frac{1}{2}1). \tag{8}$$

This can be interpreted as linear regression that estimates the desired outputs $t - \frac{1}{2}1$ from the input vectors.

3. The network with one hidden layer

Next we extend the results for the network with only one neuron to layered networks with one hidden layer. An example of the network with one hidden layer is shown in Fig. 2.

3.1. Likelihood

For this network output vector $z_p = (z_{p1}, \ldots, z_{pK})^T$ is computed for an input vector $x_p = (x_{p1}, \ldots, x_{pI})^T$ as

$$y_{pj} = f(\zeta_{pj}) = \frac{exp(\zeta_{pj})}{1 + exp(\zeta_{pj})}, \quad \zeta_{pj} = \sum_{i=1}^{I} a_{ji} x_{pi}, \tag{9}$$

$$z_{pk} = f(\eta_{pk}) = \frac{exp(\eta_{pk})}{1 + exp(\eta_{pk})}, \quad \eta_{pk} = \sum_{j=1}^{J} b_{kj} y_{pj}, \tag{10}$$

where a_{ji} is the weight from the i-th input unit to the j-th hidden unit and b_{kj} is the weight from the j-th hidden unit to the k-th output unit.

Let the set of learning samples be $\{< x_p, t_p > | p = 1, \ldots, P\}$. We also assume that the teacher's vector t_p is given as K-dimensional binary vector in which one element is unity corresponding to the correct class and all others are zero. If we assume that the each element of output vector of the network is conditionally independent, then the log-likelihood of the network for the set of learning samples is given by

$$l = \sum_{p=1}^{P} \sum_{k=1}^{K} \{t_{pk} \ln z_{pk} + (1 - t_{pk}) \ln(1 - z_{pk})\}. \tag{11}$$

Usually minus of this measure is called the cross-entropy cost function [3].

3.2. Fisher information

Here we calculate Fisher information for the network with one hidden layer explicitly.

The first and the second order derivatives of the log-likelihood (11) are given as

- The first order derivatives

$$\frac{\partial l}{\partial a_{ji}} = \sum_{p=1}^{P} \sigma_{pj} \nu_{pj} x_{pi}, \qquad \frac{\partial l}{\partial b_{kj}} = \sum_{p=1}^{P} \delta_{pk} y_{pj}, \tag{12}$$

  where $\sigma_{pj} = \sum_{k=1}^{K} \delta_{pk} b_{kj}$, $\nu_{pj} = y_{pj}(1 - y_{pj})$, and $\delta_{pk} = z_{pk}(1 - z_{pk})$.

- The second order derivatives

$$\frac{\partial^2 l}{\partial a_{ml} \partial a_{ji}} = \begin{cases} \sum_{p=1}^{P} x_{pl} \nu_{pj}(1 - 2y_{pj}) \delta_p x_{pi} \; - \sum_{p=1}^{P} x_{pl} \nu_{pj} \chi_{pjj} \nu_{pj} x_{pi} & \text{if } m = j \\ - \sum_{p=1}^{P} x_{pl} \nu_{pm} \chi_{pmj} \nu_{pj} x_{pi} & \text{otherwise} \end{cases} \tag{13}$$

$$\frac{\partial^2 l}{\partial a_{ml} \partial b_{kj}} = \begin{cases} \sum_{p=1}^{P} x_{pl} \nu_{pj} \delta_{pk} \; - \sum_{p=1}^{P} x_{pl} \nu_{pj} b_{kj} \omega_{pk} y_{pj} & \text{if } m = j \\ - \sum_{p=1}^{P} x_{pl} \nu_{pm} b_{km} \omega_{pk} y_{pj} & \text{otherwise} \end{cases} \tag{14}$$

$$\frac{\partial^2 l}{\partial b_{nm} \partial a_{ji}} = \begin{cases} \sum_{p=1}^{P} x_{pi} \nu_{pj} \delta_{pn} \; - \sum_{p=1}^{P} y_{pj} \omega_{pn} b_{nj} \nu_{pj} x_{pi} & \text{if } m = j \\ - \sum_{p=1}^{P} y_{pm} \omega_{pn} b_{nj} \nu_{pj} x_{pi} & \text{otherwise} \end{cases} \tag{15}$$

$$\frac{\partial^2 l}{\partial b_{nm} \partial b_{kj}} = \begin{cases} - \sum_{p=1}^{P} y_{pm} \omega_{pk} y_{pj} & \text{if } n = k \\ 0 & \text{otherwise} \end{cases} \tag{16}$$

  where $\delta_p = \sum_{k=1}^{K} \delta_{pk}$ and $\chi_{pmj} = \sum_{k=1}^{K} b_{km} \omega_{pk} b_{kj}$.

From the relation on log-likelihood

$$0 = \mathrm{E}\left(\frac{\partial l_p}{\partial z_{pk}}\right) = \frac{\mathrm{E}(t_{pk}) - z_{pk}}{z_{pk}(1 - z_{pk})}, \tag{17}$$

Fisher information for the weights $A = [a_{ji}]$ and $B = [b_{kj}]$ is given by

$$F_{a_{ml} a_{ji}} = \sum_{p=1}^{P} x_{pl} \nu_{pm} \chi_{pmj} \nu_{pj} x_{pi}, \qquad F_{a_{ml} b_{kj}} = \sum_{p=1}^{P} x_{pl} \nu_{pm} b_{km} \omega_{pk} y_{pj} \tag{18}$$

$$F_{b_{nm} a_{ji}} = \sum_{p=1}^{P} y_{pm} \omega_{pn} b_{nj} \nu_{pj} x_{pi}, \qquad F_{b_{nm} b_{kj}} = \begin{cases} \sum_{p=1}^{P} y_{pm} \omega_{pk} y_{pj} & \text{if } n = k \\ 0 & \text{otherwise} \end{cases} \tag{19}$$

Thus Fisher information for the weights of the network with one hidden layer is given as a weighted covariance matrix of inputs $\{x_p\}$ and outputs of hidden units $\{y_p\}$.

3.3. Iterative weighted least squares algorithm

Next we consider learning algorithms of the weights which maximizes the log-likelihood (11).

One of the simplest optimization method is steepest decent method which use information on the first order derivatives of objective function. Let the current estimates of the weights be $\theta = (a_{11}, \ldots, a_{JI}, b_{11}, \ldots, b_{KJ})^T$. Then the estimate is repeatedly updated by

$$\theta^* = \theta - \alpha \nabla l. \tag{20}$$

If we want to update the weights to each learning sample instead of the set of learning samples, this can be modified to

$$\theta^* = \theta - \alpha \nabla l_p, \tag{21}$$

where ∇l_p is the first order derivative to the sample $< \boldsymbol{x}_p, \boldsymbol{t}_p >$ and the relation $\nabla l = \sum_{p=1}^{P} \nabla l_p$ holds. In the following we call this SD (steepest decent) method.

Similar to the case of the neural networks with only one unit, we can develop Fisher's scoring algorithm which uses Fisher information. The estimate is repeatedly changed by

$$\theta^* = \theta + \delta\theta, \tag{22}$$

where the increment $\delta\theta$ is determined by solving the linear equation

$$F\delta\theta = \nabla l. \tag{23}$$

In the following we call this FS (Fisher's scoring) method.

Since there is possibility that Fisher information matrix F becomes singular, in the subsequent experiments we used $F + \beta I$ instead of F itself, where β is a constant and I is the unit matrix.

3.4. Unitwise iterative weighted least squares algorithm

In FS method it is necessary to compute Fisher information F $((IJ + JK) \times (IJ + JK))$ and solve the linear equation (23) with $IJ + JK$ unknown parameters at each step of learning. This is not easy for large networks.

Here we propose an algorithm which uses only block diagonal elements of Fisher information and neglects the other elements.

Fisher information related with the weights from input units to the unit j in the hidden layer is given by

$$F_{Aj} = \left[\sum_{p=1}^{P} x_{pl} \nu_{pj} \chi_{pjj} \nu_{pj} x_{pi} \right] = X^T W_{Aj} X, \tag{24}$$

where

$$X = [\boldsymbol{x}_1, \ldots, \boldsymbol{x}_P]^T \tag{25}$$

$$W_{Aj} = \mathrm{diag}(\nu_{pj} \chi_{pjj} \nu_{pj}). \tag{26}$$

This is a weighted covariance matrix of input vectors. From the equations (22) and (23), the normal equation of weighted least squares estimation to obtain the current estimates of the weights $\theta^*_{Aj} = (a_{j1}, \ldots, a_{jI})^T$ is given by

$$X^T W_{AJ} X \theta^*_{Aj} = X^T W_{AJ} (\zeta_j + W_{Aj}^{-1} \delta_{Aj}) \tag{27}$$

where

$$\zeta_j = (\zeta_{1j}, \ldots, \zeta_{Pj})^T \tag{28}$$
$$\delta_{Aj} = (\sigma_{1j}\nu_{1j}, \ldots, \sigma_{Pj}\nu_{Pj})^T. \tag{29}$$

Similarly, Fisher information related with the weights from hidden units to the unit k in the output layer and the corresponding normal equation are given by

$$F_{Bk} = \left[\sum_{p=1}^{P} y_{pm}\omega_{pk}y_{pj}\right] = Y^T W_{Bk} Y \tag{30}$$

and

$$Y^T W_{Bk} Y \theta^*_{Bk} = Y^T W_{Bk}(\eta_k + W_{Bk}^{-1}\delta_{Bk}) \tag{31}$$

where

$$Y^T = [y_1, \ldots, y_P] \tag{32}$$
$$W_{Bk} = \mathrm{diag}(\omega_{pk}) \tag{33}$$
$$\eta_k = (\eta_{1k}, \ldots, \eta_{Pk})^T \tag{34}$$
$$\delta_{Bk} = (\omega_{1k}, \ldots, \omega_{Pj})^T. \tag{35}$$

The estimates $\{\theta_{Aj} | j = 1, \ldots, J\}$ and $\{\theta_{Bk} | k = 1, \ldots, K\}$ are repeatedly updated by solving these normal equations. We call this UFS (Unitwise Fisher's scoring) method. In this algorithm each unit of the network estimates its own weights by the iterative weighted least squares algorithm which is similar with the one derived for the network with only one neuron in section 2.3.

Next we will consider the recursive formulas to solve these normal equations for the weighted least squares. This reveals the close relation between the proposed UFS algorithm and the back-propagation learning algorithm.

Consider a set of earlier learning samples $\{< x_p, t_p > | p = 1, \ldots, N-1\}$ and a new learning sample $< x_N, t_N >$. Then one can estimate the parameters $\theta_{Aj}^{(N)}$, which is optimal with respect to the set of earlier learning samples and the new learning samples, using the estimates $\theta_{Aj}^{(N-1)}$ for the set of earlier learning samples. The recursive formula is given by

$$\theta_{Aj}^{(N)} = \theta_{AJ}^{(N-1)} + Q_{Aj}^{(N)} x_N \sigma_{Nj} \nu_{Nj}, \tag{36}$$

where the matrix $Q_{Aj}^{(N)}$ gives the estimates of the inverse of the weighted covariance matrix ${}^{(N)}X^T\, {}^{(N)}W_{Aj}^{(N)} X$ and its recursive formula is given by

$$Q_{Aj}^{(N)} = Q_{Aj}^{(N-1)} - \frac{\nu_{Nj}\chi_{Njj}\nu_{Nj} Q_{Aj}^{(N-1)} x_N x_N^T Q_{Aj}^{(N-1)}}{1 + \nu_{Nj}\chi_{Njj}\nu_{Nj} x_N^T Q_{Aj}^{(N-1)} x_N}. \tag{37}$$

Similarly the recursive formula for the estimates θ_{Bk} is given by

$$\theta_{Bk}^{(N)} = \theta_{Bk}^{(N-1)} + Q_{Bk}^{(N)} y_N \delta_{Nj} \tag{38}$$
$$Q_{Bk}^{(N)} = Q_{Bk}^{(N-1)} - \frac{\omega_{Nj} Q_{Bk}^{(N-1)} y_N y_N^T Q_{Bk}^{(N-1)}}{1 + \omega_{Nj} y_N^T Q_{Bk}^{(N-1)} y_N}. \tag{39}$$

By comparing the equations (36) and (38) with the back-propagation algorithm in which the cross-entropy cost function is used as the cost function, one can notice that the estimates Q_{Aj} or Q_{Bk} of the inverse of the covariance matrices are used instead of the constant learning rate of the back-propagation algorithm.

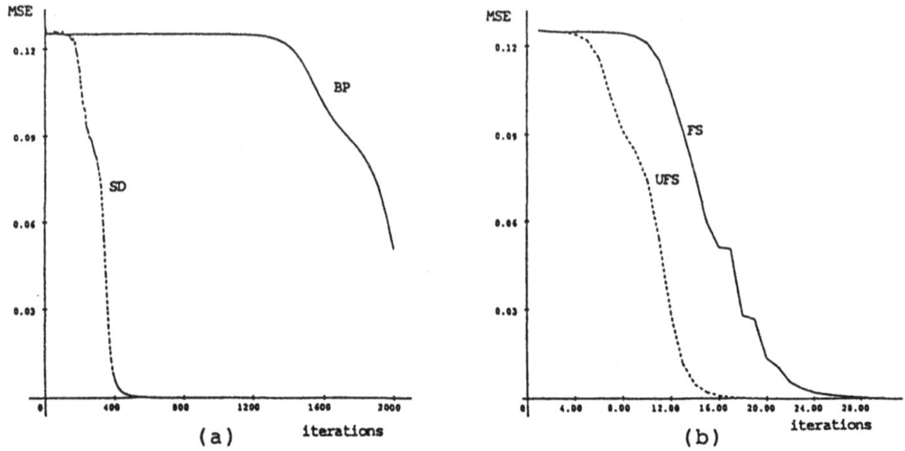

Figure 3. Results of learning of XOR problem.

4. Experiment

The learning algorithms described in this paper are compared with usual BP algorithm which uses square-error criterion.

4.1. XOR problem

The problem used in this experiment is the XOR problem. The network has 2 inputs, 2 hidden units and 1 output unit. Initial weights of the network were randomly generated within the interval $[-0.5, 0.5]$. For all algorithms the same initial weights were used. In BP and SD methods, the learning rate α was set to 0.25. The results of learning are shown in Fig. 3 (a) and (b). It is noticed that SD is faster than BP for this problem. This result agrees with the result of Holt [9]. FS and UFS methods converged with less than 30 iteration for this problem.

4.2. Pattern Recognition Problem

Experiment for classification of Fisher's Iris data was also performed. In this case the network has 4 inputs, 2 hidden units and 3 output units. Each of input features was normalized to zero mean and unit variance. Initial weights of the network were also randomly generated. In BP and SD methods, the learning rate α was set to $\frac{1}{150}$. The results of learning are shown in Fig. 4 (a) and (b). It is noticed that results are similar to the XOR problem.

5. Conclusion

This paper discussed learning algorithms of layered neural networks from the standpoint of maximum likelihood estimation. For the network with only one neuron Fisher information was explicitly calculated. It can be interpreted as a weighted covariance matrix of input vectors. A learning algorithm was presented on the basis of Fisher's scoring method. The algorithm can be interpreted as iterations of weighted least square method. Then those results were extended to the network with one hidden layer. For this network Fisher information was also

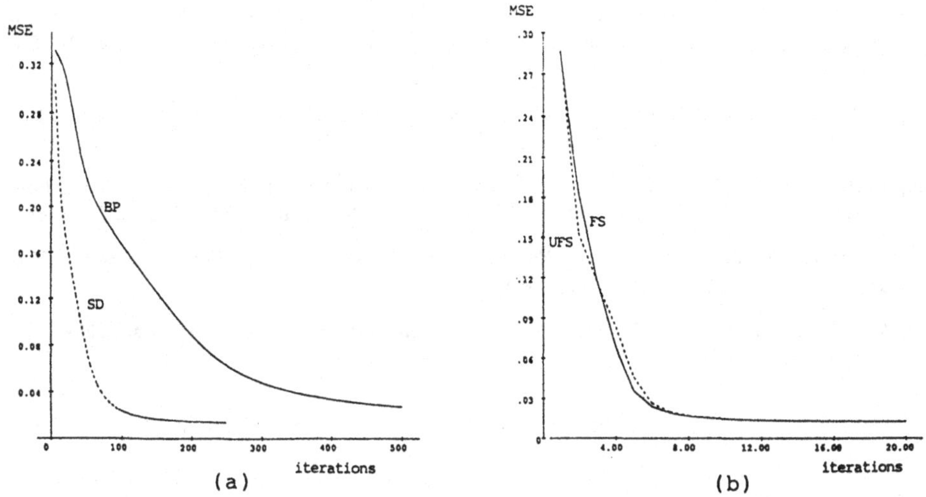

Figure 4. Results of learning of a pattern recognition problem.

a weighted covariance matrix of inputs and outputs of hidden units. Tow new algorithms were proposed by utilizing this information. The algorithms converged with fewer iterations than usual BP algorithm. Especially UFS (unitwise Fisher's scoring) method reduces to the algorithm in which each unit estimates its own weights by a weighted least squares method. In this paper we considered learning algorithms of neural networks for pattern classification. The same approach can be applied for neural networks for function approximation.

References

[1] Rumelhart,D.E., Hinton,G.E., and Williams,R.J. : Learning representations by back-propagating errors, Nature, Vol.323-9, pp.533-536 (1986).

[2] Rumelhart,D.E., Hinton,G.E., and Williams,R.J. : Learning internal representations by error propagation, in *Parallel Distributed Processing* Volume 1, McCleland,J.L., Rumelhart,D.E., and The PDP Research group, Cambridge, MA: MIT Press, 1986.

[3] Richard,M.D. and Lippmann,R.P. : Neural network classifiers estimate Bayesian *a posteriori* probabilities, Neural Computation, Vol.3, No.4, pp.461-483 (1991).

[4] Baum,E.B. and Wilczek,F. : Supervised learning of probability distributions by neural networks, In *Neural Information Processing Systems*, D.Anderson, ed.,pp.52-61. American Institute of Physics, New York (1988).

[5] Hinton,G.E. : Connectionist learning procedures, Artificial Intelligence 40, 185-234 (1989).

[6] Bridle,J.S. :Training stochastic model recognition algorithms as networks can lead to maximum mutual information estimation of parameters. In *Neural Information Processing Systems 2*, David S.Touretzky, ed., pp.211-217, Morgan Kaufmann (1990).

[7] Gish,H. : A probabilistic approach to the understanding and training of neural network classifiers. In Proceedings of IEEE Conference on Acoustics Speech and Signal Processing, pp.1361-1364 (1990).

[8] Hampshire,J.B. and Waibel,A.H. : A novel objective function for improved phoneme recognition using time-delay neural networks, IEEE Trans. on Neural Networks, Vol.1, No.2, pp.216-228 (1990).

[9] Holt,M.J.J. and Semanani,S. : Convergence of back propagation in neural networks using a log-likelihood cost function, Electronics Letters, Vol.26, No.23 (1990).

[10] Kurita,T. : A Method to Determine the Number of Hidden Units of Three Layered Neural Networks by Information Criteria, Trans. of IEICE Japan, J73-D-II, 1872-1878, 1990 (in Japanese).

[11] Kurita,T. : On Maximum Likelihood Estimation of Feed-Forward Neural Net Parameters, IEICE Tech. Report, NC91-36, 1991 (in Japanese).

[12] McCullagh,P. and Nelder FRS,J.A. : *Generalized Linear Models*, Chapman and Hall, 1989.

DOMAINS OF ATTRACTION IN AUTOASSOCIATIVE MEMORY NETWORKS FOR CHARACTER PATTERN RECOGNITION

Koichi Niijima
Department of Control Engineering and Science
Kyushu Institute of Technology
Iizuka 820, Japan

Abstract

An autoassociative memory network is constructed by storing character pattern vectors whose components consist of a small positive number ε and $1-\varepsilon$. Although its connection weights and threshold values can not be determined only by this storing condition, it is proved that the output function of the network is contractive in a region around each stored pattern, if ε is sufficiently small. This implies that the region is a domain of attraction in the network. The shape of the region is clarified in our contraction mapping analysis. In addition to this region, larger domains of attraction are also found. Any noisy pattern vector in such domains, which may have real valued components, can be recognized as one of the stored patterns. Moreover, an autoassociative memory model having large domains of attraction is proposed. This model has symmetric connection weights and is successfully applied to character pattern recognition.

1 Introduction

In the study of pattern recognition by neural networks, it is important to investigate their recalling ability of stored patterns. The ability can be estimated by measuring the size of domains of attraction in the networks. This paper concerns the domains of attraction in autoassociative memory networks. Several papers treat the regions of attraction of patterns stored in associative memory networks. McEliece, Posner, Rodemich, and Venkatesh [4] evaluated the Hamming radius of attraction of the Hopfield associative memory. In Amari and Maginu [2], it was observed by computer simulation that the basins of attraction of an autocorrelation associative model have strange shapes. Cottrell [3] estimated the sizes of domains of attraction in an associative memory by computing the radius of Hamming sphere contained in the domains. Amari [1] found the asymptotic radius of one-step recalling region in the sparsely encoded associative memory. However, all these results are in 1 and 0 (or 1 and -1) world, that is, all the stored and noisy patterns treated therein are restricted to be binary valued vectors. It can occur that noisy character patterns have real valued components. In this paper, we deal with an autoassociative memory network which permits such noisy character patterns as objects of recognition.

Our network has n input layer nodes and the same number of output layer units. When an n-dimensional pattern vector is input from the input layer, linear combinations of the vector with connection weights are computed. Each output layer unit receives one of them from which a threshold value is subtracted. The resulting quantity is output through a monotone and differentiable sigmoid function.

We prepare m n-dimensional reference character patterns. Assume that $m < n$. Components of the reference patterns consist of a small positive number ε and $1-\varepsilon$. The reason why we use ε and $1-\varepsilon$ instead of 0 and 1 lies in the form of the sigmoid function. It is later seen that ε plays an important role in our analysis. On connection weights and threshold values in our network, we impose the condition that when each reference pattern is input, the same pattern is output. When a network satisfies this condition, we say that the m reference patterns are stored or memorized in the network. Since $m < n$, the connection weights and the threshold values can not be determined only by the above condition. However, it holds that if ε is small enough, then the output function of the network becomes a contraction mapping in a region around each stored pattern. This implies that the region is a domain of attraction having the stored pattern as an attractor. It is shown that the region is a convex polyhedral domain in the Euclidean space R^n, whose boundary consists of hyperplanes to be derived in our contraction mapping analysis. In addition to this region, larger domains of attraction are also found. Any noisy pattern in such domains can be recognized as one of the memorized patterns.

An autoassociative memory network having large domains of attraction is derived using the hyperplanes defining the original domain of attraction. It is achieved by determining connection weights and threshold values of our network so as to maximize the distances from each stored pattern to the hyperplanes. It is proved that the model obtained has symmetric connection weights.

Numerical simulations are carried out using this model. As reference patterns, we give ten capital letters from "A" to "J". Some noisy patterns are presented and it is checked which domain of attraction they belong to.

2 Domains of attraction

We consider the following neural network:

$$y_i = f(\sum_{j=1}^{n} w_{ij}x_j - \theta_i), \qquad i = 1, 2, ..., n, \tag{2.1}$$

where w_{ij} denote connection weights, and θ_i threshold values. The function f represents a sigmoid function of the form

$$f(t) = \frac{1}{1+\exp(-t)}. \tag{2.2}$$

We put $W_i = {}^t(w_{i1}, w_{i2}, ..., w_{in})$ and $x = {}^t(x_1, x_2, ..., x_n)$, where ^{t} is the transpose symbol. With these notations, (2.1) may be written as

$$y_i = f(W_i \cdot x - \theta_i), \qquad i = 1, 2, ..., n, \tag{2.3}$$

where $\cdot$ is the inner product symbol. We further put

$$\varphi_i(x) = f(W_i \cdot x - \theta_i),$$

$\varphi = {}^t(\varphi_1, \varphi_2, ..., \varphi_n)$ and $y = {}^t(y_1, y_2, ..., y_n)$. Then, (2.3) can be written in the operational form

$$y = \varphi(x).$$

We see that φ is a mapping from R^n into R^n. Let x^ν, $\nu = 1, 2, ..., m$, be reference character patterns, where $m < n$. Suppose that their components consist of sufficiently small positive number ε and $1 - \varepsilon$. When the weights w_{ij} and the thresholds values θ_i satisfy

$$x^\nu = \varphi(x^\nu) \tag{2.4}$$

for all $\nu = 1, 2, ..., m$, we say that the m patterns x^ν are stored in the network. The equation (2.4) implies that x^ν, $\nu = 1, 2, ..., m$, are fixed points of φ. The i-th component of (2.4) can be written as

$$f(W_i \cdot x^\nu - \theta_i) = x_i^\nu, \qquad \nu = 1, 2, ..., m.$$

Using the inverse function f^{-1} of f, we have

$$W_i \cdot x^\nu - \theta_i = f^{-1}(x_i^\nu), \qquad \nu = 1, 2, ..., m.$$

Since $x_i^\nu = \varepsilon$ or $1 - \varepsilon$, it follows by an easy calculation that $f^{-1}(x_i^\nu) = \text{sgn}(x_i^\nu - \frac{1}{2})\ln\frac{1-\varepsilon}{\varepsilon}$, where sgn indicates the sign function. Therefore, the change of variables

$$W_i = V_i \ln\frac{1-\varepsilon}{\varepsilon}, \qquad \theta_i = \eta_i \ln\frac{1-\varepsilon}{\varepsilon}$$

yields

$$V_i \cdot x^\nu - \eta_i = \text{sgn}(x_i^\nu - \frac{1}{2}), \qquad \nu = 1, 2, ..., m. \tag{2.5}$$

From now on, we call V_i and η_i the normalized weight vector and the normalized threshold value, respectively. We note that V_i and η_i are almost independent of ε, since the component of x^ν is almost 0 or 1. We also note that V_i and η_i can not be determined only by (2.5), because the number of unknown variables is $n + 1$, but that of the equations m. However, we have the following theorem.

Theorem 1. Suppose that the normalized weight vector V_i satisfies (2.5). For any fixed ρ, $0 < \rho < 1$ and each stored pattern x^ν, we define a domain $D_\rho(x^\nu)$ in R^n by

$$D_\rho(x^\nu) = \{x \in R^n \mid \max_{i=1,2,...,n} |V_i \cdot (x - x^\nu)| \leq \rho\}.$$

Then, if ε is sufficiently small for the above ρ, we have, for x, $\tilde{x} \in D_\rho(x^\nu)$,

$$\|\varphi(x) - \varphi(\tilde{x})\| \leq \kappa \|x - \tilde{x}\|$$

with κ satisfying

$$\kappa = \sqrt{\sum_{i=1}^{n} \|V_i\|^2 \, \varepsilon^{1-\rho} \ln\frac{1}{\varepsilon}} < 1,$$

where $\|\cdot\|$ denotes the Euclidean norm.

Proof. By the definition of φ_i,

$$\varphi_i(x) - \varphi_i(\bar{x}) = f(W_i \cdot x - \theta_i) - f(W_i \cdot \bar{x} - \theta_i).$$

Using the mean value theorem and the relation $f'(z) = f(z)(1 - f(z))$, it follows that

$$\varphi_i(x) - \varphi_i(\bar{x}) = f(z_i)(1 - f(z_i))W_i \cdot (x - \bar{x}), \tag{2.6}$$

where

$$z_i = \lambda(W_i \cdot x - \theta_i) + (1 - \lambda)(W_i \cdot \bar{x} - \theta_i), \qquad 0 < \lambda < 1.$$

Note that z_i can be rewritten as

$$\begin{aligned} z_i &= W_i \cdot x^\nu - \theta_i + W_i \cdot (\lambda(x - x^\nu) + (1 - \lambda)(\bar{x} - x^\nu)) \\ &= \operatorname{sgn}(x_i^\nu - \frac{1}{2}) \ln\frac{1 - \varepsilon}{\varepsilon} + W_i \cdot (\lambda(x - x^\nu) + (1 - \lambda)(\bar{x} - x^\nu)). \end{aligned}$$

Since $D_\rho(x^\nu)$ is a convex set, $\lambda(x - x^\nu) + (1 - \lambda)(\bar{x} - x^\nu)$ is in $D_\rho(x^\nu)$. Therefore,

$$\begin{aligned} |W_i \cdot (\lambda(x - x^\nu) + (1 - \lambda)(\bar{x} - x^\nu))| &\leq \lambda|W_i \cdot (x - x^\nu)| + (1 - \lambda)|W_i \cdot (\bar{x} - x^\nu)| \\ &\leq \rho \ln\frac{1 - \varepsilon}{\varepsilon}. \end{aligned}$$

Putting here

$$z_i^- = \left(\operatorname{sgn}(x_i^\nu - \frac{1}{2}) - \rho\right) \ln\frac{1 - \varepsilon}{\varepsilon}, \tag{2.7}$$

$$z_i^+ = \left(\operatorname{sgn}(x_i^\nu - \frac{1}{2}) + \rho\right) \ln\frac{1 - \varepsilon}{\varepsilon}, \tag{2.8}$$

it follows that $z_i^- \leq z_i \leq z_i^+$ and hence

$$f(z_i^-) \leq f(z_i) \leq f(z_i^+)$$

because of the monotonicity of f.

We consider the case $x_i^\nu = \varepsilon$. Then $f(z_i^+) \leq 1/2$ holds, since $z_i^+ \leq 0$ from (2.8). Hence,

$$f(z_i)(1 - f(z_i)) \leq f(z_i^+)(1 - f(z_i^+)).$$

Denoting the right hand side by d_+, we get

$$d_+ = \frac{\exp(-z_i^+)}{(1 + \exp(-z_i^+))^2}.$$

Since $z_i^+ = (-1 + \rho) \ln\frac{1-\varepsilon}{\varepsilon}$, we have

$$\begin{aligned} d_+ &= \frac{\exp((1 - \rho)\ln\frac{1-\varepsilon}{\varepsilon})}{(1 + \exp((1 - \rho)\ln\frac{1-\varepsilon}{\varepsilon}))^2} \\ &= \frac{\varepsilon(1 - \varepsilon)\exp(-\rho\ln\frac{1-\varepsilon}{\varepsilon})}{(\varepsilon + (1 - \varepsilon)\exp(-\rho\ln\frac{1-\varepsilon}{\varepsilon}))^2}. \end{aligned}$$

Using here the inequality

$$\frac{1}{\varepsilon+(1-\varepsilon)\exp(-t)} \leq \exp(t)$$

for $t \geq 0$, d_+ is bounded as

$$\begin{aligned} d_+ &\leq \varepsilon(1-\varepsilon)\exp(\rho\ln\frac{1-\varepsilon}{\varepsilon}) \\ &\leq \varepsilon^{1-\rho}. \end{aligned}$$

Therefore, when $x_i^\nu = \varepsilon$, we have

$$f(z_i)(1-f(z_i)) \leq f(z_i^+)(1-f(z_i^+)) \leq \varepsilon^{1-\rho}. \tag{2.9}$$

We next consider the case $x_i^\nu = 1-\varepsilon$. Then $f(z_i^-) \geq 1/2$ holds, since $z_i^- \geq 0$ from (2.7). Hence,

$$f(z_i)(1-f(z_i)) \leq f(z_i^-)(1-f(z_i^-)).$$

The right hand side is also bounded by $\varepsilon^{1-\rho}$ using the same technique as in the case $x_i^\nu = \varepsilon$. Therefore, when $x_i^\nu = 1-\varepsilon$, we have

$$f(z_i)(1-f(z_i)) \leq f(z_i^-)(1-f(z_i^-)) \leq \varepsilon^{1-\rho}. \tag{2.10}$$

Combining (2.9) and (2.10) with (2.6) gives

$$\begin{aligned} |\varphi_i(x)-\varphi_i(\tilde{x})| &\leq \varepsilon^{1-\rho}\ln\frac{1-\varepsilon}{\varepsilon}\,|V_i\cdot(x-\tilde{x})| \\ &\leq \|V_i\|\,\varepsilon^{1-\rho}\ln\frac{1}{\varepsilon}\,\|x-\tilde{x}\|. \end{aligned}$$

Summing up both sides from $i=1$ until $i=n$ after squared, we get the first inequality. The inequality $\kappa < 1$ follows from the fact that $\varepsilon^{1-\rho}\ln\frac{1}{\varepsilon}$ approaches to 0 when ε tends to 0.

The next result follows immediately from Theorem 1.

Corollary 1. Let ρ and κ be defined in Theorem 1. Then we have
(i) $x = \varphi(x)$ has a unique solution in $D_\rho(x^\nu)$,
(ii) if $x^{(1)}$ is in $D_\rho(x^\nu)$, then the sequence $\{x^{(\ell)}\}$ generated by

$$x^{(\ell+1)} = \varphi(x^{(\ell)}), \qquad \ell = 1,2,... \tag{2.11}$$

converges to x^ν,
(iii) $D_\rho(x^\nu)$, $\nu = 1,2,...,m$, are mutually disjoint.

The result (ii) asserts that any noisy version x in $D_\rho(x^\nu)$ can be recognized as the stored pattern x^ν, that is, $D_\rho(x^\nu)$ is a domain of attraction of x^ν. It should be noted that $D_\rho(x^\nu)$ is a candidate among domains of attraction of x^ν.

We next study the domain $D_\rho^k(x^\nu)$ of the vector x such that its k-step recall $\varphi^k(x)$ is in $D_\rho(x^\nu)$:

$$D_\rho^k(x^\nu) = \{x \in R^n \mid \max_{i=1,2,...,n} |V_i\cdot(\varphi^k(x)-x^\nu)| \leq \rho\ \}.$$

This domain can also be regarded as a domain of attraction of x^ν.

Concerning these domains, we have

Theorem 2. Let ρ and κ be defined in Theorem 1. Assume that ϵ is so small as to satisfy

$$\sqrt{n} \max_{i=1,2,\ldots,n} \|V_i\| \, \kappa \leq \rho. \tag{2.12}$$

Then we have

$$D_\rho^0(x^\nu) \subset D_\rho^1(x^\nu) \subset \cdots \subset D_\rho^k(x^\nu) \subset \cdots,$$

where $D_\rho^0(x^\nu)$ indicates $D_\rho(x^\nu)$ defined in Theorem 1.

Proof. It suffices to show that $D_\rho^k(x^\nu)$ is included in $D_\rho^{k+1}(x^\nu)$ for any k. Let x be any vector in $D_\rho^k(x^\nu)$. Then, for $u = \varphi^k(x)$, it holds that

$$|V_i \cdot (u - x^\nu)| \leq \rho$$

which implies that u is in $D_\rho(x^\nu)$. Therefore, we get by Theorem 1,

$$\|\varphi(u) - \varphi(x^\nu)\| \leq \kappa \|u - x^\nu\|.$$

By this inequality and assumption (2.12), we have

$$\begin{aligned} |V_i \cdot (\varphi^{k+1}(x) - x^\nu)| &= |V_i \cdot (\varphi(u) - \varphi(x^\nu))| \\ &\leq \|V_i\| \, \|\varphi(u) - \varphi(x^\nu)\| \\ &\leq \|V_i\| \, \kappa \|u - x^\nu\| \\ &\leq \sqrt{n} \max_{i=1,2,\ldots,n} \|V_i\| \, \kappa \\ &\leq \rho, \end{aligned}$$

where we have used $0 < u, x^\nu < 1$. This implies that x is in $D_\rho^{k+1}(x^\nu)$, which finishes the proof.

For these domains, we have the following result.

Corollary 2. Suppose that the same conditions as in Theorem 2 are satisfied. Then, for any integer $p, q \geq 0$ and $\nu \neq \mu$, we have

$$D_\rho^p(x^\nu) \cap D_\rho^q(x^\mu) = \phi.$$

This is an extension of (iii) of Corollary 1. It is very hard to clarify the structure of the asymptotic region $\lim_{p\to\infty} D_\rho^p(x^\nu)$.

3 An autoassociative memory model

As described in the previous section, the normalized weight vector V_i and the normalized threshold value η_i can not be determined only by (2.5). We propose here a method for determining them, thereby an autoassocitive memory model is obtained. The method concerns the shape of the domain $D_\rho(x^\nu)$. As shown in Fig. 1, the boundary of the domain consists of the hyperplanes $|V_i \cdot (x - x^\nu)| = \rho, \ \ i = 1, 2, \ldots, n$.

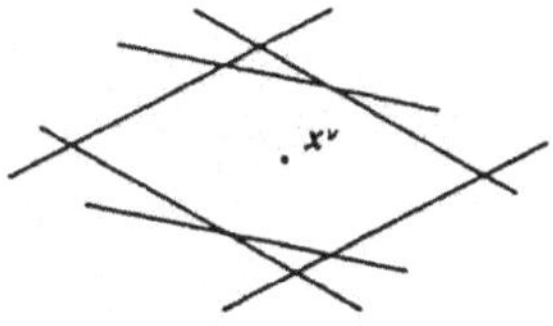

Fig. 1. The domain $D_\rho(x^\nu)$.

We shall maximize the distance from the stored pattern x^ν to each hyperplane. This approach is valuable in a sense that for each i, the layer lying between $V_i \cdot (x - x^\nu) = -\rho$ and $V_i \cdot (x - x^\nu) = \rho$ is made thick. The distance from x^ν to the i-th hyperplane is easily computed to get $\rho / \|V_i\|$ as far as $\|V_i\| \neq 0$. This idea yields the following minimization problem:

$$\|V_i\|^2 \quad \longrightarrow \quad \min \tag{3.1}$$

subject to

$$V_i \cdot x^\nu - \eta_i = \mathrm{sgn}(x_i^\nu - \frac{1}{2}), \qquad \nu = 1, 2, ..., m. \tag{3.2}$$

Concerning this problem, we have the following result.

Theorem 3. Assume that the stored patterns x^ν, $\nu = 1, 2, ..., m$, are linearly independent. Let G be the $m \times m$ Gram matrix whose (ν, μ)-element is the inner product (x^ν, x^μ), and let $\boldsymbol{a}$ be the m-dimensional vector ${}^t(2, 2, ..., 2)$. We further put $\xi_i = {}^t(\xi_{i1}, \xi_{i2}, ..., \xi_{im})$ and $r_i = {}^t(r_{i1}, r_{i2}, ..., r_{im})$, where $r_{i\nu} = -2\,\mathrm{sgn}(x_i^\nu - \frac{1}{2})$. Then the equation

$$\begin{pmatrix} G & \boldsymbol{a} \\ {}^t\boldsymbol{a} & 0 \end{pmatrix} \begin{pmatrix} \xi_i \\ \eta_i \end{pmatrix} = \begin{pmatrix} r_i \\ 0 \end{pmatrix} \tag{3.3}$$

has a unique solution $(\xi_i,\ \eta_i)$, and the solution V_i of the problem (3.1) and (3.2) is given by

$$V_i = {}^t(v_{i1}, v_{i2}, ..., v_{in}),$$

where

$$v_{ij} = -\frac{1}{2} \sum_{\mu=1}^{m} \xi_{i\mu} x_j^\mu,$$

and $\xi_{i\mu}$ denotes the μ-component of the solution ξ_i of (3.3).

Proof. We introduce the Lagrange multipliers $\xi_{i\mu}$ and put

$$J = \|V_i\|^2 + \sum_{\mu=1}^{m} \xi_{i\mu}(V_i \cdot x^\mu - \eta_i - \mathrm{sgn}(x_i^\mu - \frac{1}{2})).$$

By $\frac{\partial J}{\partial \xi_{i\nu}} = 0$, we get (3.2). Next we have from $\frac{\partial J}{\partial v_{ij}} = 0$,

$$2v_{ij} + \sum_{\mu=1}^{m} \xi_{i\mu} x_j^\mu = 0. \tag{3.4}$$

We further have by $\frac{\partial J}{\partial \eta_i} = 0$,

$$\sum_{\mu=1}^{m} \xi_{i\mu} = 0. \tag{3.5}$$

Solving (3.4),

$$v_{ij} = -\frac{1}{2}\sum_{\mu=1}^{m} \xi_{i\mu} x_j^{\mu}. \tag{3.6}$$

Substituting this into (3.2), we get

$$-\frac{1}{2}\sum_{\mu=1}^{m}(x^{\nu}, x^{\mu})\xi_{i\mu} - \eta_i = \text{sgn}(x_i^{\nu} - \frac{1}{2}), \qquad \nu = 1, 2, ..., m.$$

We combine this with (3.5) to get (3.3).

It remains to show the solvability of (3.3). Since x^{ν}, $\nu = 1, 2, ..., m$, are linearly independent, the Gram matrix G is nonsingular and the following equation holds:

$$\begin{pmatrix} G & \boldsymbol{a} \\ {}^t\boldsymbol{a} & 0 \end{pmatrix}\begin{pmatrix} E & -G^{-1}\boldsymbol{a} \\ {}^t\mathbf{o} & 1 \end{pmatrix} = \begin{pmatrix} G & \mathbf{o} \\ {}^t\boldsymbol{a} & -{}^t\boldsymbol{a}G^{-1}\boldsymbol{a} \end{pmatrix},$$

where $\mathbf{o}$ denotes m-dimensional zero vector and E denotes the $m \times m$ unit matrix. Therefore, putting

$$A = \begin{pmatrix} G & \boldsymbol{a} \\ {}^t\boldsymbol{a} & 0 \end{pmatrix},$$

we have

$$\det(A) = -{}^t\boldsymbol{a}G^{-1}\boldsymbol{a} \det(G).$$

The matrix G is positive definite, and so G^{-1} is. We already know that $\det(G) \neq 0$. Therefore, $\det(A) \neq 0$. This finishes the proof.

The connection weights w_{ij} and the threshold values θ_i are computed using the relations

$$w_{ij} = v_{ij} \ln\frac{1-\varepsilon}{\varepsilon}, \qquad \theta_i = \eta_i \ln\frac{1-\varepsilon}{\varepsilon},$$

where v_{ij} and η_i have been obtained in Theorem 3.

Thus we can obtain the desired autoassociative memory model. Next, we show that the model has the symmetric connection weights w_{ij}.

Theorem 4. The connection weights w_{ij} of the proposed model are symmetric, that is, it holds that

$$w_{ij} = w_{ji}$$

for $1 \leq i,\ j \leq n$.

Proof. It suffices to prove that $v_{ij} = v_{ji}$ holds, because $w_{ij} = v_{ij}\ln\frac{1-\varepsilon}{\varepsilon}$. Remember that

$$v_{ij} = -\frac{1}{2}\sum_{\mu=1}^{m} \xi_{i\mu} x_j^{\mu}.$$

Using two relations $x_j^{\mu} = (\frac{1}{2} - \varepsilon)\text{sgn}(x_j^{\mu} - \frac{1}{2}) + \frac{1}{2}$ and $\sum_{\mu=1}^{m} \xi_{i\mu} = 0$, we get

$$v_{ij} = -\frac{1}{2}(\frac{1}{2} - \varepsilon)\sum_{\mu=1}^{m} \xi_{i\mu}\text{sgn}(x_j^{\mu} - \frac{1}{2}).$$

By the definition of r_j, that is, $r_j = -2\,{}^t(\mathrm{sgn}(x_j^1 - \frac{1}{2}), \mathrm{sgn}(x_j^2 - \frac{1}{2}), ..., \mathrm{sgn}(x_j^m - \frac{1}{2}))$, the quantity v_{ij} can be expressed as

$$v_{ij} = (\frac{1}{2} - \varepsilon)(\xi_i, r_j)$$

using the inner product symbol $(\cdot, \cdot)$. On the other hand, ξ_i may be written as $\xi_i = Br_i$ using an $m \times m$ symmetric matrix B. Therefore,

$$\begin{aligned} v_{ij} &= (\frac{1}{2} - \varepsilon)(Br_i, r_j) \\ &= (\frac{1}{2} - \varepsilon)(r_i, Br_j) \\ &= v_{ji}. \end{aligned}$$

This completes the proof.

By virtue of Theorem 4, we see that there exists an energy function for the present model. Therefore, the stored patterns x^ν, $\nu = 1, 2, ..., m$, become local minima of the energy function and the domains of attraction of the model are the basins of attraction. Of course, the energy function may have local minima except the stored patterns.

4 Numerical simulations

We perform numerical simulations using the proposed model in Section 3. Stored patterns in this model are the capital letters from "A" to "J" below, each of which is represented by the 10×10 grid matrix.

Fig. 2. Stored capital letters.

The dimension n of the pattern is 100 and the number m of the stored patterns is 10. Black and white squares indicate $1 - \varepsilon$ and ε, respectively. As ρ appearing in Theorem 1, we choose $\rho = 0.998$ for which $\varepsilon = \exp(-10000)$ is selected. The normalized weight vectors V_i in the model were computed using Theorem 3. The result shows that $\|V_i\|$ are zero for 22 indexes i. This means that the output layer unit for such i is not connected with any input layer node. As described in Section 3, the distance from x^ν to the hyperplanes $|V_i \cdot (x - x^\nu)| = 0.998$ for nonzero $\|V_i\|$ is $0.998/\|V_i\|$. Therefore, the size of the domain attraction $D_{0.998}(x^\nu)$ can be estimated by computing $0.998/\|V_i\|$ for nonzero $\|V_i\|$. So, it is worthy to evaluate the maximum value Max, the minimum value Min, the mean value $\bar{V}$, and the variance σ^2 of nonzero $\|V_i\|$:

$$Max = 1.020, \qquad Min = 0.467, \qquad \bar{V} = 0.664, \qquad \sigma^2 = 0.021.$$

We show that κ in Theorem 1 is less than 1, that is, the output function of the present model is contractive. Since $(\sum_{i=1}^n \|V_i\|^2)^{1/2} = 6.0$ and $\varepsilon^{1-\rho}\ln\frac{1}{\varepsilon} = \exp(-20)$, we have $\kappa = 1.237 \times 10^{-4} < 1$. We next check the condition (2.12) in Theorem 2. The left hand side of (2.12) is equal to 0.00126 which is less than the present ρ. Therefore, $D_{0.998}^k(x^\nu)$ for all $k = 0, 1, 2, ...,$ and $\nu = 1, 2, ..., 10$, become domains of attraction in the model.

The first noisy capital letter x is

Fig. 3. The first noisy capital letter.

The area of each square is proportional to the component of x. To find out a domain of attraction to which the noisy pattern x belongs, we compute $\max_{i=1,2,...,100} |V_i \cdot (x - x^\nu)|$ for $\nu = 1, 2, ..., 10$:

ν	1	2	3	4	5	6	7	8	9	10
	2.050	2.392	2.054	1.903	2.392	2.018	2.392	2.018	1.706	0.993

Table 1. The values of $\max_{i=1,2,...,100} |V_i \cdot (x - x^\nu)|$.

Table 1 shows that the pattern x in Fig.3 is in $D_{0.998}(x^{10})$, that is, this pattern is recognized as "J" pattern.

The second noisy capital letter x is

Fig. 4. The second noisy capital letter.

We first try to compute $\max_{i=1,2,...,100} |V_i \cdot (x - x^\nu)|$ for $\nu = 1, 2, ..., 10$. The computed values are

ν	1	2	3	4	5	6	7	8	9	10
	1.928	1.728	1.943	1.854	1.268	1.715	1.943	1.950	1.571	1.875

Table 2. The values of $\max_{i=1,2,...,100} |V_i \cdot (x - x^\nu)|$.

All the values in Table 2 are larger than 0.998, which implies that the pattern x in Fig.4 is not in any $D_\rho(x^\nu)$. The next task is to check whether or not this x is in $D^1_\rho(x^\nu)$. For this purpose, we compute $\max_{i=1,2,\ldots,100} |V_i \cdot (\varphi(x) - x^\nu)|$ for $\nu = 1, 2, \ldots, 10$:

ν	1	2	3	4	5	6	7	8	9	10
	2.085	1.932	2.005	1.991	0.606	1.731	2.085	2.072	2.052	2.052

Table 3. The values of $\max_{i=1,2,\ldots,100} |V_i \cdot (\varphi(x) - x^\nu)|$.

Since the fifth value is less than 0.998, the noisy pattern x in Fig.4 belongs to $D^1_{0.998}(x^5)$ and this pattern is recognized as "E" pattern.

The third noisy capital letter x is given by

Fig. 5. The third noisy capital letter.

As a computational result, we see that this x is not in any $D^k_\rho(x^\nu)$ for $0 \leq k \leq 2$. However, this pattern belongs to $D^3_{0.998}(x^6)$ as shown below, that is, x in Fig.5 is recognized as "F" pattern:

ν	1	2	3	4	5	6	7	8	9	10
	2.022	2.081	2.038	1.992	2.081	0.146	2.081	2.029	2.038	2.033

Table 4. The values of $\max_{i=1,2,\ldots,100} |V_i \cdot (\varphi^3(x) - x^\nu)|$.

In the last, we give the following noisy capital letter x:

Fig. 6. The last noisy capital letter.

A computational result shows that this x is not in any $D^k_\rho(x^\nu)$ for $0 \leq k \leq 3$. However, as easily seen from the table below, this pattern belongs to $D^4_{0.998}(x^1)$, that is, x in Fig.6 is recognized as "A" pattern:

ν	1	2	3	4	5	6	7	8	9	10
	0.325	2.191	2.102	2.121	2.191	2.121	2.191	2.121	2.049	2.116

Table 5. The values of $\max_{i=1,2,\ldots,100} |V_i \cdot (\varphi^4(x) - x^\nu)|$.

5 Conclusions

In this paper, we constructed an autoassociative memory network by a recording algorithm and sought its domains of attraction based on the contraction mapping principle. Regions of attraction containing such domains were also found. We further designed an autoassociative memory network by using the shape of the original domain of attraction.

There are many problems to be studied in the future. The model in Section 3 seems to be closely related to the associative memory networks obtained by the correlation recording and the generalized-inverse recording. It is needed to compare their recalling ability with ours. The domains of attraction found in this paper are still restricted and their sizes are independent of the stored patterns. We need to develop the present approach to get larger domains of attraction depending on the memorized patterns.

References

[1] Amari,S., Characteristics of Sparsely Encoded Associative Memory, *Neural Networks*,Vol.2, 1989, pp.451-457.

[2] Amari,S. and Maginu,K., Statistical Neurodynamics of Associative Memory, *Neural Networks*,Vol.1, 1988, pp.63-73.

[3] Cottrell,M., Stability and Attractivity in Associative Memory Networks, *Biological Cybernetics*,Vol.58, 1988, pp.129-139.

[4] McEliece,R.J., Posner,E.C., Rodemich,E.R., and Venkatesh,S.S., The Capacity of the Hopfield Associative Memory, *IEEE Transactions on Information Theory*,Vol.33, 1987, pp.461-482.

REGULARIZATION LEARNING OF NEURAL NETWORKS FOR GENERALIZATION

Shotaro Akaho

Mathematical Informatics Section
Electrotechnical Laboratory
1-1-4 Umezono, Tsukuba 305, Japan

Abstract

In this paper, we propose a learning method of neural networks based on the regularization method and analyze its generalization capability. In learning from examples, training samples are independently drawn from some unknown probability distribution. The goal of learning is minimizing the expected risk for future test samples, which are also drawn from the same distribution. The problem can be reduced to estimating the probability distribution with only samples, but it is generally ill-posed. In order to solve it stably, we use the regularization method. Regularization learning can be done in practice by increasing samples by adding appropriate amount of noise to the training samples. We estimate its generalization error, which is defined as a difference between the expected risk accomplished by the learning and the truly minimum expected risk. Assume p-dimensional density function is s-times differentiable for any variable. We show the mean square of the generalization error of regularization learning is given as $Dn^{-2s/(2s+p)}$, where n is the number of samples and D is a constant dependent on the complexity of the neural network and the difficulty of the problem.

1 Introduction

Generalization is one of the most important problems of neural network learning. As it is, even if the learned network is adapted to a given set of training samples well, it often does not fit unknown test samples. In this paper, in order to get the network with small generalization error, we propose a learning method of neural networks based on the regularization method, and analyze its effectiveness.

Neural network learning can be formulated as learning from examples: Assume that training samples are given from environment and each sample z is independently generated according to some unknown probability distribution $P(z)$. The goal of learning is to minimize a loss function $Q(z, \alpha)$ not only for training samples but also for test samples, where $Q(z, \alpha)$ belongs to a set of function parameterized by α. For example, in pattern recognition problem, each z denotes a pair of input pattern x and class y. A loss function is provided as $(y - q(x, \alpha))^2$, where $q(x, \alpha)$ denotes a function parameterized by α. Suppose test samples are generated according to the same probability distribution as training samples, the problem is to minimize the *expected risk* defines as

$$I(\alpha) = \int Q(z, \alpha) P(z)\, dz, \tag{1}$$

when the density function $P(z)$ is unknown but random independent samples are given[9].

It is known that neural networks with enough number of units has a capability to approximate arbitrary functions[4]. However, because the density function $P(z)$ is assumed to be unknown, we cannot minimize the expected risk $I(\alpha)$ just with a small number of training samples. Usually, instead of $I(\alpha)$, *empirical risk* defined as

$$I_{\text{emp}}(\alpha) = \frac{1}{n}\sum_{i=1}^{n} Q(\underset{(i)}{Z}, \alpha). \tag{2}$$

is minimized, where $\underset{(1)}{Z}, \ldots, \underset{(n)}{Z}$ denote training samples.

Two minimizations of (1) and (2) above do not agree with each other in general. As the capability or complexity of a network is getting higher, the capability to describe unknown data becomes lower, because the learned function becomes unstable and overfits just training samples (*over-learning* or *lack of generalization capability*). One method to avoid this problem is to limit a set of realizable functions by restricting the size or structure of a network (*statistical model selection*). Though many methods are proposed from a statistical point of view, they have problems as follows:

1. $I_{\text{emp}}(\alpha)$ is not so good estimate of $I(\alpha)$.

2. Model selection is done for a discrete set of functions, and each function is structurally different from others. Thus usually, we cannot apply a method like the steepest descent method, and model selection needs two steps: first, we train several networks and estimate their generalization capability, and next, we select the best one among them.

3. The method based on AIC and MDL (information criteria) [10][5] is not so robust, because they are essentially parametric.

4. The method based on VC dimension (function complexity) [9][2][1] cannot be used in practice, because the bound of the estimated generalization error is not so tight.

In this paper, we try to avoid model selection procedure. For this purpose, instead of minimizing empirical risk, we consider the minimization of expected risk itself, using $P(z)$ estimated with training samples. It is known that the nonparametric estimation of density function from finite number of data is an *ill-posed* problem and the solution is unstable. *Regularization method* is used for stabilizing the solution.

In section 2, we describe the density approximation method based on the regularization method. Then in section 3, we propose the learning method and analyze its effectiveness. Main theoretical result is theorem 1 in section 3, which valuates the expected error of the estimation of $I(\alpha)$.

2 Nonparametric density approximation

2.1 Regularization method

As noticed in the previous section, the problem of density estimation is ill-posed. The *regularization method* is a general method to solve ill-posed problems nonparametrically[8] and for the estimation of density function $f(t)$ it is formalized as follows.

Consider the problem to solve the following integral equation from n empirical data.

$$Af(t) \equiv \int_{-\infty}^{\infty} u(x-t)f(t)\,dt = F(x), \tag{3}$$

where $u(x)$ is a unit step function. Though we can estimate the cumulative distribution function $F(x)$ stably, only an unstable solution $f(x)$ can be derived by putting it in above equation directly. Now, instead of solving the equation (3), consider the functional minimization problem below:

$$R_\gamma(\hat{f}, F) = \rho^2(A\hat{f}, F) + \gamma\Omega(\hat{f}), \tag{4}$$

where ρ denotes a distance between two functions, Ω denotes a functional called stabilizer that satisfies several conditions such as compactness and positiveness, and γ is a positive parameter. It is known the problem above can be solved stably: $\hat{f}$ converges to the true solution of $Af = F$, when the value γ is decreasing slowly enough in proportion to the approximation accuracy of $F(x)$.

2.2 Regularization method and Parzen method

It is known that the regularization method is equivalent to *Parzen method* (or kernel type density estimation method) for the problem of density estimation[9].

Parzen method is described as follows. Let $\underset{(1)}{X}, \ldots, \underset{(n)}{X}$ be n empirical data from a density function $f(x)$. The approximation of $f(x)$ is given as follows (one dimensional case).

$$f_n(x, a_n) = \frac{a_n}{n} \sum_i K(a_n(x - \underset{(i)}{X})), \tag{5}$$

where $K(x)$ is a function that satisfies several conditions and a_n is a constant.

Remark that $K(x)$ and a_n correspond to stabilizer and regularization parameter respectively in the regularization method.

Roughly speaking, it is a method to shade with $K(x)$ around each sample point. In this method, it is important to determine the shading parameter and to estimate the accuracy of approximation. If we cannot deal with them well, over-learning or over-shading eventually occurs.

In Parzen method, an asymptotically optimum value of a_n can be determined from empirical data [6]. Its detail result is shown in appendix A. The essential points are

- Assume that the density function (p dimensional) is s times partially continuously differentiable.
- Pick $K(x)$ that satisfies some appropriate conditions.
- Then the mean square error of the approximated function can be estimated ((14),(15), theorem 2).
- We can determine an asymptotically optimum value of parameter a_n from that estimation ((22)).

3 Learning algorithm via regularization method

In this section, we propose a learning algorithm using approximated density function, and estimate its capability.

3.1 Estimation of generalization capability

In order to minimize expected risk $I(\alpha)$, we consider the minimization of $\hat{I}(\alpha)$ defined below, instead of $I(\alpha)$.

$$\hat{I}(\alpha) = \int Q(z,\alpha)\hat{P}(z)\,dz, \tag{6}$$

where $\hat{P}(\alpha)$ is a density approximated by Parzen method. Since $\hat{I}(\alpha)$ depends upon just empirical data, this minimization is able in theory. Let us estimate how close this value is to the true expected risk under the condition that this minimization has been succeeded. In the next section, we state how this minimization can be done.

We shall say the network has generalization property if the following equation holds for an arbitrary given δ,

$$\mathrm{E}[I(\hat{\alpha}_0) - I(\alpha_0)]^2 < \delta, \tag{7}$$

where $\hat{\alpha}_0$ denotes the α that minimizes $\hat{I}(\alpha)$, and α_0 denotes the α that minimizes $I(\alpha)$.

From theorem 3 in appendix A, we can bound the left hand side of (7) and the following theorem is valid. (see appendix B about its proof).

Theorem 1 *Let p-dimensional density function $P(z)$ be s times differentiable for any variable, and let the domain region X be bounded. Then, for large n,*

$$\mathrm{E}[I(\hat{\alpha}_0) - I(\alpha_0)]^2 < \overline{D}_1 D_2 n^{-2s/(2s+p)}, \tag{8}$$

holds, where E *denotes the expectation*

$$\mathrm{E}[\,\cdot\,] = \int \cdot \prod_j P(\underset{(j)}{Z})\,d\underset{(j)}{Z},$$

$\overline{D}_1$ denotes

$$\overline{D}_1 = \sup_\alpha D_1(\alpha), \qquad D_1(\alpha) = \int_X Q(z,\alpha)^2\,dz, \tag{9}$$

and the value D_2 depends on $P(z), p$ and s (see appendix B).

Accordingly, the value D_2 corresponds to the difficulty of the problem itself, and the value $\overline{D}_1$ corresponds to the complexity of the set of functions.

For example, consider the case that $Q(z) = (y - q(x,\alpha))^2$ and x_i, y, q are bounded between 0 and 1, which is a typical assumption of neural networks. In this case, $D_1(\alpha)$ is also bounded by 1 for any α.

3.2 A method to minimize the estimated expected risk

Here we consider how to minimize $\hat{I}(\alpha)$. Since $\hat{I}(\alpha)$ includes the calculation of integration and sum, it seems difficult to minimize directly. One simple idea is to increase the number of data by generating samples from the estimated density function $\hat{P}(z)$, and to train a network with them. This process can be repeated incrementally. We call these generated samples as "quasi-samples".

The following algorithm is a possible implementation of this idea. (it is described in one-dimensional notation for simplicity).

1: Take n samples $(\underset{(1)}{X}, \ldots, \underset{(n)}{X})$.

2: Repeat I),II):

I) Generate some quasi-samples; Each of them is generated by i), ii):

i) Select one sample $\underset{(i)}{X}$ arbitrary.

ii) Generate a random sample according to the density $a_n K(a_n(x - \underset{(i)}{X}))$

II) Train the network slightly with the quasi-samples.

Remark. $a_n K(a_n(x - \underset{(i)}{X}))$ does not always satisfy conditions for density function. In this case, it becomes a little more complex to generate quasi-samples and to train the network. Let $f(x)$ be $a_n K(a_n(x - \underset{(i)}{X}))$ for arbitrary selected X_i, and separate $f(x)$ into the positive part and the negative part, namely,

$$f(x) = f^+(x) - f^-(x), \qquad f^+(x), f^-(x) \geq 0. \tag{10}$$

Generate two quasi-samples x^+ according to $f^+(x)/\overline{f^+}$ and x^- according to $f^-(x)/\overline{f^-}$, where $\overline{f^+}$ and $\overline{f^-}$ are the normalization parameters for $f^+(x)$ and $f^-(x)$ respectively. Using x^+ and x^-, train the network to minimize

$$\overline{f^+} Q(x^+, \alpha) - \overline{f^-} Q(x^-, \alpha). \tag{11}$$

This learning becomes equivalent to minimizing $\hat{I}(\alpha)$, as the number of quasi-samples increases.

In the learning for quasi-samples, it is desirable not to learn so much in order to avoid the problem of converging to local minima.

3.3 Experiments

We show the results of a simple numerical experiment.

Samples: Both input x and output y are one-dimensional, x is generated from uniform distribution on $[0, 1]$, and y is from constant function $y = 0.5$ with additive normal noise of variance 0.01.

Network size: Feedforward three-layered network with 1,10,1 units in the input layer, the hidden layer, and the output layer respectively.

Learning algorithm: In the regularization learning, we generated 1000 quasi-samples from all training samples and trained the network 100 steps with batch-type back propagation. We call this procedure a "learning-step" and we repeated 50 learning-steps.

In the learning with just training samples, we trained the network (1000×100/ Sample size) steps with batch-type back propagation for each learning-step of regularization learning.

Back propagation used here is the simplest one with no acceleration, and a coefficient of the steepest descent is taken 0.5.

Error estimation: MSE(mean square error) for both training samples and 10000 test samples.

Result: Figure 1 shows the MSE plots for one learning process, where the number of samples is 14.

Figure 2 shows the error plots for several sample sizes. Each error is averaged for the last 20 learning-steps, and further it is averaged for 10 such experiments.

Since each sample includes a noise of variance 0.01, there remains about 0.01 mean square error for test data, even if learning could converge to the optimum function.

In every cases, the learning with just training samples does not seem to have generalization property, overfitting to training samples, while regularization learning vibrates around 0.01 both for training samples and test samples (the reason of vibration is thought to be generating new quasi-samples one after another).

4 Concluding remarks

We proposed a learning method using regularization and analyzed its generalization capability. This method overcomes some drawbacks of model selection, and it also includes another measure of the complexity that does not change so much for the selection of a function class.

The method to add some noise to training samples has been used as a heuristic method in order to reduce generalization error. The present paper provides this kind of methods with theoretical foundation in a general framework, and it can also answer questions such as how large the noise should be.

The field of smoothing spline is deeply related to our method[7][3]. In the present paper, we mainly treated the estimation of density function rather than the estimation of regression function and our main concern is the analysis of generalization capability rather than estimation itself.

The estimation in theorem 1 and theorem 3 is getting worse as the dimension of data increase when differentiability is not high. Thus for using this method, it is important to reduce dimensionality of input data previously by extracting efficient low dimensional features from raw data.

The criterion to estimate the generalization error in this paper is different from Vapnik's one. To show the regularization learning is superior to the learning with just training samples, we must estimate it on the same criterion and solve some real world problems in practice (The experiment we've shown is too simple to show the effectiveness of our method). These tasks are left in future.

Acknowledgment

The author would like to thank K.Tamura, Director of Information Science Division of ETL, for affording an opportunity of this study. He is also deeply indebted to N.Otsu, Director of Machine Understanding Division, H.Asoh, T.Kurita and all other members of Mathematical Informatics Section for their helpful discussions.

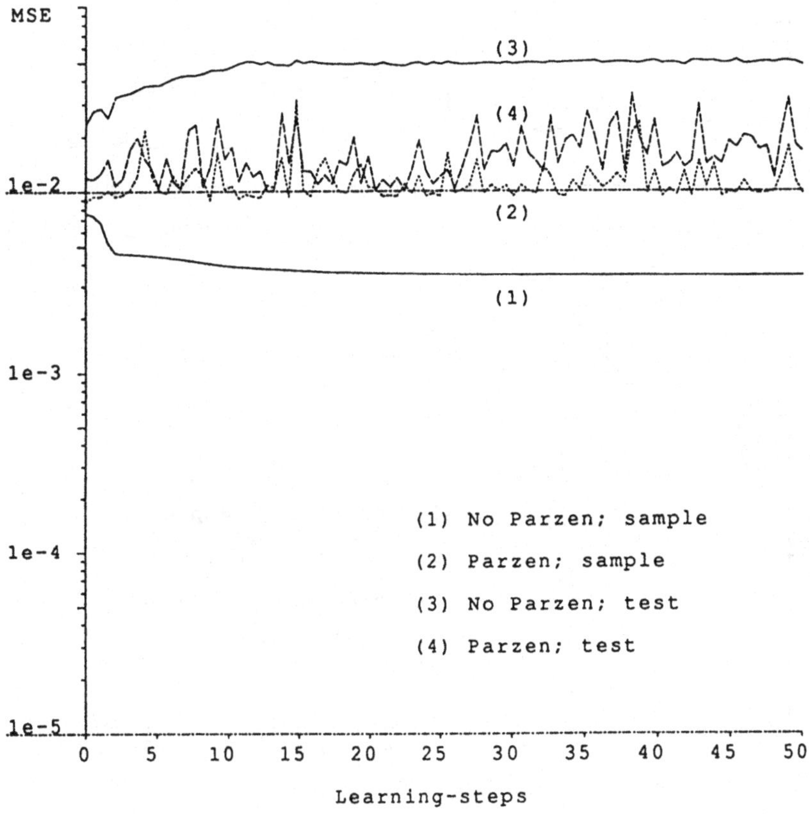

Figure 1: MSE in one learning process both for training samples (#:14) and for test samples(#:10000).

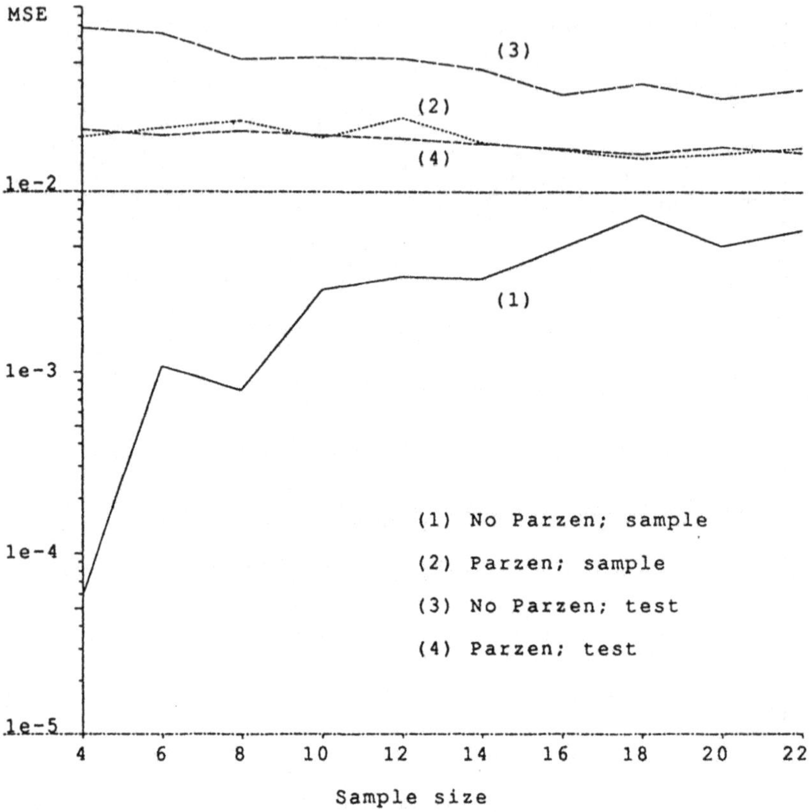

Figure 2: MSE for several sample sizes, each of which is averaged for the last 20 learning steps and further averaged for 10 such experiments.

Appendix A Optimum parameter in Parzen method

In parzen method, an asymptotically optimum a_n can be obtained from empirical data[6]. We summarize the result below.

Consider the density function of p-dimensional random variable $\boldsymbol{x} = (x_1, \ldots, x_p)$. Here we assume the density function $f(\boldsymbol{x})$ belongs to G_s which is a set of functions that is s times differentiable for all x_i. (Multi-dimensional) Parzen method is defined as approximating $f(\boldsymbol{x})$ from n sample data $\underset{(1)}{X}, \ldots, \underset{(n)}{X}$ with

$$f_n(\boldsymbol{x}, a_n) = \frac{1}{n} \sum_{j=1}^{n} \prod_{i=1}^{p} a_n K(a_n(x_i - \underset{(j)}{X_i})). \tag{12}$$

Now consider the estimation of the following expression,

$$U(\boldsymbol{x}, a_n) = \mathrm{E}[f_n(\boldsymbol{x}, a_n) - f(\boldsymbol{x})]^2, \tag{13}$$

where E denotes the expectation: $\mathrm{E}[\,\cdot\,] = \int \cdot \prod_j f(\underset{(j)}{\boldsymbol{X}})\, d\underset{(j)}{\boldsymbol{X}}$.

A class of kernel function is assumed to belong to H_s that is a set of functions $K(x)$ which satisfy (14), (15) below

$$K(x) = K(-x), \qquad \int K(x)\, dx = 1, \qquad \sup |K(x)| < \infty, \tag{14}$$

$$\int x^i K(x)\, dx = 0 \quad (i = 1, \cdots, s-1), \qquad \alpha = \int x^s K(x)\, dx \neq 0, \qquad \int x^s |K(x)|\, dx < \infty. \tag{15}$$

For example, normal distribution function belongs to H_2.

The following theorem is valid.

Theorem 2 *If $f(x) \in G_s$ and $K(x) \in H_s$,*

$$U(\boldsymbol{x}, a_n) \simeq \frac{a_n^p}{n} \|K\|^{2p} f(\boldsymbol{x}) + a_n^{-2s} \frac{\alpha^2}{(s!)^2} (\sum_{i=1}^{p} \partial_i^s f(\boldsymbol{x}))^2, \tag{16}$$

where

$$\partial_i^s f(\boldsymbol{x}) = \frac{\partial^s f(\boldsymbol{x})}{\partial {x_i}^s}, \qquad \|K\|^2 = \int K(x)^2\, dx, \tag{17}$$

and $a \simeq b$ denotes $a/b \to 1$ $(n \to \infty)$.

From this theorem, the optimum $a_n (= a_n^0)$ that minimizes $U(\boldsymbol{x}, a_n)$ is obtained as below, which depends upon $f(\boldsymbol{x})$.

$$a_n^0 = C(\boldsymbol{x}) n^\gamma, \qquad \gamma = \frac{1}{2s+p}, \qquad C(\boldsymbol{x}) = \left(\frac{2s\alpha^2}{(s!)^2 p \|K\|^{2p}} \frac{(\sum_i \partial_i^s f(\boldsymbol{x}))^2}{f(\boldsymbol{x})} \right)^\gamma. \tag{18}$$

Substituting a_n with this value, we have $U(\boldsymbol{x}, a_n^0) = O(n^{-2s/(2s+p)})$.

Since $f(\boldsymbol{x})$ is assumed to be unknown, we must estimate $C(\boldsymbol{x})$ in some way. The method applied here is to approximate $f(\boldsymbol{x})$ with $f_n(\boldsymbol{x})$. For this purpose, consider two sequences $\{\tau_{n,i}\}, \{b_n\}$.

$$\{\tau_{n,0}\} : O(n^\gamma) \qquad \{\tau_{n,i}\} : O(n^{\gamma^2}) \quad (i = 1, \cdots, p), \tag{19}$$

$$\{b_n\} : b_n \to 0 \quad (n \to \infty), \quad nb_n \geq C > 0. \tag{20}$$

We approximate $C(\boldsymbol{x})$ with those sequences as follows.

$$\hat{C}(\boldsymbol{x}) = \left(\frac{2s\alpha^2}{(s!)^2 p \|K\|^{2p}} \frac{(\sum_i \partial_i^s f_n(\boldsymbol{x}, \tau_{n,i}))^2 + b_n}{|f_n(\boldsymbol{x}, \tau_{n,0})| + b_n} \right)^\gamma. \tag{21}$$

Then an approximation of a_n^0 is obtained as below,

$$\hat{a_n^0}(\boldsymbol{x}) = \hat{C}(\boldsymbol{x}) n^\gamma, \tag{22}$$

and the following theorem is obtained.

Theorem 3 *Under the condition of theorem 2 and moreover let $K(x)$ possess bounded, integrable s-th derivatives and*

$$\int x^s K^{(s)}(x)\, dx < \infty. \tag{23}$$

If $K^*(x)$ below admits nonincreasing and integrable majorant $K_0(x)$ on $[0, \infty)^p$,

$$K^*(x) \equiv p \prod_{j=1}^{p} K(x_j) + \sum_{i=1}^{p} (\prod_{\substack{j=1 \\ j \neq i}}^{p} K(x_j)) K^{(1)}(x_i) x_i, \tag{24}$$

then the following equation is valid for $\hat{a_n^0}$ defined as (22).

$$U(\hat{a_n^0}) \simeq U(a_n^0) = O(n^{-2s/(2s+p)}). \tag{25}$$

Appendix B Proof of theorem 1

The following lemma is valid.

Lemma 1 *Let p-dimensional density function $P(z)$ be s times differentiable for any variable and let the domain region X be bounded. Then for large n,*

$$\mathrm{E}[\hat{I}(\alpha) - I(\alpha)]^2 \leq \frac{1}{4} D_1(\alpha) D_2 n^{-2s/(2s+p)}, \tag{26}$$

is valid, where E denotes the expectation

$$\mathrm{E}[\,\cdot\,] = \int \cdot \prod_j P(\underset{(j)}{Z})\, d\underset{(j)}{Z},$$

and $D_1(\alpha)$ denotes

$$D_1(\alpha) = \int_X Q(z, \alpha)^2\, dz, \tag{27}$$

and the value D_2 depends on $P(z)$, p and s.

Proof of lemma 1 Let $P(z)$ have a bounded domain region, namely, a set $\{ z \mid P(z) > 0\}$ belongs to a bounded set X. Denote

$$V(\alpha) = \mathrm{E}[\hat{I}(\alpha) - I(\alpha)]^2 = \mathrm{E}[\int_X Q(z,\alpha)(\hat{P}(z) - P(z))dz]^2. \tag{28}$$

Then, from Cauchy-Schwartz's inequality and exchanging operation of integrations,

$$V(\alpha) \le \int_X Q(z,\alpha)^2\,dz \int_X \mathrm{E}[\hat{P}(z) - P(z))]^2\,dz. \tag{29}$$

The second factor of right hand side can be estimated by (18), theorem 2 and theorem 3. Thus

$$V(\alpha) \le \frac{1}{4} D_1(\alpha) D_2 n^{-2s/(2s+p)} \tag{30}$$

is obtained, where

$$D_1(\alpha) = \int_X Q(z,\alpha)^2\,dz \tag{31}$$

$$D_2 = 4(p+2s)\int_X (\frac{\|K\|^{2p}}{2s} P(z))^{2s\gamma} (\frac{1}{s!\sqrt{p}} \sum_i \partial_i^s P(z))^{2p\gamma}\,dz. \tag{32}$$

Q.E.D.

Proof of theorem 1 It is easy to show the inequality

$$|I(\hat{\alpha}_0) - I(\alpha_0)| < |I(\hat{\alpha}_0) - \hat{I}(\hat{\alpha}_0)| + |I(\alpha_0) - \hat{I}(\alpha_0)|, \tag{33}$$

where $\hat{\alpha}_0$ is the α that minimizes $\hat{I}(\alpha)$ and α_0 is the α that minimizes $I(\alpha)$.
On the other hand, if the inequality $0 \le x < y+z$ is satisfied, $x^2 < (y+z)^2 \le 2(y^2+z^2)$ holds in general. Thus the following inequality is valid.

$$\mathrm{E}[I(\hat{\alpha}_0) - I(\alpha_0)]^2 \le 2\left(\mathrm{E}[\hat{I}(\alpha_0) - I(\alpha_0)]^2 + \mathrm{E}[\hat{I}(\hat{\alpha}_0) - I(\hat{\alpha}_0)]^2\right). \tag{34}$$

Consequently from lemma 1, we have

$$\mathrm{E}[I(\hat{\alpha}_0) - I(\alpha_0)]^2 \le 4 \sup_\alpha V(\alpha). \tag{35}$$

The theorem is proved. Q.E.D.

In lemma 1, we assumed that the domain region is bounded. Let us consider the case that it is not bounded. Even in this case, we can always take a bounded X which satisfies for any β,

$$\int_{X^c} |\hat{P}(z)| + P(z)\,dz = O(n^\beta). \tag{36}$$

Since X depends on n in general, the value

$$\int_X Q(z,\alpha)^2\,dz \tag{37}$$

becomes $g(n;\beta)$, which is a function of n and β. Take X so that $g(n;\beta)$ is as small as possible. Let us assume Q is bounded on X^c. Then

$$\begin{aligned} V(\alpha) &\le \mathrm{E}[\int_X Q(z)|P(z) - \hat{P}(z)|\,dz + Q_{\max} n^\beta]^2 \\ &\le 2\,\mathrm{E}[\int_X Q(z)|P(z) - \hat{P}(z)|]^2\,dz + 2Q_{\max}^2 n^{2\beta} \qquad (38) \\ &= O(g(n) n^{-2s/(2s+p)} + n^{2\beta}) \qquad (39) \end{aligned}$$

As β is decreasing, $g(n;\beta)$ decreases. Hence, let β_0 be a solution of the following equation,

$$2\beta = \log_n g(n;\beta) - \frac{2s}{2s+p}, \tag{40}$$

$V(\alpha)$ is bounded by

$$V(\alpha) \leq O(n^{2\beta_0}). \tag{41}$$

References

[1] E.B. Baum and D. Haussler: What size net gives valid generalization? *Neural Computation*, Vol. 1, pp. 151–160, 1989.

[2] A. Blumer, A. Ehrenfeucht, D. Haussler, and M.K. Warmuth: Learnability and the Vapnik-Chervonenkis dimension. *J. of the Assoc. for Comp. Machinery*, pp. 929–965, 1989.

[3] P. Craven and G. Wahba: Smoothing noisy data with spline functions. *Numerische Mathematik*, Vol. 31, pp. 377–403, 1979.

[4] K. Hornik, M. Stinchcombe, and H. White: Multilayer feedforward networks are universal approximators. *Neural Networks*, Vol. 2, pp. 359–366, 1989.

[5] T. Kurita: An attempt on model selection for neural networks. In *IEICE Technical Report* PRU89-16, 1989. In Japanese.

[6] E.A. Nadaraya: *Nonparametric estimation of probability densities and regression curves.* Kluwer Academic Publishers, 1989.

[7] T. Poggio: Networks for approximation and learning. *Proc. IEEE*, Vol. 78, No. 9, pp. 1481–1496, 1990.

[8] A.N. Tikhonov and V.Ya. Arsenin: *Solutions of Ill-posed Problems.* Winston, Washington, 1977.

[9] V.A. Vapnik: *Estimation of Dependences Based on Empirical Data.* Springer-Verlag, 1984.

[10] K. Yamanishi: Learning non-parametric-densities using finite-dimensional parametric hypotheses. In *Proc. of ALT '91*, pp. 175–186, 1991.

COMPETITIVE LEARNING BY ENTROPY MINIMIZATION

Ryotaro Kamimura
Information Science Laboratory
Tokai University
1117 Kitakaname Hiratsuka Kanagawa 259-12, Japan

Abstract

In this paper, we present a method of entropy minimization for competitive learning with winner-take-all activation rule. In the competitive learning, only one unit is turned on as a winner, while all the other units are off as losers. Thus, the learning is mainly considered to be a process of entropy minimization. If entropy in competitive layer is minimized, only one unit is on, while all the other units are turned off. If entropy is maximized, all the units are equally activated.

We applied this method of entropy minimization to two problems: autoencoder as feature detector and the organization of internal representation: the estimation of well-formedness of English sentences. For an autoencoder, we observed that networks with entropy method could classify four input patterns into two categories clearly. For a sentence well-formedness problem, a feature of input patterns was explicitly seen in competitive hidden layer. In other words, explicit internal representation could be obtained. In two cases, multiple inhibitory connections were observed to be produced. Thus, entropy minimization method is completely equivalent to competitive learning approaches through mutual inhibition. Entropy minimization method is more simple and easy to calculate. In the formulation and experiments, supervised learning (autoencoder) was used. However, the entropy method can be extended to fully unsupervised learning, which may replace ordinary competitive learning with winner-take-all activation rule.

1 Introduction

1.1 Competitive Learning as Entropy Minimization

Competitive learning approaches have been used as feature detector or pattern classifiers. One of the simplest forms was developed by Rumelhart and Zipser[11]. Into several other learning algorithms, for example, Kohonen's feature map[5], ART[3], counter propagation networks[6], competitive learning is incorporated as a substructure.

In the learning, one of the most essential characteristics is the winner-take-all activation rule in which only one unit or at least one unit of a group of units is turned on as a winner, while all the other units are turned off as losers. In this context, competitive learning is considered to be a process of entropy minimization. For example, suppose that an entropy function is formulated for competitive layer. If only one competitive unit (a unit of competitive layer) can appropriately be turned on, entropy is minimized. On the other hand, if entropy of competitive layer is maximized, all the units are equally activated. Thus, it is necessary to formulate an appropriate entropy function for competitive layer.

1.2 Formulation of Entropy Function

Several methods to formulate entropy function may be possible. Our approach is formulated as follows. Suppose that a network consists of input, competitive, and output units. In this case, we are only concerned with competitive units among which a method must be presented to choose a winner. Let an ith activity of a competitive unit be v_i, then the activity can be normalized by an equation: $p_i = \frac{v_i}{\sum_j v_j}$, where the summation is taken over all the competitive units. By using this normalized activity, entropy for competitive layer is defined by an equation: $H = -\sum_i p_i \log p_i$. Thus, the problem in competitive learning to be solved is to minimize this entropy function. If this entropy function is minimized, only one competitive unit is turned on, while all the other competitive units are turned off. On the other hand, if the entropy is maximized, all the competitive units are equally activated.

Competitive learning is one of the typical examples of unsupervised learning. However, we have applied competitive learning as entropy minimization to supervised learning. For problems of unsupervised learning, we have used a method of autoencoder in which all the activities of input units must exactly be reproduced at output unit through competitive hidden layer. Fully unsupervised learning with entropy method is now under investigation [9].

1.3 Paper Outline

In section 2, we formulate entropy minimization procedure and apply it to recurrent back-propagation. In section 3, we present two experimental results. First, an autoencoder must classify four input data into two classes. Second, networks with entropy minimization are applied to a sentence well-formedness problem. In this problem, network must classify English sentences into two categories: well-formed and ill-formed sentences. It is shown that in competitive hidden layer networks try to detect features of English sentences. In other words, we can obtain explicit internal representation by using entropy method.

2 Theory and Computational Methods

2.1 Entropy Minimization for Competitive Layer

To compute the information entropy for a competitive layer, an activity of a competitive unit(v_i), that is, a unit in competitive layer, must be normalized, for example, as

$$p_i = \frac{v_i}{\sum_{j=1}^{M} v_j}, \tag{1}$$

where M is the number of competitive units and v_i is an activity of ith competitive unit.

By using the normalized activity(p_i), the information entropy is defined by an equation

$$H = -\sum_{i=1}^{M} p_i \log p_i. \tag{2}$$

If this entropy is minimized, uncertainty for the activities of units is zero. This means that only one unit is on, while all the other units are turned off.

Using this entropy function, total function(F) to be minimized in learning is defined by

$$F = \eta E + \lambda H, \tag{3}$$

where E is a quadratic error function, η is a learning rate and λ is a parameter to determine the strength of the effect of entropy function.

Thus, a modified learning rule is defined by

$$\begin{aligned} \frac{dw_{ij}}{dt} &= -\frac{\partial F}{\partial w_{ij}} \\ &= -\eta\frac{\partial E}{\partial w_{ij}} - \lambda\frac{\partial H}{\partial w_{ij}}. \end{aligned} \tag{4}$$

2.2 Application to Recurrent Back-propagation

In this section, we apply entropy minimization to recurrent back-propagation developed by Pineda[10].

Suppose that a network architecture is composed of input units (Ω_I), competitive hidden units (Ω_C) and output units (Ω_O). The activity of network is governed by the following differential equations:

$$\frac{dv_i}{dt} = -v_i + f(u_i) + I_i, \qquad i = 1, ..., N, \tag{5}$$

where v_i is an activity of ith unit of the network, w_{ij} is a connection strength from jth unit to ith unit, f is any differentiable function, N is the number of the units in the network, and u_i is a total net output defined by an equation:

$$u_i = \sum_j w_{ij} v_j.$$

In this paper, a logistic function is used for producing the output value of an unit, that is,

$$f(u_i) = \frac{1}{1 + e^{-u_i}}. \tag{6}$$

An external bias I_i is given by

$$I_i = \begin{cases} \xi_i, & \text{if } i \in \Omega_I, \\ 0, & \text{otherwise,} \end{cases}$$

where ξ_i is an ith element of an input pattern. Fixed points(v_i^*) are solutions of equations:

$$v_i^* = f(\sum_j w_{ij} v_j^*) + I_i. \tag{7}$$

Let us define a quadratic error function as

$$E = \frac{1}{2}\sum_{i=1}^{N} J_i^2, \tag{8}$$

where

$$J_i = \begin{cases} \tau_i - v_i^*, & \text{if } i \in \Omega_O, \\ 0, & \text{otherwise,} \end{cases}$$

and where τ_i is a target value at ith output unit.

Then, information entropy of a competitive layer is defined by

$$H = -\sum_{i=1}^{M} p_i \log p_i. \tag{9}$$

Using these terms, total function(F) to be minimized in learning is defined by

$$F = \eta E + \lambda H, \tag{10}$$

where η is a learning rate, and λ is a parameter to determine the magnitude of the effect of entropy function.

Connection weights are updated by the following equation:

$$\begin{aligned} \frac{dw_{ij}}{dt} &= -\frac{\partial F}{\partial w_{ij}} \\ &= -\eta\frac{\partial E}{\partial w_{ij}} - \lambda\frac{\partial H}{\partial w_{ij}}. \end{aligned} \tag{11}$$

For connections into competitive units, we have

$$\begin{aligned} -\frac{\partial H}{\partial w_{ij}} &= -\frac{\partial H}{\partial v_i}\frac{\partial v_i}{\partial w_{ij}} \\ &= (\log p_i + 1)\frac{\sum_r v_r^* - v_i^*}{(\sum_r v_r^*)^2} f'(u_i)v_j^*, \end{aligned} \tag{12}$$

where a subscript i is over all the competitive units and a subscript j is over all the units.

Thus, an update rule is summarized as follows. For connections into all the units except competitive units, weights are updated by an equation

$$\begin{aligned} \frac{dw_{ij}}{dt} &= -\eta\frac{\partial E}{\partial w_{ij}} \\ &= \eta f'(u_i)z_i^* v_j^* \end{aligned} \tag{13}$$

where z_i is a solution of the following equations

$$\frac{dz_i}{dt} = -z_i + f(u_i) + J_i. \tag{14}$$

For connections into competitive units, we have

$$\begin{aligned} \frac{dw_{ij}}{dt} &= -\eta\frac{\partial E}{\partial w_{ij}} - \lambda\frac{\partial H}{\partial w_{ij}}. \\ &= \eta f'(u_i)z_i^* v_j^* \\ &\quad +\lambda(\log p_i + 1)\frac{\sum_r v_r^* - v_i^*}{(\sum_r v_r^*)^2} f'(u_i)v_j^*, \end{aligned} \tag{15}$$

where a subscript i is only over competitive units.

2.3 Information Entropy and Redundancy

In this section, we present how to compute redundancy from entropy to see easily the meaning of tables in Section 3.

Since entropy is dependent upon the number of competitive units, entropy should be normalized to see easily the effect of entropy function. To compute the information entropy, an activity of a competitive unit(v_i) for kth input pattern must be normalized, for example, as

$$p_i^k = \frac{v_i^k}{\sum_{j=1}^{M} v_j^k}, \tag{16}$$

where M is the number of competitive units and v_i^k is an activity of ith competitive unit for kth input pattern.

By using the normalized activity(p_i^k), the information entropy is defined by an equation

$$H_k = -\sum_{i=1}^{M} p_i^k \log p_i^k. \tag{17}$$

Since the information entropy is dependent upon the number of competitive units, it is useful to normalize the entropy by maximum entropy.

As well known, entropy function is bounded within

$$0 \leq H_k \leq \log M. \tag{18}$$

Thus, using this maximum entropy, the information entropy is normalized as

$$H_k^{nrm} = \frac{H_k}{H_{max}}. \tag{19}$$

The redundancy is defined by

$$R_k = 1 - H_k^{nrm}. \tag{20}$$

Redundancy is bounded within

$$0 \leq R_k \leq 1. \tag{21}$$

By averaging the redundancy over all the patterns, we have,

$$< R_k >= \frac{1}{U} \sum_{k=1}^{U} R_k, \tag{22}$$

where U is the number of input patterns.

Redundancy reaches its maximum(unity), when the information entropy(H) is minimum(zero). Thus, if the redundancy is maximum, uncertainty for the activities of competitive units is zero. This means that only one competitive unit is on, while all the other competitive units are turned off. If the redundancy is minimum(zero), uncertainty of the activities of competitive units is maximum (maximum information entropy). In this case, the activities of competitive units are uniformly distributed.

3 Results and Discussion

3.1 Pattern Classifier by Autoencoder

3.1.1 Data and Network Architecture

Networks were trained to classify input patterns with different Hamming distances into two categories. In Table II, we can see four input patterns. Hamming distance between a pattern No.1(101) and No.2(100) is only one, meaning that two pattern can be grouped into one same category. On the other hand, Hamming distance between a pattern No.3(010) and No.4(011) is also one, and can be classify into one category. For further details, see Dayhoff[4].

In the experiments, we used recurrent neural networks in which input patterns must be reproduced at output units through competitive layer. See Figure 1. Initial values for connection matrix were uniformly distributed and ranged from -0.4 to 0.4. Initial values for activity values of units were also uniformly distributed, ranging between 0.1 and 0.9. The learning was performed with the so-called *batch learning*, meaning that connection weights were updated after all the input patterns were processed. A parameter(η) was first set to unity and then another parameter(λ) was increased for networks to have maximum information theoretical redundancy.

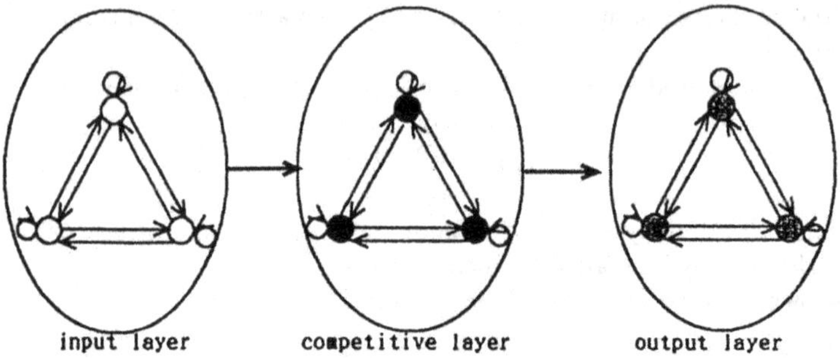

Figure 1: An example of autoencoder, which performs an identity mapping. The number of input, competitive and output units are all three. In this network, input patterns must exactly be reproduced at output units.

3.1.2 Redundancy

In this section, we try to show that redundancy is more than three times higher by using entropy method. This means that in competitive layer, only a small number of competitive units are turned on.

Table I shows values of redundancy, when a parameter λ ranges from 0.000 (standard error function) to 0.073. Above this point of 0.073, networks could not finish the learning. As can be seen in the table, values of redundancy increase approximately linearly as λ increases. When λ is 0.073, redundancy reaches its highest value of 0.765. This means that redundancy computed with entropy function ($\lambda = 0.073$) is about 3.2 times higher than that computed with standard error function.

As already mentioned in Section 2, redundancy is bounded between zero and one. The highest value of redundancy (0.765) is sufficiently close to one. Thus, it can be expected and was actually observed that only one competitive unit is on, while all other units are off.

3.2 Feature Extraction

To extract features from input patterns, we had to overcome several difficult problems. One of the most troublesome problems is that networks tend to learn input patterns excessively, meaning that networks tend to differentiate all the input patterns. In other words, all the patterns are considered to be different patterns and differently classified. To overcome this problem, we chose a method to stop the learning as early as possible. One of the easiest way to do so is to take a loose criterion for finishing the learning. In the experiments, the learning was considered to be finished, when the difference between outputs from output units and target values is greater than or equal to 4.5 for all the output units.

Table II shows the number of active competitive hidden units, when a parameter λ is 0.073. A competitive unit with an activation value greater than or equal to 0.5 is considered to be an active hidden unit. As can be seen in the table, in the case of standard error

Table I: Redundancy, computed with entropy function for an autoencoder.

λ	$< R_k >$	Ratio
0.000	0.236	1.000
0.010	0.261	1.106
0.020	0.252	1.068
0.030	0.247	1.047
0.040	0.250	1.059
0.050	0.323	1.369
0.060	0.524	2.220
0.070	0.528	2.237
0.071	0.693	2.936
0.072	0.454	1.924
0.073	0.765	3.242

Table II: The number of active competitive units for an autoencoder. An activation value for an active competitive unit is greater than or equal to 0.5. A symbol □ means an active unit.

	Standard			Entropy		
Pattern	Unit 1	Unit 2	Unit 3	Unit 1	Unit 2	Unit 3
101	□	□	□	□		
100	□	□	□	□		
010					□	
011		□			□	

function, all the three competitive units are used for two patterns: 101 and 100. For the other two patterns, no unit or only one unit is turned on. Thus, it seems that network classify input patterns into two categories.

This tendency is more clearly observed for entropy function. For the first two input patterns, only the first competitive unit is on, while the second unit is on for the rest of input patterns. The final third unit is not used for the classification. This means that networks succeed in classifying input patterns into two categories and a redundant unit is not used.

Finally we note that the activities of competitive units are somewhat randomly distributed by using standard quadratic error function. Thus, if we use severer criteria for finishing the learning, for example, more than 0.6, activity patterns of competitive units are somewhat random. On the other hand, the activity patterns are same for entropy function, even if we use much severer criteria.

3.2.1 Competitive layer

We observed that entropy was decreased by increasing the number and the strength of inhibitory connections. Figure 2 shows a state of competitive layer obtained after finishing the learning. As can be seen in the figure, entropy minimization is performed by using inhibitory connection. Except two connections (self-connections for unit No.1 and No.3), all the con-

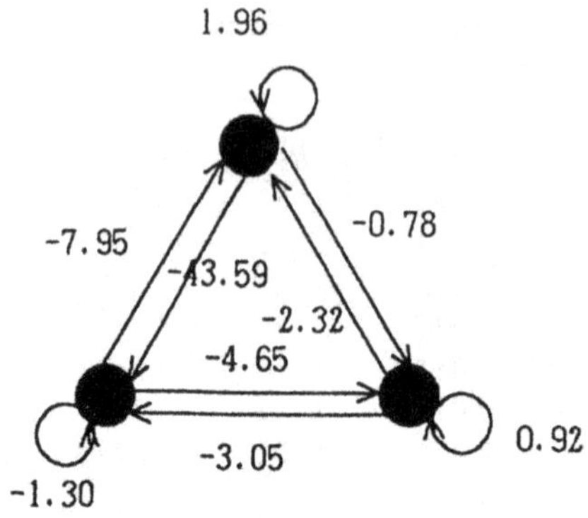

Figure 2: A final state of competitive layer for the autoencoder after finishing the learning.

nections are negative. This means that entropy decreases through inhibitory connections, which is completely compatible with the competitive learning through inhibition.

3.3 Sentence Well-formedness Problem

3.3.1 Data and Network Architecture

We applied entropy method to a sentence well-formedness problem to examine to what extent entropy method can organize internal representation. In this problem, networks must infer the well-formedness (grammaticality) of English sentences. For example, a sentence: "the man is tall" must be estimated to be well-formed. On the other hand, another sentence: "the is man tall" must be inferred to be ill-formed.

We made twenty-five well-formed sentences, and twenty-five ill-formed sentences, composed of ten distinct English words. Then, networks were trained to produce a correct answer to a given sentence. Networks for experiments consisted of 100 input units and two output units and ten competitive hidden units. See Figure 3. The learning was considered to be finished when the difference between outputs from output units and target values was less than or equal to 0.3 for all the output units. We used the so-called *pattern learning* to speed up the learning. Thus, weights were updated for each input pattern. For further details about this problem, see [8].

3.3.2 Redundancy

In this section, it is shown that redundancy of competitive hidden units computed with entropy function is much higher than that for standard error function.

Table III shows redundancy computed with entropy function, when a parameter λ ranges from 0.00 to 0.11. Above this point, networks could not finish the learning. Redundancy increases gradually as the parameter λ increases. When λ is 0.09, the redundancy reaches its highest value of 0.483. In that case, redundancy with entropy function ($\lambda = 0.09$) is about 1.75 times higher than the redundancy with standard error function.

Table IV shows the number of patterns making a given competitive unit active, computed with standard error function for a sentence well-formedness problem. A unit is considered

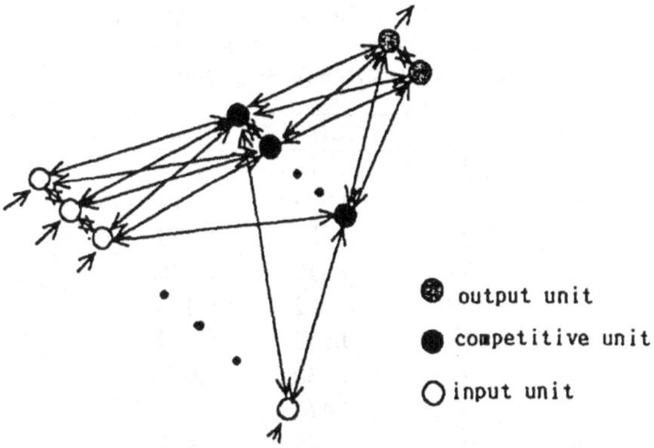

Figure 3: An example of recurrent network for a sentence well-formedness problem. The network consists of 100 input, ten hidden and two output units. The network must be trained to produce *yes*, if a grammatical sentence is given to the network. On the other hand, the network must produce *no*, if an ungrammatical sentence is given to the network.

to be active when an activation value of the unit is greater than or equal to 0.5. As can be seen in the table, unit activation patterns for well-formed sentences are different from those for ill-formed sentences. For example, for well-formed sentences, unit No.3, No.4 and No.5 are active, while the units are inactive for ill-formedness sentences. Similarly, unit No.9 and No.10 are also active only for well-formed sentences.

This tendency can more clearly be seen in Table V, which shows the number of patterns activating a given competitive unit, computed with entropy function. For well-formed sentences, only unit No.3, No.4 and No.10 are clearly active, while for ill-formed sentences unit No.1, No.2 and No.7 are active. The total number of active units is only six. Theses results shows that by using entropy function, internal representation is more clearly organized and a small number of competitive hidden units are only used.

3.4 Discussion

In this section, we will discuss a problem with competitive learning by autoencoder, generalization performance, and a possibility of a unified approach to the learning by entropy method.

First, we have applied the method to a pattern classifier by autoencoder. This autoencoder could successfully classify input patterns into distinct two categories. However, this network architecture (autoencoder) is not efficient, when the size of networks is larger. Thus, traditional competitive learning must be incorporated into entropy method[9], in which in addition to winner-take-all activation rule, the difference between input patterns and connections into a winning unit must be decreased.

Generalization performance must carefully be examined, when entropy function is introduced. We think that internal representation by entropy method (for example, sentence well-formedness problem) can be simple and explicit. Thus, it can be expected that gener-

Table III: Redundancy, computed with entropy function for a sentence well-formedness problem.

λ	$< R_k >$	Ratio
0.00	0.276	1.000
0.01	0.324	1.174
0.02	0.332	1.203
0.03	0.316	1.145
0.04	0.412	1.493
0.05	0.436	1.580
0.06	0.460	1.667
0.07	0.475	1.721
0.08	0.409	1.482
0.09	0.483	1.750
0.10	0.453	1.641
0.11	0.427	1.547

Table IV: The number of patterns activating a given competitive unit, computed with standard error function for a sentence well-formedness problem. An activation value for an active unit is greater than or equal to 0.5.

	Number of Active Units	
Unit	Well-formed	Ill-formed
1	0	2
2	6	25
3	11	0
4	18	0
5	22	0
6	2	19
7	6	25
8	0	23
9	14	0
10	7	0

Table V: The number of patterns activating a given competitive unit, computed with entropy function for a sentence well-formedness problem. An activation value for an active unit is greater than or equal to 0.5.

	Number of Active Units	
Unit	Well-formed	Ill-formed
1	0	24
2	0	23
3	9	0
4	22	0
5	0	0
6	0	0
7	0	23
8	0	0
9	0	0
10	22	0

alization performance may be increased. For related experimental results, see [9].

Finally, it must be pointed out that compared with competitive leanings with winner-take-all methods, entropy method is more general. Under a simple principle that an entropy must be minimized, multiple other learning rules may be included in entropy method [9]. For example, similar method of entropy minimization was applied to the problem of *minimum entropy coding* by Barlow *et al.*[1]. To increase the generalization performance, and to obtain optimal minimum networks, entropy minimization method was applied to the elimination of weights [12] and [13].

4 Conclusion

In the present paper, we have presented a method of entropy minimization for competitive learning. In competitive learning, only one competitive unit is on, while all other competitive units are off. Thus, competitive learning is considered to be a process of entropy minimization at competitive layer. We have formulated a procedure of entropy minimization in competitive layer and applied it to recurrent back-propagation. By applying entropy minimization to an autoencoder, we have observed that networks can detect a salient feature of input patterns automatically. Then, applied to supervised learning of classifying English sentences, we have also observed that in competitive hidden layer, network can detect a feature of input English sentences clearly.

Finally, competitive learning is developed especially for unsupervised learning. Thus, it is necessary to develop a learning algorithm of entropy minimization for fully unsupervised learning.

References

[1] H. B. Barlow, T. P. Kaushal and G. J. Mitchison, "Finding minimum entropy codes," *Neural Computing*, vol.1, pp.412-423, 1989.

[2] S. Becker, "Unsupervised learning procedures for neural networks," *International Journal of Neural Systems*, Vol.2, No.1, pp.17-33, 1991.

[3] G. A. Carpenter and S. Grossberg, "The ART of adaptive pattern recognition by a self-organizing neural network," *Computer*, March, pp.77-88, 1988.

[4] J. E. Dayhoff, *Neural Network Architectures*, New York: Van Nostrand Reinhold, 1990.

[5] T. Kohonen, *Self-Organization and Associated Memory*, New York: Springer-Verlag, 1988.

[6] R. Hecht-Nielsen, *Neurocomputing*, Addison-Wesley Publishing Company, 1989.

[7] J. Hertz, A. Krough and R. G. Palmer, *Introduction to the Theory of Neural Computation*, Redwood City, CA: Addison-Wesley, 1991.

[8] R. Kamimura, "Acquisition of the grammatical competence with recurrent neural networks," in *Artificial Neural Networks*, T. Kohonen, K. Makisara, O. Simula, and J. Kangas, Ed, Amsterdam, The Netherlands, Vol.1, 1991, pp.903-908.

[9] R. Kamimura, "Minimum entropy method in neural networks," Research Report, Information Science Laboratory, Tokai University, ISL-RR-92-05, 1992.

[10] F.J. Pineda, "Generalization of back-propagation to recurrent neural networks," *Physical Review Letters*, Vol.59, No.19, pp. 2229-2232, 1987.

[11] D. E. Rumelhart and D. Zipser, "Feature discovery by competitive learning," in *Parallel Distributed Processing*, D. E. Rumelhart, J. L. McClelland, and the PDP Research Group, Cambridge, Massachusetts: the MIT press, Vol.1, pp.318-362, 1986.

[12] H. Uchida and M. Ishikawa, "A structural learning of neural networks based on entropy criterion"(in Japanese), IEICE Technical Report, The Institute of Electronics, Information and Communication Engineers, Vol.91, No.530, pp.161-167, 1992.

[13] W. Zhang, A. Hasegawa, K. Itoh and Y. Ichioka, "Error back propagation with minimum-entropy weights: A technique for better generalization of 2-D Shift-Invariant NNs," in *Proceedings of International Joint Conference on Neural Networks*, (Seattle, WA), Vol.1, 1991, pp.645-648.

Inductive Inference

INDUCTIVE INFERENCE WITH BOUNDED MIND CHANGES

YASUHITO MUKOUCHI†

Department of Information Systems,
Kyushu University 39, Kasuga 816, Japan

Abstract. In this paper, we deal with inductive inference for a class of recursive languages with a bounded number of mind changes. We introduce an n-bounded finite tell-tale and a pair of n-bounded finite tell-tales of a language, and present a necessary and sufficient condition for a class to be inferable with bounded mind changes, when the equivalence of any two languages in the class is effectively decidable. We also show that the inferability of a class from positive data strictly increases, when the allowed number of mind changes increases. In his previous paper, Mukouchi gave necessary and sufficient conditions for a class of recursive languages to be finitely identifiable, that is, to be inferable without any mind changes from positive or complete data. The results we present in this paper are natural extensions of the above results.

1. Introduction

Inductive inference is a process of hypothesizing a general rule from examples. As a correct inference criterion for inductive inference of formal languages and models of logic programming, we have mainly used Gold's identification in the limit[5]. In this criterion, an inference machine is allowed to change its guesses finitely many times, and the guesses are required to converge to a correct guess. Angluin[1], Wright[11] and Sato&Umayahara[6] discussed conditions for a class of formal languages to be inferable from positive data. Shinohara[7, 8] also discussed inductive inferability from positive data in more general setting and exhibited that inductive inference from positive data is much more powerful than it has been believed.

Considering ordinary learning process of human beings, the criterion of identification in the limit seems to be natural. However, we can not decide in general whether a sequence of guesses from an inference machine converges or not at a certain time, and the results of the inference necessarily involve some risks. In his previous paper[10], Mukouchi gave necessary and sufficient conditions for a class of recursive languages to be finitely identifiable, that is, to be inferable without any mind changes from positive or complete data. We use the phrase 'mind change' to mean that an inference machine changes its guess.

†Mailing address: Research Institute of Fundamental Information Science, Kyushu University 33, Fukuoka 812, Japan.

E-mail: mukouchi@rifis.sci.kyushu-u.ac.jp

In this paper, we deal with inductive inference for a class of recursive languages with a bounded number of mind changes. The results we present in this paper are natural extensions of the above results concerning finite identification.

Note that Case&Smith[3] discussed inductive inference of a class of recursive functions from view point of anomalies and mind changes, and showed that there is a natural hierarchy. Case&Lynes[4] also showed that an anomaly hierarchy exists even in case of a class of recursive languages.

In Section 2, we prepare some necessary concepts for our discussions. We also recall the results on finite identification from positive and complete data. In Section 3, we discuss conditions for a class to be inferable with bounded mind changes from positive data. Angluin[1] introduced the notion of a finite tell-tale of a language to discuss inferability of formal languages from positive data, and showed that a class is inferable from positive data if and only if there is a recursive procedure to enumerate all elements in the finite tell-tale of any language of the class. In this paper, we introduce an n-bounded finite tell-tale of a language, and present a necessary and sufficient condition for a class to be inferable with bounded mind changes, when the equivalence of any two languages in the class is effectively decidable. We also exhibit a concrete class of recursive languages which is inferable with at most n mind changes but not inferable with at most $n-1$ mind changes, and show that the inferability of a class strictly increases, when the allowed number of mind changes increases. Case&Smith[3] showed similar results for a class of recursive functions. In Section 4, we give a necessary and sufficient condition for a class to be inferable with bounded mind changes *from complete data*, which is analogous to the above condition concerning positive data.

2. Preliminaries

Let U be a recursively enumerable set to which we refer as a *universal set*. Then we call $L \subseteq U$ a *language*. We do not consider the empty language in this paper.

Definition 2.1. *A class of languages $\Gamma = L_1, L_2, \cdots$ is said to be an indexed family of recursive languages if there exists a computable function $f : N \times U \to \{0,1\}$ such that*

$$f(i,w) = \begin{cases} 1, & \text{if } w \in L_i, \\ 0, & \text{otherwise.} \end{cases}$$

From now on, we assume a class of languages is an indexed family of recursive languages without any notice.

Definition 2.2. *A positive presentation of a language L is an infinite sequence $w_1, w_2, \cdots$ of elements of U such that $\{w_1, w_2, \cdots\} = L$.*

A complete presentation of a language L is an infinite sequence $(w_1,t_1),(w_2,t_2),\cdots$ of elements of $U \times \{0,1\}$ such that $\{w_i \mid t_i = 1, i \geq 1\} = L$ and $\{w_j \mid t_j = 0, j \geq 1\} = U - L$.

Positive or complete presentations are denoted by σ, δ, the finite sequence which consists of first $n \geq 0$ data in σ by $\sigma[n]$ and the finite set by $\sigma(n)$.

For a finite sequence $\sigma[n]$ and a sequence δ, the sequence which is obtained by concatenating $\sigma[n]$ with δ is denoted by $\sigma[n] \cdot \delta$.

Definition 2.3. *An n-bounded inference machine (abbreviated to IM_n; $n \geq 0$ or $n = *$) is an effective procedure that requests inputs from time to time and produces outputs from time to time, where if $n \geq 0$, it produces at most $n+1$ outputs, and if $n = *$, it produces at most finitely many outputs.*

The outputs produced by the machine are called guesses.

For a finite sequence $\sigma[m] = w_1, w_2, \ldots, w_m$, we denote by $M(\sigma[m])$ the last guess produced by an IM_n M which is successively fed $w_1, w_2, \ldots, w_m$ on its input requests.

The inference machines we are dealing with in this paper may not produce a guess after reading a datum until requesting a next datum.

Definition 2.4. *A class $\Gamma = L_1, L_2, \cdots$ is said to be EX_n identifiable from positive data (resp., complete data) if there exists an IM_n M satisfying the following ($n \geq 0$ or $n = *$): For any language L_i of Γ and for any positive presentation (resp., complete presentation) σ of L_i, the last guess k of M which is successively fed σ's data satisfies $L_k = L_i$.*

A class Γ is said to be finitely identifiable (resp., identifiable in the limit) if it is EX_0 identifiable (resp., EX_* identifiable).

A class Γ is also said to be EX-TXT_n identifiable (resp., EX-INF_n identifiable) if it is EX_n identifiable from positive data (resp., complete data). By the same notation EX-TXT_n (resp., EX-INF_n), we also denote the set of the classes that are EX-TXT_n identifiable (resp., EX-INF_n identifiable).

In this paper, a finite-set-valued function F is said to be *computable* if there exists an effective procedure that produces all elements in $F(x)$ and then halts uniformly for any argument x.

Mukouchi[10] presented necessary and sufficient conditions for an indexed family of recursive languages to be finitely identifiable.

Definition 2.5 (Mukouchi[10]). *A set S_i is said to be a definite finite tell-tale of L_i if*

(1) *S_i is a finite subset of L_i, and*
(2) *$S_i \subseteq L_j$ implies $L_j = L_i$ for any index j.*

Theorem 2.1 (Mukouchi[10]). *A class Γ is finitely identifiable from positive data if and only if a definite finite tell-tale of L_i is uniformly computable for any index i, that is, there exists an effective procedure that on input i produces all elements of a definite finite tell-tale of L_i and then halts.*

Definition 2.6 (Mukouchi[10]). *A language L is said to be consistent with a pair of sets $\langle T, F\rangle$ if $T \subseteq L$ and $F \subseteq U - L$.*

A pair of sets $\langle T_i, F_i\rangle$ is said to be a pair of definite finite tell-tales of L_i if

(1) *T_i is a finite subset of L_i,*
(2) *F_i is a finite subset of $U - L_i$, and*
(3) *if L_j is consistent with the pair $\langle T_i, F_i\rangle$, then $L_j = L_i$.*

Theorem 2.2 (Mukouchi[10]). *A class Γ is finitely identifiable from complete data if and only if a pair of definite finite tell-tales of L_i is uniformly computable for any index i.*

The following corollary shows a necessary condition for a class to be finitely identifiable.

Corollary 2.3. *If a class Γ is finitely identifiable from positive or complete data, then whether $L_i = L_j$ or not is effectively decidable for any indices i, j.*

Proof: Clearly, if Γ is finitely identifiable from positive data, then Γ is also finitely identifiable from complete data. Therefore, it suffices to show the case of complete data.

Suppose Γ is finitely identifiable from complete data. Fix arbitrary indices i, j. To begin with, compute a pair of definite finite tell-tales of L_i, and set it to $\langle T_i, F_i\rangle$. We can effectively compute this pair by Theorem 2.2. Then check whether L_j is consistent with $\langle T_i, F_i\rangle$. We can effectively check this, because T_i and F_i are explicitly given finite sets. If L_j is not consistent with $\langle T_i, F_i\rangle$, then we conclude $L_i \neq L_j$, because L_i is consistent with $\langle T_i, F_i\rangle$. Otherwise, we conclude $L_i = L_j$ by Definition 2.6. ∎

3. Inductive Inference with Bounded Mind Changes from Positive Data

First of all, we give a necessary condition for a class Γ to be $EX\text{-}TXT_n$ identifiable.

Proposition 3.1. *For any $n \geq 1$, if a class Γ contains languages $L_{i_0}, L_{i_1}, \ldots, L_{i_n}$ such that $L_{i_0} \subsetneq L_{i_1} \subsetneq \cdots \subsetneq L_{i_n}$, then Γ is not $EX\text{-}TXT_{n-1}$ identifiable.*

Proof: Suppose that Γ contains languages $L_{i_0}, L_{i_1}, \ldots, L_{i_n}$ such that $L_{i_0} \subsetneq L_{i_1} \subsetneq \cdots \subsetneq L_{i_n}$ and that Γ is $EX\text{-}TXT_*$ identifiable by an IM_* M. For simplicity, put $L'_j = L_{i_j}$ $(0 \leq j \leq n)$. We show that M needs to change its guesses more than n times to identify L'_n from a certain positive presentation of L'_n. Let σ_j be an arbitrary positive presentation of L'_j. We recursively define c_j and δ_j as follows:

Stage 0:

Let $c_0 := 0$ and $\delta_0 := \sigma_0$. Goto Stage 1.

Stage $m (1 \leq m \leq n)$:

Let

$$c_m := \min\{c > c_{m-1} \mid \exists k \text{ s.t. } M(\delta_{m-1}[c]) = k \wedge L'_{m-1} = L_k\} \quad \text{and}$$
$$\delta_m := \delta_{m-1}[c_m] \cdot \sigma_m.$$

Such an integer c_m exists, because δ_{m-1} is a positive presentation of L'_{m-1} and Γ is $EX\text{-}TXT_*$ identifiable by M. Note that the above δ_m becomes a positive presentation of L'_m, because $L'_{m-1} \subsetneq L'_m$.

Goto Stage $m + 1$.

Stage $n + 1$:

Let

$$c_{n+1} := \min\{c > c_n \mid \exists k \text{ s.t. } M(\delta_n[c]) = k \wedge L'_n = L_k\}.$$

When we feed a positive presentation δ_n of L'_n successively to M, it should output guesses after reading c_1-th, c_2-th, $\ldots$, c_{n+1}-th datum, and so it can not identify L'_n within n mind changes. ∎

Before proceeding to the next corollary, we briefly recall a pattern and a pattern language. (For more details, see Angluin[2] or Mukouchi[9].)

Fix a finite alphabet with at least two constant symbols. A pattern is a nonnull finite string of constant and variable symbols. The pattern language $L(\pi)$ generated by a pattern π is the set of all strings obtained by substituting nonnull strings of constant symbols for the variables of π. Since two patterns that are identical except for renaming of variables generate the same pattern language, we do not distinguish one from the other. We can enumerate all patterns recursively and whether $w \in L(\pi)$ or not for any w and π is effectively decidable. Therefore, we can consider the class of pattern languages as an indexed family of recursive languages, where the pattern itself is considered to be an index.

Corollary 3.2. *For any $n \geq 0$, a class of pattern languages is not $EX\text{-}TXT_n$ identifiable.*

Proof: By Proposition 3.1, it suffices to show that there exist patterns $\pi_0, \pi_1, \ldots, \pi_m$ such that $L(\pi_0) \subsetneq L(\pi_1) \subsetneq \cdots \subsetneq L(\pi_m)$ for any $m \geq 1$.

In fact, let $\pi_0 = x_1 x_2 \cdots x_{m+1}, \pi_1 = x_1 x_2 \cdots x_m, \ldots, \pi_m = x_1$. Then $L(\pi_i)$ is the set of all constant strings of length more than $m - i$, and clearly

$$L(\pi_0) \subsetneq L(\pi_1) \subsetneq \cdots \subsetneq L(\pi_m). \quad \blacksquare$$

Angluin[1] showed that the class of pattern languages is inferable from positive data in the limit, that is, it is $EX\text{-}TXT_*$ identifiable.

Definition 3.1. *Let $\Gamma = L_1, L_2, \cdots$. A set S_i is said to be a 0-**bounded finite tell-tale** (abbreviated to FT_0) in Γ of L_i if S_i is a definite finite tell-tale of L_i, that is,*

(1) S_i is a finite subset of L_i, and

(2) $S_i \subseteq L_j$ implies $L_j = L_i$ for any index j.

*A set S_i is said to be an n-**bounded finite tell-tale** (abbreviated to FT_n ; $n \geq 1$) in Γ of L_i if*

(1) S_i is a finite subset of L_i, and

(2) if $L_j \neq L_i$ and $S_i \subseteq L_j$, then there exist an FT_{n-1} in Γ of L_j.

Intuitively, an FT_n in Γ of L_i is a tell-tale which validates producing the guess i, when the inference machine is allowed to produce another $n - 1$ guesses.

We can easily prove by induction on n that if a certain finite set S is an FT_n in Γ of L_i, then S is also an FT_{n+1} in Γ of L_i ($n \geq 0$).

Definition 3.2. *An FT_0 in Γ of L_i is said to be **recurrently computable** if a certain FT_0 S_i in Γ of L_i is computable.*

An FT_n in Γ of L_i is said to be recurrently computable ($n \geq 1$) if

(1) a certain FT_n S_i in Γ of L_i is computable, and

(2) for any index j, if $L_j \neq L_i$ and $S_i \subseteq L_j$, then an FT_{n-1} in Γ of L_j is recurrently computable.

*An FT_n of Γ is said to be **recurrently constructible** if an FT_n in Γ of L_i is recurrently computable for any index i ($n \geq 0$).*

Lemma 3.3. *Suppose whether $L_i = L_j$ or not is effectively decidable for any indices i, j.*

For any $n \geq 0$, if a class Γ is $EX\text{-}TXT_n$ identifiable, then an FT_n of Γ is recurrently constructible.

Proof: The proof of the lemma is done by induction on n.

(I) In case of $n = 0$, it is clear from Theorem 2.1.

(II) Suppose the assertion is valid in case of $n < m$, and we consider the case of $n = m$. Suppose Γ is $EX\text{-}TXT_m$ identifiable by an IM_m M, and we would like to construct an FT_m of L_h recurrently for any index h. Fix an arbitrary index h, and let σ_h be an arbitrary positive presentation of L_h. When we feed σ_h successively to M, M outputs k such that $L_k = L_h$ by the assumption. Let $d = \min\{l \mid \exists k \text{ s.t. } M(\sigma_h[l]) = k \wedge L_h = L_k\}$ and $S = \sigma_h(d)$.

We consider the subclass $\Gamma_S = L'_1, L'_2, \cdots$ of Γ defined as follows:

$$L'_i = \begin{cases} L_i, & \text{if } S \subseteq L_i \\ L_b, & \text{otherwise} \end{cases} \qquad (i \geq 1),$$
$$b = \min\{j \mid S \subseteq L_j\}.$$

Since a set S is an explicitly given finite set, the subclass Γ_S becomes an indexed family of recursive languages.

The class Γ_S is $EX\text{-}TXT_{m-1}$ identifiable by the following IM_{m-1}· M' which is constructible using M, where we are going to feed a positive presentation δ to M':

Procedure M';
begin
  Simulate M with feeding $\sigma_h[d] \cdot \delta$ successively on its input requests;
  While the above simulation, if M produces guesses, then output them except for the first one;
end.

In fact, suppose the subclass Γ_S is not $EX\text{-}TXT_{m-1}$ identifiable by this M'. Then we can easily show that the class Γ is not $EX\text{-}TXT_m$ identifiable by M, which contradicts the assumption.

From the induction hypothesis, an FT_{m-1} of Γ_S is recurrently constructible.

For a while, we show by induction on n' that the following claim is valid for any index i ($n' \leq m-1$):

Claim: If $S \subseteq L_i$ and a set D is an $FT_{n'}$ in Γ_S of $L'_i (= L_i)$, then the set $D \cup S$ is an $FT_{n'}$ in Γ of L_i.

(i) In case of $n' = 0$. Let $S \subseteq L_i$ and D be an FT_0 in Γ_S of L'_i. From the definition of an FT_0, $D \subseteq L'_j$ implies $L'_j = L'_i$. Since $S \subseteq L_j$ implies $L'_j = L_j$, it follows that $D \cup S \subseteq L_j$ implies $L_i = L'_i = L'_j = L_j$, that is, $L_i = L_j$ for any index j. Therefore, the set $D \cup S$ is an FT_0 in Γ of L_i.

(ii) Suppose the assertion is valid in case of $n' < m'$, and we consider the case of $n' = m'$. Suppose $S \subseteq L_i$ and D is an $FT_{m'}$ in Γ_S of L'_i. For any index j, if $L'_j \neq L'_i$ and $D \subseteq L'_j$, then an $FT_{m'-1}$ in Γ_S of L'_j is computable, and set it E_j. For any index j, if $D \cup S \subseteq L_j$, then $S \subseteq L_j$, and it follows that $L_j = L'_j$. Therefore, for any index j, if $L_j \neq L_i$ and $D \cup S \subseteq L_j$, then the set $E_j \cup S$ becomes an $FT_{m'-1}$ in Γ of L_j by the induction hypothesis. Thus the set $D \cup S$ is an $FT_{m'}$ in Γ of L_i.

By (i) and (ii), the above claim is valid for any $n' \leq m-1$.

Clearly by the above claim, for any index i such that $S \subseteq L_i$, an FT_{m-1} in Γ of L_i is recurrently computable. Therefore, the set S is an FT_m in Γ of L_h, and it follows that an FT_m in Γ of L_h is recurrently computable. Thus an FT_m of Γ is recurrently constructible. ∎

Lemma 3.4. *Suppose whether $L_i = L_j$ or not is effectively decidable for any indices i, j.*

For any $n \geq 0$, if an FT_n of a class Γ is recurrently constructible, then Γ is EX-TXT_n identifiable.

Proof: Suppose an FT_n of Γ is recurrently constructible. We denote by $FT_m(i)$ the result of computation of an FT_m in Γ of L_i ($m \geq 0$). We consider the following procedure:

```
Procedure M;
begin
    m := n;    k := 0;
    S := φ;    T := φ;
    for j := 1 to ∞ do begin
        read a next datum and add it to T;        /* Note that ♯T = j */
        for i := 1 to j do
            if (k = 0) ∨ (L_k ≠ L_i ∧ S ⊆ L_i) then
                if FT_m(i) ⊆ T then begin
                    output i;
                    if m = 0 then stop;
                    S := FT_m(i);    k := i;
                    m := m − 1;
                end;
    end;
end.
```

Clearly, this procedure produces at most $n+1$ outputs. Suppose we are going to feed a positive presentation σ successively to the procedure on its input requests.

(1) This procedure produces at least one guess. In fact, suppose this procedure never produces a guess. When it reaches the case

$$j = \max\{h, \min\{l \mid FT_n(h) \subseteq \sigma(l)\}\} \quad \text{and} \quad i = h,$$

this procedure should produce the guess h, which contradicts the assumption.

(2) Suppose the last guess, say g, produced by this procedure is not correct.

(i) In case of $m = 0$, when the procedure produced the last guess. It contradicts the definition of an FT_0.

(ii) Otherwise, note that $S \subseteq T$ and $L_g \neq L_h$. When it reaches the case

$$j \geq \max\{h, \min\{l \mid FT_n(h) \subseteq \sigma(l)\}\} \quad \text{and} \quad i = h,$$

this procedure should produce the next guess h, which contradicts the assumption. ∎

We obtain the following Theorem 3.5 by Lemma 3.3 and Lemma 3.4.

Theorem 3.5. *Suppose whether $L_i = L_j$ or not is effectively decidable for any indices i, j.*

For any $n \geq 0$, a class Γ is EX-TXT_n identifiable if and only if an FT_n of Γ is recurrently constructible.

Note that in case of $n = 0$, the above theorem is equivalent to Theorem 2.1 by Corollary 2.3.

Example 3.1. *Fix an arbitrary number $n \geq 0$. We consider the following set:*

$$C_n = \left\{ (q_1, q_2, \cdots, q_{n+1}) \in N^{n+1} \;\middle|\; \begin{array}{c} q_1, q_2, \cdots, q_{n+1} \text{ are prime numbers with} \\ q_1 < q_2 < \cdots < q_{n+1} \end{array} \right\}.$$

Fix an arbitrary computable bijection from C_n to N, and we denote it by $\langle\langle \; \rangle\rangle$. We consider the following class:

$$\Gamma = L_1, L_2, \cdots,$$

where

$$L_{\langle\langle q_1, q_2, \ldots, q_{n+1} \rangle\rangle} = \{m \in N \mid m \text{ is a multiple of } q_j \text{ for some } j \ (1 \leq j \leq n+1)\}.$$

Clearly, this class Γ is an indexed family of recursive languages. This class is also finitely identifiable. In fact, there exists a computable FT_0 of $L_{\langle\langle q_1, q_2, \ldots, q_{n+1} \rangle\rangle}$ such as

$$\{q_1, q_2, \cdots, q_{n+1}\}.$$

Note that if $n \geq 1$, this class Γ does not have the property of so-called finite thickness[1, 11], which is a sufficient condition to be inferable from positive data in the limit.

Example 3.2. *Fix an arbitrary number $n \geq 0$. We consider the following set:*

$$D_n = \left\{ (q_1, q_2, \cdots, q_k) \in N^k \;\middle|\; \begin{array}{c} 1 \leq k \leq n+1, \\ q_1, q_2, \cdots, q_k \text{ are prime numbers with} \\ q_1 < q_2 < \cdots < q_k \end{array} \right\}.$$

Fix an arbitrary computable bijection from D_n to N, and we denote it by $[\![\;]\!]$. We consider the following class:

$$\Gamma' = L_1, L_2, \cdots,$$

where

$$L_{[\![q_1, q_2, \ldots, q_k]\!]} = \{m \in N \mid m \text{ is a multiple of } q_j \text{ for some } j \ (1 \leq j \leq k)\} \quad (1 \leq k \leq n+1).$$

Clearly, this class Γ' is an indexed family of recursive languages. This class is also $EX\text{-}TXT_n$ identifiable. In fact, there exists a computable FT_m $(n - k + 1 \leq m \leq n)$ of $L_{[\![q_1, q_2, \ldots, q_k]\!]}$ such as

$$FT_{n-k+1} = \cdots = FT_n = \{q_1, q_2, \cdots, q_k\},$$

and it follows that an FT_n of Γ' is recurrently constructible.

On the other hand, this class is shown to be not $EX\text{-}TXT_{n-1}$ identifiable by Proposition 3.1 if $n \geq 1$.

Note that this class Γ' does not have the property of finite thickness if $n \geq 1$.

From Corollary 3.2, Example 3.2 and the fact that the class of pattern languages is $EX\text{-}TXT_*$ identifiable but not $EX\text{-}TXT_n$ identifiable for any $n \geq 0$, we see that there exists a hierarchy such as

$$EX\text{-}TXT_0 \subsetneq EX\text{-}TXT_1 \subsetneq \cdots \subsetneq EX\text{-}TXT_n \subsetneq \cdots \subsetneq EX\text{-}TXT_*.$$

4. Inductive Inference with Bounded Mind Changes from Complete Data

In this section, we give the results concerning complete data, which are analogous to the results in Section 3 concerning positive data.

The following Definition 4.1 and Theorem 4.1 form a remarkable contrast to Definition 3.1 and Theorem 3.5 concerning positive data.

Definition 4.1. *Let $\Gamma = L_1, L_2, \cdots$ A pair $\langle T_i, F_i \rangle$ is said to be a pair of 0-bounded finite tell-tales (abbreviated to PFT_0) in Γ of L_i if $\langle T_i, F_i \rangle$ is a pair of definite finite tell-tales of L_i, that is,*

(1) T_i is a finite subset of L_i,
(2) F_i is a finite subset of $U - L_i$, and
(3) for any index j, if L_j is consistent with the pair $\langle T_i, F_i \rangle$, then $L_j = L_i$.

A pair $\langle T_i, F_i \rangle$ is said to be a pair of n-bounded finite tell-tales (abbreviated to PFT_n ; $n \geq 1$) in Γ of L_i if

(1) T_i is a finite subset of L_i,
(2) F_i is a finite subset of $U - L_i$, and
(3) for any index j, if $L_j \neq L_i$ and L_j is consistent with $\langle T_i, F_i \rangle$, then there exists a PFT_{n-1} in Γ of L_j.

We can easily prove by induction on n that if a certain pair of finite sets $\langle T, F \rangle$ is a PFT_n in Γ of L_i, then $\langle T, F \rangle$ is also a PFT_{n+1} in Γ of L_i $(n \geq 0)$.

Similarly to Definition 3.2, we define the recurrent computability and the recurrent constructibility of a PFT_n.

We can also prove the following theorem in a similar way to the proofs of Lemma 3.3 and Lemma 3.4, where we take the pair $\langle D_T \cup S_T, D_F \cup S_F \rangle$ as the union of $\langle D_T, D_F \rangle$ and $\langle S_T, S_F \rangle$.

Theorem 4.1. *Suppose whether $L_i = L_j$ or not is effectively decidable for any indices i, j.*

For any $n \geq 0$, a class Γ is EX-INF_n identifiable if and only if a PFT_n of Γ is recurrently constructible.

Note that in case of $n = 0$, the above theorem is equivalent to Theorem 2.2 by Corollary 2.3.

5. Concluding Remarks

We have investigated some characterization theorems on language learning with a bounded number of mind changes, and exhibited necessities of mind changes.

We have also recognized again the importance of paying attention to a characteristic subset of each language in the class, when we consider language learning.

Acknowledgements

The author wishes to thank Setsuo Arikawa for many suggestions, valuable comments and productive discussions.

References

[1] Angluin, D., *Inductive inference of formal languages from positive data*, Information and Control **45** (1980), 117–135.

[2] Angluin, D., *Finding patterns common to a set of strings*, Proc. 11th Annual Symposium on Theory of Computing (1979), 130–141.

[3] Case, J. and Lynes, C., *Machine inductive inference and language identification*, Proc. of the International Colloquium on Automata, Languages and Programming(ICALP) (1982), 107–115.

[4] Case, J. and Smith, C., *Comparison of identification criteria for machine inductive inference*, Theoretical Computer Science **25** (1983), 193–220.

[5] Gold, E.M., *Language identification in the limit*, Information and Control **10** (1967), 447–474.

[6] Sato, M. and Umayahara, K., *Inductive inferability for formal languages from positive data*, Proc. 2nd Workshop on Algorithmic Learning Theory (1991), 84–92.

[7] Shinohara, T., *Inductive inference from positive data is powerful*, Proc. 3rd Workshop on Comput. Learning Theory (1990), 97–110.

[8] Shinohara, T., *Inductive inference of monotonic formal systems from positive data*, Proc. 1st Workshop on Algorithmic Learning Theory (1990), 339–351.

[9] Mukouchi, Y., *Characterization of pattern languages*, Proc. 2nd Workshop on Algorithmic Learning Theory (1991), 93–104.

[10] Mukouchi, Y., *Characterization of finite identification*, to appear in Proc. 3rd International Workshop on Analogical and Inductive Inference (1992).

[11] Wright, K., *Identification of unions of languages drawn from an identifiable class*, Proc. 2nd Workshop on Comput. Learning Theory (1989), 328–333.

EFFICIENT INDUCTIVE INFERENCE OF PRIMITIVE PROLOGS FROM POSITIVE DATA

Hiroki Ishizaka† hiro@iias.flab.fujitsu.co.jp

Hiroki Arimura‡ arim@ai.kyutech.ac.jp

Takeshi Shinohara‡ shino@ai.kyutech.ac.jp

† FUJITSU LABORATORIES, IIAS
140, Miyamoto, Numazu, Shizuoka 410-03, Japan

‡ Department of Artificial Intelligence,
Kyushu Institute of Technology
680-4, Kawazu, Iizuka 820, Japan

Abstract

This paper is concerned with the problem of efficient inductive inference of primitive Prologs from only positive data. The class of primitive Prologs is a proper subclass of one of linear Prologs that is known to be inferable from only positive data. In this paper, we discuss on the consistent and conservative polynomial update time inferability of the subclass. We give a consistent and conservative polynomial update time inference algorithm that, when it is given the base case of the target program as a hint, identifies the subclass in the limit. Furthermore, we give a consistent but not conservative polynomial update time inference algorithm for the subclass using a 2-mmg (minimal multiple generalization) algorithm. The notion of 2-mmg is a natural extension of Plotkin's least generalization. The second inference algorithm uses the 2-mmg algorithm to infer the heads of clauses in a target program.

1 Introduction

In this paper, we consider the problem of efficient inductive inference of primitive Prologs from positive data. A primitive Prolog is a very restricted logic program. It has at most two clauses that consist of one unary predicate symbol. The class of primitive Prologs is a subclass of *linear Prologs with at most k clauses* that is known to be inferable from only positive data but not proven to be polynomial update time inferable [Shi90]. One aim of this study is to investigate what subclass of linear Prologs is inferable in polynomial update time.

On the other hand, Arimura[ASO91b, ASO91a] gives a notion of minimal multiple generalization (*mmg*, for short) that is a natural extension of Plotkin's least generalization (*lg*, for short). While a *lg* covers a given set of terms (or atoms) precisely by one term (or atom), an *mmg* covers the set minimally by several terms (or atoms). For example, suppose that

$$M = \{app([],[],[]), app([a],[];[a]), app([],[b],[b]), app([a],[b],[a,b]), app([a,b],[c],[a,b,c]), \ldots\}$$

is the least Herbrand model of a program for appending lists $app(X, Y, Z)$. Then the lg of M is $app(X, Y, Z)$. On the other hand, a pair $\{app([], X, X), app([X|Y], Z, [X|W])\}$ of atoms is an mmg of M. Note that the pair corresponds to the heads of clauses in normal append program.

In some inductive inference algorithms for logic programs such as GEMINI [Ish88] or CIGOL [MB88], lg plays very important role to infer heads of clauses. However, in general, a program consists of several clauses. In order to infer several heads using lg, inference algorithm has to divide a given set of positive examples (finite subset of the least Herbrand model of a target program) into several appropriate subsets at first then it can get candidates for heads of clauses by computing a lg of each subset. This process, that is, dividing a set of positive examples appropriately then generalizing each obtained subset of examples, exactly corresponds to mmg calculation. Hence, it seems that mmg is more useful than lg in inductive inference of logic programs. Another aim of this study is to investigate effective application of mmg to inductive inference of logic programs.

We also introduce the notion of a *most specific version* of a logic program P ($msv(P)$). An $msv(P)$ consists of most specific instances of clauses in P that cover the same set of atoms covered by the original clauses. Also, an $msv(P)$ has the same Herbrand model as original P's. In general, an lg of a set of ground atoms covered by a clause in P is more specific than the head of the clause. Hence, the notion of most (or more) specific version is important for constructing an inference algorithm that infers the heads of program clauses using lg or mmg. In fact, we show that, for any primitive Prolog P, the pair of heads of clauses in $msv(P)$ is an mmg of the least Herbrand model $M(P)$ of P.

2 Preliminary

In this paper, the reader assumed to be familiar with rudiments of logic programs [Llo84]. In what follows, we assume that a first order language $\mathcal{L}$, that has only one unary predicate symbol and finitely many function symbols (we regard a constant symbol as a 0-ary function symbol), is given. An atom, a term, a clause, a logic program (program, for short) and related notions are defined over $\mathcal{L}$. We denotes the set of function symbols of $\mathcal{L}$ by Σ. A *word* is an atom or a term. A set of variables appearing in a word t is denoted by $var(t)$. A term containing mutually distinct variables $x_1, \ldots, x_m$ is denoted by $t[x_1, \ldots, x_m]$, that is $\{x_1, \ldots, x_m\} \subseteq var(t[x_1, \ldots, x_m])$. A term obtained from $t[x_1, \ldots, x_m]$ by replacing each occurrence of the variable x_i by a term R_i ($1 \leq i \leq m$) is denoted by $t[R_1, \ldots, R_m]$. For a word t, $L(t)$ denotes the set of all ground instances of t. If a substitution θ binds only variables occurring in a term t and $t\theta$ is ground, then θ is called *ground substitution* for t. The amount of the number of occurrences of symbols appearing in a word t is denoted by $size(t)$. For a set S of words, $size(S)$ is defined as $\sum_{t \in S} size(t)$ and $|S|$ as the number of elements of S.

A sequence of integers of the form $\langle i_1, \ldots, i_n \rangle$ is called an *index*. For a subword t of W, the index I of t in W is defined as follows:

- if $W = t$, then $I = \langle\, \rangle$
- otherwise, let W be of the form $\varphi(t_1, \ldots, t_n)$ and $\langle j_1 \ldots, j_m \rangle$ be the index of t in t_i, $I = \langle i, j_1 \ldots, j_m \rangle$.

$W(I)$ denotes the subword of W which has an index I in W.

A *primitive Prolog* P is a program that satisfies the following conditions (a)–(d):

(a) Only one unary predicate symbol appears in P.

(b) P consists of at most two clauses.

(c) If P consists of two clauses, then both heads of clauses have no common instance.

(d) Atoms appearing in the body of a clause are most general atoms as $p(x)$.

(e) Variables appearing in the body of a clause are mutually distinct and also appear in the head of the clause.

In a word, a primitive Prolog is a program of the form:

$$p(t[x_1,\ldots,x_m]) \leftarrow p(x_1),\ldots,p(x_m).$$
$$p(s).$$

where $L(t[x_1,\ldots,x_m]) \cap L(s) = \emptyset$ (empty set).

For a program P, M_P (or $M(P)$) denotes the least Herbrand model of P. Let α be a ground atom, M be a set of ground atoms and $C = A \leftarrow B_1,\ldots,B_n$ be a clause. We say that C *covers* α in M if there exists a substitution θ such that $A\theta = \alpha$ and $B_i\theta \in M$ $(1 \le i \le n)$. $C(M)$ denotes the set of all ground atoms that are covered by C in M.

3 Minimal multiple generalization

For words s and t, we define a relation $\succeq$ over words as $s \succeq t$ (read 's is a *generalization* of t') iff $s\theta = t$ for some substitution θ. We write $s \succ t$ when $s \succeq t$ but $t \not\succeq s$ and $s \equiv t$ when $s \succeq t$ and $t \succeq s$. When $s \equiv t$, we say that s (t) is a *variant* of t (s). For a set S of words, we say that t is a *generalization* of S iff $t \succeq \alpha$ for any $\alpha \in S$. Let t be a generalization of S. We say that t is a *least generalization* of S (denoted by $lg(S)$) iff $t' \succeq t$ for any generalization t' of S. For any set S of words, an $lg(S)$ is unique up to variable renaming [Plo70]. Furthermore, the following proposition follows from the similar argument as on lg[1] calculation for an infinite set of words in [JLMM88].

Proposition 1 *Let S be a set of words. Then there exists a finite subset S' of S such that $lg(S) \equiv lg(S')$.*

Let k be a fixed integer and S be a set of words. We say that a set of words $\{t_1,\ldots,t_m\}$ $(m \le k)$ is a *k-minimal multiple generalization* (*k-MMG*, for short) of S iff $S \subseteq L(t_1) \cup \cdots \cup L(t_m)$ and, for any $\{s_1,\ldots,s_n\}$ $(n \le k)$ such that $S \subseteq L(s_1) \cup \cdots \cup L(s_n)$, $L(s_1) \cup \cdots \cup L(s_n) \not\subset L(t_1) \cup \cdots \cup L(t_m)$. The property shown in Theorem 2 is called *compactness* of unions of tree pattern languages [ASO91a]. The property is very useful for proving not only the following Theorem 3 but also a lot of theorems given in this paper.

Theorem 2 ([ASO91a]) *Suppose that $|\Sigma| \ge k+1$. Let $t, s_1, s_2, \ldots, s_k$ be words. If $L(t) \subseteq L(s_1) \cup L(s_2) \cup \cdots \cup L(s_k)$, then $L(t) \subseteq L(s_i)$ for some $1 \le i \le k$.*

Theorem 3 ([ASO91a]) *Let k be any fixed integer and S be a set of words. If $|\Sigma| \ge k+1$, then a k-MMG of S is computable in polynomial time of size(S).*

In general, a *k-MMG* of S is not unique. Arimura's algorithm [ASO91b] to compute a *k-MMG* of S is not exahusitive for all *k-MMG*'s of S. However, it is ensured that when the algorithm is give a set S of words for its input, it finds out at least one *k-MMG* of S in time polynomial in *size*(S) and it outputs only *k-MMG*'s of S. We denotes the set of outputs of the algorithm for an input S by *k-mmg*(S).

From the way to calculate 2-*mmg*(S), Proposition 1 and Theorem 2, we can prove the following lemma.

Lemma 4 *Suppose that $|\Sigma| \ge 3$. Then, for any primitive Prolog $P = \{C_0, C_1\}$, there exists a finite set $S \subseteq M_P$ that satisfies the following two claims for any S_Δ such that $S \subseteq S_\Delta \subseteq M_P$.*

[1]In [JLMM88], lg is called as lca (least common anti-instance).

(1) $\{lg(C_0(M_P)), lg(C_1(M_P))\} \in 2\text{-}mmg(S_\Delta)$.

(2) *For any pair of atoms* $\{q_0, q_1\} \in 2\text{-}mmg(S_\Delta)$, *either*

(a) $q_0 \succeq lg(C_0(M_P))$ *and* $q_1 \preceq lg(C_1(M_P))$ *or*

(b) $q_1 \succeq lg(C_0(M_P))$ *and* $q_0 \preceq lg(C_1(M_P))$

holds.

The above lemma ensures that, for any given enumeration of a target model, the pair of atoms $lg(C_0(M_P))$, $lg(C_1(M_P))$ can be found in the limit using the 2-*mmg* algorithm. In the next section, we show that $lg(C_0(M_P))$ and $lg(C_1(M_P))$ correspond to the heads of clauses in a *most specific version* of P. The most specific version of P is a program that has the same least Herbrand model as P's. Hence, the above lemma suggests applicability of the 2-*mmg* algorithm to infer heads of clauses of the target program.

4 Most specific versions

Let $P = \{C_1, \ldots, C_n\}$ be a program, $C \in P$ be a clause and θ be a substitution. A *more specific version* C' of C for P, denoted $C \succeq_P C'$, is a clause such that $C' = C\theta$ and $C'(M_P) = C(M_P)$. A *more specific version* P' of P, denoted $P \succeq_P P'$, is a program which consists of clauses C'_i $(1 \le i \le n)$ where C'_i is a more specific version of C_i for P. $SV(P)$ denotes the set of all more specific version of P. The *most specific version* P' of P, denoted $msv(P)$, is a program such that $P \succeq_P P'$ and, for any $P'' \in SV(P)$, $P'' \succeq_{P''} P'$.

The above definition for the notion of more specific version slightly differ from the original one by Marriott et al. [MNL88]. For example, let P consist of clauses:

$$C_1 = p(x) \leftarrow q(x,y),$$
$$C_2 = q(f(x),y) \leftarrow q(x,y),$$
$$C_3 = q(f(a),y).$$

The program P' which obtained from P by replacing C_1 with $C'_1 = p(f(x)) \leftarrow q(f(x),a)$ is a more specific version of P by the above definition, but not by the Marriott's one. However, if the original program P is variable bounded[2], then both definitions are identical. Since primitive Prologs are variable bounded, the following proposition holds.

Proposition 5 ([MNL88]) *Let P be a primitive Prolog. Then, for any $P' \in SV(P)$, it holds that $M(P') = M(P)$. Furthermore, P has a most specific version $msv(P)$ that is unique up to a variable renaming of its clauses.*

The inference algorithms presented in this paper try to infer the heads of clauses as a pair $\{lg(C_0(M_P)), lg(C_1(M_P))\}$ of least generalizations. As shown in Proposition 6 and Lemma 8, the pair corresponds to the heads of $msv(P)$. Thus the algorithms, in fact, take the class of most specific versions of primitive Prologs as its hypothesis space. The above proposition ensures that the sift of hypothesis space is acceptable. In the following, we show several other properties of most specific versions of primitive Prologs that are useful for constructing our inference algorithms.

Let $P = \{C_0, C_1\}$ be a primitive Prolog, where

$$C_0 = p(s),$$
$$C_1 = p(t[x_1, \ldots, x_m]) \leftarrow p(x_1), \ldots, p(x_m).$$

[2]A program is variable bounded iff it consists of clauses in which every variable appearing in the body appears in the head.

Proposition 6 *Suppose that* $|\Sigma| \geq 2$. *Then, for any* $P' \in SV(P)$, P' *is of the form* $\{C_0, C_1\theta\}$ *where* $\theta = \{x_1/R_1, \ldots, x_m/R_m\}$.

Let $\theta = \{x_1/R_1, \ldots, x_m/R_m\}$ and $msv(P) = \{C_0, C_1\theta\}$, namely,

$$C_0 = p(s),$$
$$C_1\theta = p(t[R_1, \ldots, R_m]) \leftarrow p(R_1), \ldots, p(R_m).$$

Lemma 7 $lg(\{s, t[R_1, \ldots, R_m]\}) \succ s$ *and* $lg(\{s, t[R_1, \ldots, R_m]\}) \succ t[R_1, \ldots, R_m]$.

Proof: If $lg(\{s, t[R_1, \ldots, R_m]\}) \equiv s$, then $s \succeq t[R_1, \ldots, R_m]$. This contradicts the condition of a primitive Prolog: $L(s) \cap L(t[x_1, \ldots, x_m]) = \emptyset$. A similar argument for the alternate claim completes the proof of the lemma. □

Lemma 8 *If* $|\Sigma| \geq 2$, *then* $p(t[R_1, \ldots, R_m]) \equiv lg(C_1(M_P))$.

Proof: Along the same line of argument as in the proof of Theorem 2 in [Ish88], we can show that, for a substitution σ such that $head(C_1)\sigma = lg(C_1(M_P))$, $C_1\sigma(M_P) = C_1(M_P)$. Thus $\{C_0, C_1\sigma\} \in SV(P)$.

On the other hand, since $C_1'(M_P) = C_1(M_P)$ for any $P' = \{C_0, C_1'\} \in \dot{S}V(P)$, $head(C_1') \succeq lg(C_1(M_P))$. Hence, there exists a substitution δ such that $head(C_1'\delta) = head(C_1\sigma)$. Since C_1 is variable bounded, $C_1'\delta = C_1\sigma$. From Proposition 5, it follows that $M_P = M_{P'}$. Thus

$$C_1'(M_{P'}) = C_1'(M_P) = C_1(M_P) = C_1\sigma(M_P) = C_1\sigma(M_{P'}).$$

That is, $C_1' \succeq_{P'} C_1\sigma$ and $P' \succeq_{P'} \{C_0, C_1\sigma\}$. Hence, $\{C_0, C_1\sigma\} = msv(P)$. This completes the proof of the lemma. □

Proposition 6 and Lemma 8 ensure that if $|\Sigma| \geq 3$ then the heads of clauses in $msv(P)$ are given as a pair $\{p(s) = lg(C_0(M_P)), lg(C_1(M_P))\}$ of atoms that is in 2-$mmg(S)$ for a sufficiently large subset S of M_P.

Lemma 9 *For any* $1 \leq i \leq m$, $p(R_i) \equiv lg(M_P)$.

Proof: Assume that, for some $1 \leq i \leq m$, $p(R_i) \not\succeq lg(M_P)$. Then, there exists $p(\beta) \in M_P$ such that $p(R_i) \not\succeq p(\beta)$. On the other hand, there exists a substitution γ such that $p(t[\beta, \ldots, \beta])\gamma \in C_1(M_P)$. Since $p(R_i) \not\succeq p(\beta)$ for some $1 \leq i \leq m$, it holds that $head(C_1\theta) = p(t[R_1, \ldots, R_m]) \not\succeq p(t[\beta, \ldots, \beta])\gamma$. Hence, $p(t[\beta, \ldots, \beta])\gamma \notin C_1\theta(M_P)$. This contradicts the fact that $C_1\theta(M_P) = C_1(M_P)$. Thus $p(R_i) \succeq lg(M_P)$ for any $1 \leq i \leq m$.

Assume that $p(R_i) \succ lg(M_P)$ for some $1 \leq i \leq m$. Let $p(\tau) = lg(M_P)$ and $\sigma = \{x_i/\tau\}$. Clearly, it holds that $C_1\sigma(M_P) = C_1(M_P)$. Let $P' = \{C_0, C_1\sigma\}$. Since $C_1 \succeq_P C_1\sigma$, $P' \in SV(P)$. On the other hand, since $R_i \succ \tau$, there is no substitution γ such that $C_1\sigma\gamma = C_1\theta$, that is, $P' \not\succeq_{P'} msv(P)$. This contradicts the definition of $msv(P)$. □

The following Lemma 10 ensures that, for the pair $\{p(s), p(t')\}$ of heads of clauses in $msv(P)$, each atom appearing in the body of a clause in $msv(P)$ is of the form $lg(\{p(s), p(t')\})$. A greedy search algorithm for the body of a clause given in the next section is based on this property.

Lemma 10 *For any* $1 \leq i \leq m$, $p(R_i) \equiv lg(\{p(s), p(t[R_1, \ldots, R_m])\})$.

Proof: From Proposition 6 and Lemma 8, it follows that

$$lg(\{p(s), p(t[R_1, \ldots, R_m])\}) \equiv lg(\{lg(C_0(M_P)), lg(C_1(M_P))\}).$$

Since $C_0(M_P) \cup C_1(M_P) = M_P$, from Lemma 4 in [JLMM88], it follows that

$$lg(\{lg(C_0(M_P)), lg(C_1(M_P))\}) \equiv lg(C_0(M_P) \cup C_1(M_P)) \equiv lg(M_P).$$

Thus, from Lemma 9, it holds that $lg(M_P) \equiv p(R_i)$. □

The following propositions and lemmata are useful in proving validity of the greedy search for the body of a clause. Since the algorithm is required to find an appropriate body from only positive data, it has to avoid constructing an overgeneralized hypothesis. Proposition 12 says that a hypothesis with an appropriate body that corresponds to a subset of the body of the original $msv(P)$ is consistent with all positive examples. On the other hand, Lemma 13 and Lemma 14 ensure that, for any hypothesis with an inappropriate body, there exists a positive counter example, even if the body consists of only one atom.

Proposition 11 *For any $1 \leq i \neq j \leq m$, it holds that $var(R_i) \cap var(R_j) = \emptyset$ and $var(R_i) \cap (var(t[x_1, \ldots, x_m]) - \{x_1, \ldots, x_m\}) = \emptyset$.*

Proposition 12 *Let $P' = \{C_0, C_1'\}$ where*

$$C_1' = p(t[R_1, \ldots, R_m]) \leftarrow p(R_{i_1}), \ldots, p(R_{i_k}) \text{ and } \{R_{i_1}, \ldots, R_{i_k}\} \subseteq \{R_1, \ldots, R_m\}.$$

Then it holds that $M_P \subseteq M_{P'}$.

Lemma 13 *Suppose that $|\Sigma| \geq 3$. Let $P' = \{C_0, C_1'\}$ where $C_1' = p(t[R_1, \ldots, R_m]) \leftarrow p(R')$ and R' is a subterm of $t[R_1, \ldots, R_m]$ such that $R' \equiv R_i$ $(1 \leq i \leq m)$ and $R' \notin \{R_1, \ldots, R_m\}$. Then it holds that $M_P - M_{P'} \neq \emptyset$.*

Proof: From Proposition 11 and the condition on R', it follows that $var(R') \cap var(R_i) = \emptyset$ $(1 \leq i \leq m)$. Hence, for any ground substitution σ for R', there exists a ground substitution γ for $p(t[R_1, \ldots, R_m])\sigma$ such that $p(t[R_1, \ldots, R_m])\sigma\gamma \in M_P$.

For proving the lemma, it is sufficient to show that there exists a ground substitution σ for R' such that

$$t[R_1, \ldots, R_m] \not\succeq R'\sigma \text{ and } s \not\succeq R'\sigma. \qquad (*)$$

Since it follows that $p(R')\sigma \notin M_{P'}$ from $(*)$, $p(t[R_1, \ldots, R_m])\sigma\gamma \notin M_{P'}$ for any substitution γ. With the above fact, this implies an existence of a substitution γ such that

$$p(t[R_1, \ldots, R_m])\sigma\gamma \in M_P \text{ but } p(t[R_1, \ldots, R_m])\sigma\gamma \notin M_{P'}.$$

Since $R' \equiv R_i$, from Lemma 7, it holds that $R' \succ s$ and $R' \succ t[R_1, \ldots, R_m]$. In what follows, $t[R_1, \ldots, R_m]$ is abbreviated as t.

Since $R' \succ s$, the following two cases are possible:

(1) There is an index I such that $R'(I) = x$ and $s(I) = u$ where x is a variable and u is a non-variable term.

(2) There are indexes I_1 and I_2 such that $R'(I_1) = x$, $R'(I_2) = y$, $s(I_1) = z$ and $s(I_2) = z$ where x, y, z are mutually distinct variables.

Furthermore, since $R' \equiv lg(s, t)$, these cases can be divided into the following four cases:

(1-1) There exists an index I such that $R'(I) = x$, $s(I) = u$ and $t(I) = v$ where x is a variable and u, v are non-variable terms with different principle functors.

(1-2) There exists an index I such that $R'(I) = x$, $s(I) = u$ and $t(I) = y$ where x, y are variables and u is a non-variable term.

(2-1) There exist indexes I_1 and I_2 such that $R'(I_1) = x$, $R'(I_2) = y$, $s(I_1) = z$, $s(I_2) = z$, $t(I_1) = u$ and $t(I_2) = v$ where x, y, z are mutually distinct variables and u, v are mutually distinct terms such that at least one of them is not variable.

(2-2) There exist indexes I_1 and I_2 such that $R'(I_1) = x$, $R'(I_2) = y$, $s(I_1) = z$, $s(I_2) = z$, $t(I_1) = x_1$ and $t(I_2) = y_1$ where x, x_1, y, y_1, z are mutually distinct variables.

Case (1-1): From the assumption that $|\Sigma| \geq 3$, there exists a ground term w whose principle functor differs from neither u's nor v's. For any ground substitution σ that binds x with w, $(*)$ holds.

Case (1-2): Since $R' \succ t$, there are two possible cases:

(1') There exists an index I' such that $R'(I') = x'$ and $t(I') = u'$ where x' is a variable and u' is a non-variable term.

(2') There exist indexes I'_1 and I'_2 such that $R'(I'_1) = x'$, $R'(I'_2) = y'$, $t(I'_1) = z'$ and $t(I'_2) = z'$ where x', y', z' are mutually distinct variables.

In the case (1'), it holds that $x \neq x'$[3]. Let σ be a ground substitution that binds x with a ground term w whose principle functor differs from u's and x' with a ground term w' whose principle functor differs from u''s. Then σ satisfies $(*)$. In the case (2'), let w_1 be a ground term whose principle functor differs from u's and w_2 be a different ground term from w_1. Then $(*)$ holds for any ground substitution σ that binds x with w_1, x' with one of which w_1 or w_2 and y' with the other[4].

Case (2-1): Without loss of generality, we may assume that u is not a variable. Let w_1 be a ground term whose principle functor differs from u's and w_2 be a different ground term from w_1. Then $(*)$ holds for any ground substitution σ that binds x with w_1 and y with w_2.

Case (2-2): Similarly for the case (1-2), there are two possible cases (1') and (2'). The case (1') for the case (2-2) is same as the case (2') for the case (1-2). In the case (2'), let w and w' be different ground terms. Since $x \neq y$ and $x' \neq y'$, it suffices to consider a ground substitution that binds x, x' with one of which w_1 or w_2 and y, y' with the other. □

Lemma 14 *Let $P = \{C'_0, head(C_1\theta)\}$ where $C'_0 = p(s) \leftarrow p(R')$ and R' is a subterm of s such that $R' \equiv R_i$ $(1 \leq i \leq m)$. Then it holds that $M_P - M_{P'} \neq \emptyset$.*

5 Finding the body of a clause

In this section, we describe an algorithm that, for a set S of atoms and a ordered pair $\langle p(s), p(t)\rangle$ of atoms such that $S \subseteq L(p(s)) \cup L(p(t))$ and $L(p(s)) \cap L(p(t)) = \emptyset$, searches a program $P = \{C_0, C_1\}$ where

$$C_0 = p(s),$$
$$C_1 = p(t) \leftarrow p(r_1), \ldots, p(r_k),$$

and P satisfies the following conditions:

1. $S \subseteq M_P$:
2. Each r_i is a proper subterm of t such that $r_i \equiv lg(t, s)$ $(1 \leq i \leq k)$ and $var(r_i) \cap var(r_j) = \emptyset$ $(i \neq j)$.
3. Let $t'[x_1, \ldots, x_k]$ be the term that obtained from t by replacing each occurrence of r_i by x_i $(1 \leq i \leq k)$ where x_i $(1 \leq i \leq k)$ is a variable that does not appear in t and $x_i \neq x_j$ $(i \neq j)$. Then it holds that $L(t'[x_1, \ldots, x_k]) \cap L(s) = \emptyset$.
4. The body of C_1 is maximal one in such bodies satisfying the above conditions. That is, for any other subterm r_{k+1} of t that satisfies the condition 2 and for the clause $C = p(t) \leftarrow p(r_1), \ldots, p(r_k), p(r_{k+1})$, C does not satisfy the condition 3 or $S \not\subseteq M(\{C_0, C\})$.

[3]The term $t(I)$ is a variable and the term $t(I')$ is a non-variable. Thus, if $x = x'$, then $R' \not\succ t$.

[4]When x, x', y' differs from each other, this choice of bindings can be arbitrary. When x is same as x', bind y' with w_2. When x is same as y', bind x' with w_2.

We denote a program that satisfies the above conditions by $P(S,\langle p(s),p(t)\rangle)$. While, in general, there may exist several $P(S,\langle p(s),p(t)\rangle)$s, one of them can be found in polynomial time by simple greedy search.

Algorithm 1: A greedy search algorithm for $P(S,\langle p(s),p(t)\rangle)$

Input: A set of positive examples S and a tuple of atoms $\langle p(s),p(t)\rangle$ such that $S\subseteq L(p(s))\cup L(p(t))$ and $L(p(s))\cap L(p(t))=\emptyset$.
Output: A $P(S,\langle p(s),p(t)\rangle)$
Procedure:

Let $SUB(t)$ be the set of proper subterms of t, that are variant of $lg(\{s,t\})$,
$Body := \phi$; $Pat := t$;
$H := \{p(t) \leftarrow Body.,p(s).\}$;
for each $T \in SUB(t)$ s.t. $var(T)\cap var(Body)=\phi$ **do**
  $H := \{p(t) \leftarrow Body \cup \{p(T)\}.,p(s).\}$;
  if $S \not\subseteq M_H$ **then**
    $H := \{p(t) \leftarrow Body.,p(s).\}$;
  else Let Pat' be the term which is obtained from Pat by replacing
    the all occurrences of T in Pat with a new variable x;
    if $L(Pat')\cap L(s) \neq \phi$ **then** $H := \{p(t) \leftarrow Body.,p(s).\}$;
    else $Body := Body \cup \{p(T)\}$;
      $Pat := Pat'$;
Output H and halt;

Lemma 15 *For a set S of atoms and a ordered pair $\langle p(s),p(t)\rangle$ of atoms such that $S\subseteq L(p(s))\cup L(p(t))$ and $L(p(s))\cap L(p(t))=\emptyset$, Algorithm 1 outputs a $P(S,\langle p(s),p(t)\rangle)$ in time polynomial in $size(S\cup\{p(s),p(t)\})$.*

Let $P(S,\langle p(s),p(t)\rangle)=\{C_0,C_1\}$ be an output of Algorithm 1 where $C_0=p(s)$ and $C_1=p(t) \leftarrow p(r_1),\ldots,p(r_m)$. Let C_1' be the clause which obtained from C_1 by replacing each occurrence of r_i by x_i $(1\leq i\leq m)$ where $x_1,\ldots,x_m$ are mutually distinct variables that do not occur in C_1. Since no r_i contains other r_j as its subterm, C_1' is well-defined and obtained from C_1 in time polynomial in $size(t)$. We denotes $\{C_0,C_1'\}$ by $pr(P(S,\langle p(s),p(t)\rangle))$.

Lemma 16 *The program $pr(P(S,\langle p(s),p(t)\rangle))$ is a primitive Prolog which has the same least Herbrand model as $P(S,\langle p(s),p(t)\rangle)$'s.*

6 Polynomial update time inference of primitive Prologs

An *inference algorithm* $\mathcal{A}$ is an algorithm that iterates the process "input request $\rightarrow$ computation $\rightarrow$ output". Let $g_1,g_2,\ldots$ be a sequence of outputs of $\mathcal{A}$ for an input sequence $e_1,e_2,\ldots$. We say that $\mathcal{A}$ *converges* to g for the input sequence $e_1,e_2,\ldots$ iff there exists $n\geq 1$ such that $g_i=g$ for any $i\geq n$.

An *enumeration* of a model M is a sequence $e_1,e_2,\ldots$ of elements in M such that every atom in M occurs as e_i for some $i\geq 1$. We say that $\mathcal{A}$ *identifies* a model M *in the limit from positive data* iff $\mathcal{A}$ converges to a program P such that $M_P=M$ for any enumeration of M.

Let $P_1,P_2,\ldots$ be a sequence of outputs of $\mathcal{A}$ for an enumeration $e_1,e_2,\ldots$ of a model M and S_i be the set $\{e_1,\ldots,e_i\}$. An inference algorithm $\mathcal{A}$ is *consistent* iff $S_i\subseteq M(P_i)$ for any

i. An inference algorithm $\mathcal{A}$ is *conservative* iff $P_i = P_{i-1}$ for any i such that $e_i \in M(P_{i-1})$. An inference algorithm $\mathcal{A}$ is a *polynomial update time inference algorithm* iff there exists some polynomial f such that, for any stage i, after $\mathcal{A}$ feeds the input e_i it produces the output P_i in $f(size(S_i))$ steps.

In what follows, let $P = \{C_0, C_1\}$ be any primitive Prolog where

$$C_0 = p(s),$$
$$C_1 = p(t[x_1, \ldots, x_m]) \leftarrow p(x_1), \ldots, p(x_m),$$

and $msv(P) = \{C_0, C_1\theta\}$ be the most specific version of P where $\theta = \{x_1/R_1, \ldots, x_m/R_m\}$. For notational convenience, we abbreviate $t[x_1, \ldots, x_m]$ as t and $t[R_1, \ldots, R_m]$ as t_R. Let $e_1, e_2, \ldots$ be any enumeration of M_P and S_i be a set $\{e_1, \ldots, e_i\}$ of first i elements in the enumeration.

Theorem 17 *There exists an integer N such that, for any integer $n \geq N$, Algorithm 1 outputs $msv(P)$ for given S_n and $\langle p(s), p(t_R)\rangle$ as its inputs.*

Proof: Let $T_1, \ldots, T_k$ be all subterms of t_R except R_i's such that $p(T_i) \equiv lg(\{p(s), p(t_R)\})$. From Lemma 13, for any T_i $(1 \leq i \leq k)$, a program $P_{T_i} = \{C_0, p(t_R) \leftarrow p(T_i)\}$ has a positive counter example $c_i \in M_P$ such that $c_i \notin M(P_{T_i})$. Thus, for any input S_n such that $\{c_1, \ldots, c_k\} \subseteq S_n$, Algorithm 1 outputs a program $P = \{C_0, C_1'\}$ where $C_1' = p(t_R) \leftarrow p(R_{i_1}), \ldots, p(R_{i_j})$ and $\{R_{i_1}, \ldots, R_{i_j}\} \subseteq \{R_1, \ldots, R_m\}$. On the other hand, from Proposition 12 and maximality of the body of the clause in $P(S, \langle p(s), p(t_R)\rangle)$ (the condition 4 in the definition of $P(S, \langle p(s), p(t_R)\rangle)$), there is no case in which $\{R_{i_1}, \ldots, R_{i_j}\}$ is a proper subset of $\{R_1, \ldots, R_m\}$. □

By Theorem 17, if we assume that a unit clause $C_0 = p(s)$ is given to an inference algorithm, we can easily construct a consistent and conservative polynomial update time inference algorithm that identifies the class of models of primitive Prologs in the limit.

Algorithm 2: A consistent and conservative polynomial update time inference algorithm

Given: $p(s)$ that is the unit clause of a target program P.
Input: $e_1, e_2, \ldots$: any enumeration of M_P.
Output: $P_1, P_2, \ldots$: a sequence of primitive Prologs.
Procedure:

$S := \{\}$; $H := \{p(s)\}$;
Read the next example e; $S := S \cup \{e\}$;
if $e \in M_H$ then output H;
else $S_- := S - (S \cap L(p(s)))$;
  Compute $P(S, \langle p(s), lg(S_-)\rangle)$ using Algorithm 1;
  $H := pr(P(S, \langle p(s), lg(S_-)\rangle))$; output H;

Lemma 18 *Algorithm 2 is a consistent and conservative polynomial update time inference algorithm.*

Theorem 19 *Suppose that $|\Sigma| \geq 2$. Then, for any primitive Prolog P, Algorithm 2 identifies M_P in the limit.*

Proof: Since $S_- = S - (S \cap L(p(s))) \subseteq C_1(M_P)$ for any $S \subseteq M_P$, it holds that $lg(S_-) \preceq lg(C_1(M_P))$. On the other hand, from Lemma 8, it holds that $p(t_R) \equiv lg(C_1(M_P))$. From Proposition 1, there exists a finite subset S^* of $C_1(M_P)$ such that $lg(S^*) \equiv p(t_R)$. Hence, for any integer n such that $S^* \subseteq S_n$, it holds that $lg(S_-) = p(t_R)$. With Lemma 16 and Theorem 17, this proves the theorem. □

From the above theorem, if the inference algorithm can find a unit clause in some way, the algorithm identifies a target model efficiently. In general, there is no way to find a unit clause from arbitrary enumeration of a model in bounded finite time. However, there is a way to identify it in the limit (unbounded finite time) using the 2-*mmg* algorithm. Finally, we describe a polynomial update time inference algorithm that is consistent but not conservative.

Let $P = \{C_0, C_1\}$ be a primitive Prolog where C_0 is a unit clause. If C_1 has nonempty body, then it holds that $min\{size(\alpha) \mid \alpha \in C_0(M_P)\} < min\{size(\beta) \mid \beta \in C_1(M_P)\}$. On the other hand, from Lemma 4, for any $\{q_0, q_1\} \in 2\text{-}mmg(S)$, either $q_0 \succeq C_0$ or $q_1 \succeq C_0$ holds in the limit. In both case, C_0 is less general. Thus, to identify a unit clause C_0, an inference algorithm have only to keep a minimal size example given so far and find a less general atom, that coves the example, in atoms contained outputs produced by 2-*mmg* algorithm.

Algorithm 3: A consistent polynomial update time inference algorithm using the 2-*mmg* algorithm

Input: $e_1, e_2, \ldots$: any enumeration of M_P.
Output: $P_1, P_2, \ldots$: a sequence of primitive Prologs.
Procedure:

$S := \{\}$; $Size := +\infty$;
Read the next example e; $S := S \cup \{e\}$;
if $size(e) < Size$ **then**
  $Min := e$; $Size := size(e)$
Let H be the set of all $\{q_0, q_1\} \in 2\text{-}mmg(S)$ such that $L(q_0) \cap L(q_1) = \emptyset$;
if there exists an atom h_0' appearing in H such that
$Min \in L(h_0')$ and $h_0' \not\succ q$ for any q appearing in H **then**
  Compute $P(S, \langle h_0', h_1' \rangle)$ using Algorithm 1 where $\{h_0', h_1'\} \in H$.
  Output $pr(P(S, \langle h_0', h_1' \rangle))$
else output $lg(S)$

Theorem 20 *Suppose that $|\Sigma| \geq 3$. Then, for any primitive Prolog P, Algorithm 3 identifies M_P in the limit. Furthermore, Algorithm 3 is a consistent polynomial update time inference algorithm.*

Proof: If a target program consists of only unit clauses, then the problem is just same as one for inferring unions of tree pattern languages [ASO91b] and it is easily shown that the sequence of outputs produced by Algorithm 3 converges to a primitive Prolog H such that $M_H = M_P$.

Thus, we assume that a target program consists of a unit clause C_0 and a nonunit clause C_1. From (1) in Lemma 4, there exists a finite subset S of M_P such that, for any $S \subseteq S_\Delta \subseteq M_P$, $\{lg(C_0(M_P)), lg(C_1(M_P))\} \in 2\text{-}mmg(S_\Delta)$. Furthermore, $lg(C_0(M_P)) = C_0$ and $lg(C_1(M_P))$ is an instance of $head(C_1)$. Since $L(C_0) \cap L(head(C_1))$ is empty, $L(lg(C_0(M_P))) \cap L(lg(C_1(M_P)))$ is also empty. Thus, after the algorithm has fed all elements in S, the pair of atoms $\{lg(C_0(M_P)), lg(C_1(M_P))\}$ is always contained in H.

On the other hand, from (2) in Lemma 4, for any atom q appearing in $2\text{-}mmg(S_\Delta)$, either $q \succeq lg(C_0(M_P))$ or $q \preceq lg(C_1(M_P))$ holds. Since $L(lg(C_0(M_P))) \cap L(lg(C_1(M_P))) = \emptyset$, for any atom $q \preceq lg(C_1(M_P))$, it holds that $q \not\preceq lg(C_0(M_P))$. Hence, for any atom q appearing in $2\text{-}mmg(S_\Delta)$, it holds that $lg(C_0(M_P)) \not\succ q$.

Let α be a minimal size element of M_P. Since $\alpha \in C_0(M_P) = L(lg(C_0(M_P)))$ and $L(lg(C_0(M_P))) \cap L(lg(C_1(M_P))) = \emptyset$, it holds that $\alpha \notin L(lg(C_1(M_P)))$ and also $\alpha \notin L(q)$ for any $q \preceq lg(C_1(M_P))$.

Thus, after the algorithm has fed all elements in S and α, the atom h'_0 in the algorithm is always found and satisfies that $h'_0 \equiv lg(C_0(M_P))$. From the way to calculate 2-$mmg(S_\Delta)$ [ASO91b] where $S_\Delta \supseteq S$, for any pair $\{h'_0, h'_1\} \in$ 2-$mmg(S_\Delta)$, it necessarily holds that $h'_1 \equiv lg(C_1(M_P))$. Hence, after the algorithm has fed all elements in S and α, it continues to output $pr(P(S \cup \{\alpha\}, \langle lg(C_0(M_P)), lg(C_1(M_P))\rangle))$. With Lemma 16 and Theorem 17, this proves that the algorithm identifies M_P in the limit.

Clearly, the algorithm is consistent, because it outputs either $pr(P(S, \langle h'_0, h'_1\rangle))$ or $lg(S)$. Since $|2\text{-}mmg(S)|$ is bounded by a polynomial in $size(S)$, it is also clear that the algorithm produces each output in time polynomial in $size(S)$. □

7 Concluding Remarks

In this paper, we presented two inference algorithms which identify the class of primitive Prologs in the limit from positive data. The first is a consistent and conservative polynomial update time algorithm that, given the base case of the unknown program, identifies the class in the limit from positive data with polynomial time updating hypotheses. The second is a consistent but not conservative polynomial update time algorithm that identifies the class in the limit from positive data. The second inference algorithm employing a natural technique, that is, using the 2-mmg algorithm to infer heads of clauses in a target program. This technique is considered as an extension of the method proposed by Ishizaka in [Ish88].

For this kind of polynomial update time inference algorithms, like the second one proposed in this paper, Pitt [Pit89] pointed out the problem that if we have any exponential update time inference algorithm which identifies a class, then we can obtain an algorithm that achieves polynomial update time inference by postponing to output a conjecture until they have enough size of examples. Thus, the results presented in this paper is not satisfactory.

The difficulty of accomplishing conservativeness of the inference algorithm using 2-mmg essentially originates in non-uniqueness of 2-mmg for the entire model M_P of a primitive Prolog P. For example, consider the following primitive Prolog P.

$$p([a,b,a]).$$
$$p([b|X]) \leftarrow p(X).$$

For the least Herbrand model

$$M_P = \{p([a,b,a]), p([b,a,b,a]), p([b,b,a,b,a]), p([b,b,b,a,b,a]), \ldots\},$$

there exist two kinds of 2-mmg of $M(P)$:

$$\{p([a,b,a]), p([b,X,Y,Z|W])\} \text{ and } \{p([b,a,b,a]), p([X,b,Y|Z])\}.$$

Actually the former is an instance of the heads of the program P. For any non-empty finite subset S of M_P, it holds that

$$\begin{aligned} P(S, \langle p([a,b,a]), p([b,X,Y,Z|W])\rangle) &= \{\ p([a,b,a]). \\ &\qquad p([b,X,Y,Z|W]) \leftarrow p([X,Y,Z|W]).\} \\ P(S, \langle p([b,a,b,a]), p([X,b,Y|Z])\rangle) &= \{\ p([b,a,b,a]). \\ &\qquad p([X,b,Y|Z]).\}. \end{aligned}$$

Hence, an inference algorithm that uses the 2-mmg algorithm and Algorithm 1 as its subprocedure have a chance to meet the former correct instance of P. Since we know the target program P, we know the former is correct but the latter is overgeneralized. However, it seems difficult for the algorithm to decide which is better, because the algorithm is given only positive examples and both candidates are consistent with all of them. If the algorithm can efficiently (that is, in polynomial time) decide which of competitive hypotheses has a smaller model, then it may avoid producing an overgeneralized hypothesis and achieve consistent and conservative polynomial update time inference. However, it is still open whether a model containment problem for primitive Prologs is solved efficiently.

Contrastively, in [AIS92], we introduced another subclass of linear Prologs of which element has only one 2-*mmg* of its model and presented a consistent and conservative polynomial update time inference algorithm for the class. The class is also a subclass of *context-free transformations* that was originally introduced by Shapiro in his study on MIS [Sha81]. Although the subclass is so restrictive, it can be shown that the subclass still includes several non-trivial programs in context-free transformations such as *append*, *plus*, *prefix* etc. For the class of primitive Prologs, it is still open if there exists a consistent and conservative polynomial update time inference algorithm.

References

[AIS92] Hiroki Arimura, Hiroki Ishizaka, and Takeshi Shinohara. Polynomial time inference of a subclass of context-free transformations. In *Proceedings of 5th Workshop on Computational Learning Theory*, 1992.

[ASO91a] Hiroki Arimura, Takeshi Shinohara, and Setsuko Otsuki. A polynomial time algorithm for finite unions of tree pattern languages. In *Proc. of the 2nd International Workshop on Nonmonotonic and Inductive Logics*, 1991. To appear in LNCS.

[ASO91b] Hiroki Arimura, Takeshi Shinohara, and Setsuko Otsuki. Polynomial time inference of unions of tree pattern languages. In S. Arikawa, A. Maruoka, and T. Sato, editors, *Proc. ALT '91*, pp. 105–114. Ohmsha, 1991.

[Ish88] Hiroki Ishizaka. Model inference incorporating generalization. *Journal of Information Processing*, 11(3):206–211, 1988.

[JLMM88] J-L.Lassez, M. J. Maher, and K. Marriott. Unification revisited. In J. Minker, editor, *Foundations of Deductive Databases and Logic Programming*, pp. 587–625. Morgan Kaufmann, 1988.

[Llo84] John W. Lloyd. *Foundations of Logic Programming*. Springer-Verlag, 1984.

[MB88] Stephen Muggleton and Wray Buntine. Machine invention of first-order predicates by inverting resolution. In *Proc. 5th International Conference on Machine Learning*, pp. 339–352, 1988.

[MNL88] K. Marriott, L. Naish, and J-L. Lassez. Most specific logic programs. In *Logic Programming: Proceedings of the Fifth International Conference and Symposium*, pp. 910–923. MIT Press, 1988.

[Pit89] Leonard Pitt. Inductive inference, dfas, and computational comlexity. In K. P. Jantke, editor, *Proc. AII '89, LNAI 397*, pp. 18–44. Springer-Verlag, 1989.

[Plo70] Gordon D. Plotkin. A note on inductive generalization. In B. Meltzer and D. Michie, editors, *Machine Intelligence 5*, pp. 153–163. Edinburgh University Press, 1970.

[Sha81] Ehud Y. Shapiro. Inductive inference of theories from facts. Technical Report 192, Yale University Computer Science Dept., 1981.

[Shi90] Takeshi Shinohara. Inductive inference of monotonic fomal systems from positive data. In S. Arikawa, S. Goto, S. Ohsuga, and T. Yokomori, editors, *Proc. ALT '90*, pp. 339–351. Ohmsha, 1990.

MONOTONIC LANGUAGE LEARNING

Shyam Kapur*
Institute for Research in Cognitive Science
University of Pennsylvania
3401 Walnut Street Rm 412C
Philadelphia, PA 19104 USA
skapur@linc.cis.upenn.edu

Abstract. Learnability of families of recursive languages from positive data is studied in the Gold paradigm of inductive inference, where the learner obeys certain constraints motivated by work in inductive reasoning. Previously, various notions of monotonicity have been defined in the context of language learning. These constraints require that the learner's guess monotonically 'improves' with regard to the target language. In this paper, the ideas from inductive reasoning are instantiated in alternative ways. Links are established between the various new constraints both among themselves as well as with other well-known constraints, such as conservativeness. Exactly learnable families are characterized for prudent learners which obey various combinations of these constraints. Applications of these characterizations are also shown.

1 Introduction

In the Gold paradigm for inductive inference [3], the language learner is presented with the *text* of a language, i.e., an infinite sequence of strings made up of all and only strings from the language. This model is motivated by the well-established hypothesis that the child learns her native language from positive evidence alone. (For a discussion, see [2].) The learner is said to learn a language if, on any text for it, the learner's guess *converges* to the same language, i.e., from some point onwards, the guess coincides with the language being presented. The learner is said to learn a family of languages if it learns each language in the family.

There has been some recent work toward relating research on inductive reasoning in logic to work on algorithmic inductive inference [4, 8, 15]. (In the past, one deep connection was established in [14], where a technique to infer Prolog programs was developed.) In particular, Lange and Zeugmann [8] investigate various versions of monotonicity in the context of language learning. They define strong-monotonic, monotonic and weak-monotonic language learning and relate these notions to obtain a strong hierarchy. In [9], various classes learnable under the different constraints are characterized. One way of appreciating the significance of these ideas

*This work was supported in part by ARO grant DAAL 03-89-C-0031, DARPA grant N00014-90-J-1863, NSF grant IRI 90-16592 and Ben Franklin grant 91S.3078C-1.

is to recognize that, in general, the purpose of *identification in the limit* [3] is not only at the limit. Thus, it is natural to expect that learners which are constrained to guess in a restricted and thereby informative way are more useful for various applications. For example, consider the *monotonic* constraint which requires that the learner's guess must monotonically improve with regard to the target language. In other words, if the learner's guess at any stage correctly predicts that some string x is in the target language, then all subsequent guesses must contain x as well.

In this paper, we introduce additional constraints which are also motivated by considerations of inductive reasoning. For example, we define the *dual monotonic* constraint that also requires that the learner's guess must monotonically improve with regard to the target language, but in the following sense: if the learner's guess at any stage correctly predicts that some string x is not in the target language, then all subsequent guesses must not contain x as well. We relate these new constraints to the previously defined constraints and characterize families learnable under various combinations of these constraints. Intuitively, the main underlying question we consider is whether it is possible to infer an unknown language in such a way that the intermediate hypotheses are all monotonically better generalizations and/or specializations.

In Section 2, we set up our learning model. In Section 3, we define the various constraints which we investigate in this paper. We also relate the classes of language families learnable under different constraints. Finally, in Section 4, we develop some characterizations.

2 Background

Let Σ^* be a free monoid over Σ, a finite alphabet of symbols. Let $M_1, M_2, M_3, \ldots$ be any standard enumeration of all Turing machines over Σ. Let Z_+ be the set of positive integers. For any *index* $I \in Z_+$, let W_I denote the *language* (subset of Σ^*) accepted by the machine M_I. The complement of W_I in Σ^* is denoted as $\overline{W_I}$. An index I is *total* if the corresponding machine M_I is total and accepts a non-empty language. If $I_1, I_2, \ldots$ is a recursive enumeration of total indices, then $\mathcal{F} = W_{I_1}, W_{I_2}, \ldots$ is called an *indexed family of non-empty recursive languages* (hereafter, simply an *indexed family*). We denote by $\Delta_{\mathcal{F}}$ the set of all non-empty finite subsets of Σ^* that are contained in some language in family $\mathcal{F}$ and let Δ denote $\Delta_{\{\Sigma^*\}}$. Given classes of families $\mathcal{L}_1$ and $\mathcal{L}_2$, $\mathcal{L}_1 \# \mathcal{L}_2$ denotes that the two classes are incomparable.

A *text* is an infinite sequence of strings from Σ^*; t ranges over texts, t_n is the nth string in text t, $\bar{t}_n$ is the initial prefix of length n of text t, and $content(\bar{t}_n)$ is the set of strings in the prefix $\bar{t}_n$; t is *for* a language L if and only if the set of strings in t equals L.

An *inductive inference machine (IIM)* M is an algorithmic device whose input is a text $t_1, t_2, \ldots$ and whose output is a sequence of nonnegative integers $M(\bar{t}_1), M(\bar{t}_2), \ldots$ constrained to be either 0 or total indices. The procedure works in stages, but it can happen that a stage never gets completed. At the nth stage, t_n is input and $M(\bar{t}_n)$ is output. The interpretation is as follows: If $M(\bar{t}_n) = 0$, then the IIM makes no guess; otherwise, it guesses the language $W_{M(\bar{t}_n)}$.

An IIM M is said to learn the language L if and only if, for each text t for L, there is a k such that $W_{M(\bar{t}_k)} = L$ and, for all $n > k$, $M(\bar{t}_n) = M(\bar{t}_k)$. Intuitively, the guess converges to a total index for the input language. (This is similar to the TxtEx-identification criterion [3].) We say that M learns a family $\mathcal{F}$ if M learns each language in $\mathcal{F}$.

An IIM M is said to be *prudent* if it learns each of the languages it ever guesses. A family $\mathcal{F}$ is said to be *exactly* learned by M if M learns $\mathcal{F}$ but does not learn any superset of $\mathcal{F}$. In this paper, we only consider exact, prudent learning.

3 Variants of Monotonic Learning

In this section, we consider various definitions of monotonicity and investigate their relation to each other.

Definition 1 *[8]* An IIM M is said to learn a language L

1. strong-monotonically
2. monotonically
3. weak-monotonically

if and only if M learns L and on any text t for L and any two consecutive non-zero hypotheses, say $M(\bar{t}_n)$ and $M(\bar{t}_{n+k})$, $k \geq 1$, the following condition is satisfied:

1. $W_{M(\bar{t}_n)} \subseteq W_{M(\bar{t}_{n+k})}$
2. $W_{M(\bar{t}_n)} \cap L \subseteq W_{M(\bar{t}_{n+k})} \cap L$
3. if $content(\bar{t}_{n+k}) \subseteq W_{M(\bar{t}_n)}$, then $W_{M(\bar{t}_n)} \subseteq W_{M(\bar{t}_{n+k})}$

Let *SMON*, *MON* and *WMON* denote the class of families for which there is an IIM M which learns each language strong-monotonically, monotonically and weak-monotonically, respectively.

Based on results obtained by Lange, Zeugmann & Kapur [8, 16], the existence of a strong hierarchy is indicated by the following theorem:

Theorem 1

$$FIN \subset SMON \subset MON \subset WMON = CONSERVATIVE \subset LIM$$

(Note: FIN denotes the class of families learnable by machines that on any text for any language in the family produce only a single and correct guess. CONSERVATIVE is the class of families learnable by IIMs that exclusively perform justified mind changes. LIM is the class of learnable families.)

In order to further investigate the monotonicity requirements, we define new constraints which are natural duals of the constraints defined in Definition 1. Thus, we have alternative instantiations of the concepts of monotonicity from recursive function learning in language learning. Recall that the study of monotonicity in recursive function learning was motivated by its analogues in logic [4].

Definition 2 An IIM M is said to learn a language L

1. dual strong-monotonically
2. dual monotonically
3. dual weak-monotonically

if and only if M learns L and on any text t for L and any two consecutive non-zero hypotheses, say $M(\bar{t}_n)$ and $M(\bar{t}_{n+k})$, $k \geq 1$, the following condition is satisfied:

1. $W_{M(\bar{t}_n)} \supseteq W_{M(\bar{t}_{n+k})}$
2. $\overline{W_{M(\bar{t}_n)}} \cap \bar{L} \subseteq \overline{W_{M(\bar{t}_{n+k})}} \cap \bar{L}$
3. if $content(\bar{t}_{n+k}) \subseteq W_{M(\bar{t}_n)}$, then $W_{M(\bar{t}_n)} \supseteq W_{M(\bar{t}_{n+k})}$

Let $SMON^d$, MON^d and $WMON^d$ denote the class of families for which there is an IIM M which learns each language dual strong-monotonically, dual monotonically and dual weak-monotonically, respectively.

It should be noted that the various notions of monotonicity and those of dual monotonicity are truly duals of *each other*. To appreciate the structure of the newly defined classes, the following two propositions are helpful.

Proposition 1

(1) $WMON^d \supseteq CONSERVATIVE = WMON$

(2) $MON \# MON^d$

(3) $SMON^d = FIN \subset SMON$

Proof: (1) Obvious.

(2) We first show a family $\mathcal{F}$ in $MON^d \setminus MON$. Let $\Sigma = \{a\}$. For $k \geq 1$, let $W_{I_k} = \{a\}^* \setminus \{a^k\}$. Suppose an IIM M is claimed to learn $\mathcal{F}$ monotonically. If M is run on the input sequence $a^2, a^3, \ldots$, at some stage $k-1$, the machine must guess an index for the language W_{I_1}. From then on, continue the sequence replacing the string a^{k+1} by a. Clearly, at some stage $k+m-1$, the machine must now output an index for the language $W_{I_{k+1}}$. Continue the presentation as before, now replacing the string a^{k+m+1} by a^{k+1}. Clearly, this presentation is a text for the language $W_{I_{k+m+1}}$, but now, when the machine guesses $W_{I_{k+m+1}}$, it violates monotonicity. On the other hand, the IIM that learns $\mathcal{F}$ using the *identification by enumeration* technique [3] works in a dual monotonic fashion.

For the converse, we let $\Sigma = \{a, b, c\}$ and define $\mathcal{F}$ as follows. Let $W_{I_1} = \{a\}^*$. For $k > 1$, let the language $W_{I_k} = \{a, a^2, \ldots a^{k-1}\} \cup \{c\}^*$. For $k, m \geq 1$, let $W_{I_{k,m}} = \{a, a^2, \ldots a^k\} \cup \{c, c^2, \ldots c^m\} \cup \{b\}$. (We abuse notation here to make the structure of the family transparent.) Suppose there is a machine M that is claimed to learn $\mathcal{F}$ dual monotonically. Consider a presentation of the strings $a, a^2, \ldots$ to M. At some stage k, the machine must output an index for the language W_{I_1}. From that stage on, continue the presentation with the sequence of strings $c, c^2, \ldots$. At some stage $k+m$, the machine must now output an index for the language $W_{I_{k+1}}$. Thereafter, present the string b repeatedly. The entire presentation is a text for the language $W_{I_{k,m}}$ but the machine cannot guess this language without violating its dual monotonic nature. On the other hand, consider the IIM M that guesses the language W_{I_1} as long as only strings from the set $\{a\}^*$ have been seen. If any string from the set $\{c\}^*$ is seen, the machine begins to guess the least language W_{I_k} containing all the evidence. If the string b is seen, the machine guesses the least language $W_{I_{k,m}}$ containing all the evidence. Clearly, M behaves in a monotonic fashion. It can easily be shown that M learns $\mathcal{F}$.

(3) Trivially, $FIN \subseteq SMON^d$. Further, if an IIM which learns $\mathcal{F}$ dual strong-monotonically shrinks its guess, the first guess cannot be learned. Hence, the machine must make only one guess and it must be correct. (Notice that a machine that learns strong-monotonically can make an inconsistent guess, i.e., a guess such that it does not contain all the evidence seen up to that point, while a machine that learns dual strong-monotonically cannot. However, that is not where the extra power of strong-monotonicity over its dual lies [8].) The second part of the assertion follows from Theorem 1. ∎

Proposition 2

(1) $MON^d \subset WMON^d$

(2) $SMON^d \subset MON^d$

Proof: (1) By a combination of parts (1) and (2) of Proposition 1 and Theorem 1, we have $WMON^d \setminus MON^d \neq \emptyset$. Consider a machine M that learns a family $\mathcal{F}$ dual monotonically. We claim that M must learn dual weak-monotonically as well. Suppose this is not the case. This means that, at some stage, M changes its guess, say L, on consistent data to something that is not a subset of L. Clearly, if the presentation is now extended to make up a text for L itself, the machine cannot revert to guessing L without violating dual monotonicity.

(2) Clearly, a machine that behaves dual strong-monotonically also behaves dual monotonically. Using part (3) of Proposition 1 and observing that the family $\mathcal{F}$ shown to be in $MON^d \setminus MON$ in the proof of part (2) of that proposition is not in FIN, we have the required result. ∎

Notice the progression of the relationship of the dual constraints to their counterparts. For the strong-monotonic case, its dual is more restrictive; for the monotonic case, the dual is incomparable, while for the weak-monotonic situation, its dual is at least as powerful.

Some well-known classes can be obtained by combining constraints in Definition 1 and Definition 2. For example, if we insist that a machine learn a family strong-monotonically as well as dual strong-monotonically, then it is essentially equivalent to asking the machine to make only one guess which must be correct. Likewise, a machine that learns a family weak-monotonically as well as dual weak-monotonically is essentially equivalent to a machine that learns conservatively. It is also worthwhile to observe that a family consisting of all finite sets is learnable both monotonically and dual monotonically by the algorithm that always guesses the language equal to the evidence seen. Motivated by these considerations, we next define three new classes that arise by combining the monotonicity and dual monotonicity constraints in the following way:

Definition 3 Let $SMON^{\&}$ represent the class of families learnable by some IIM that behaves strong-monotonically as well as dual strong-monotonically. Let $MON^{\&}$ and $WMON^{\&}$ be analogously defined.

The following proposition is easy to establish.

Proposition 3

(1) $FIN = SMON^{\&} \subset MON^{\&} \subset WMON^{\&} = CONSERVATIVE$

(2) $SMON \subset MON^{\&}$

While the classes $SMON^{\&}$ and $WMON^{\&}$ are equivalent to $SMON^{d}$ and $WMON$, respectively, it is not obvious whether $MON^{\&}$ is exactly the intersection of MON and MON^{d}. This relationship as well as the exact relationship between MON^{d} and $CONSERVATIVE$, $CONSERVATIVE$ and $WMON^{d}$, and $WMON^{d}$ and LIM has recently been obtained by Lange, Zeugmann & Kapur [10].

4 Characterizations

We have defined a number of variants of monotonicity and related them to each other and to other well-known classes. In this section, we characterize some of them.

Previously, the classes LIM and $CONSERVATIVE$ have been characterized. We repeat these characterizations below. We begin with the following important definition.

Definition 4 *[1]* A finite set T is a *tell-tale* subset of L in $\mathcal{F}$ if $T \subseteq L$ and

$$(\forall L' \in \mathcal{F})(L' \subset L \Rightarrow T \not\subseteq L').$$

Angluin's [1] characterized LIM in the following way:

Theorem 2 *An indexed family $\mathcal{F}$ is in LIM if and only if there is an effective procedure that, given as input a total index I such that $W_I \in \mathcal{F}$, recursively enumerates a tell-tale subset T_I of W_I.*

Various sufficient conditions, useful to show that a family is in LIM, can be found in [1, 7, 13]. The notion of *least upper bound* (l.u.b.) is a dual to that of tell-tale subsets. Thus, a language $L \in \mathcal{F}$ is an l.u.b. of a finite set T if and only if T is a tell-tale subset of L. We use the following definitions in the subsequent characterizations.

Definition 5 A partial recursive function $f : \Delta \times Z_+ \mapsto Z_+ \cup \{0\}$ is a *good* function for an indexed family $\mathcal{F}$ if, for all $S \in \Delta_{\mathcal{F}}$ and $k \in Z_+$, $f(S,k)$ is defined and is either 0 or a total index such that $W_{f(S,k)} \in \mathcal{F}$. The function f is *full-range* if each L in $\mathcal{F}$ has a subset S such that, for some $k \in Z_+$, $W_{f(S,k)} = L$.

Definition 6 *[5, 6]* A partial recursive function $f : \Delta \times Z_+ \mapsto Z_+ \cup \{0\}$ is a *least upper bound (l.u.b.)* function of *type* 0 for an indexed family $\mathcal{F}$ if it is good and if, for any S and k, $S \subseteq W_{f(S,k)}$, then $W_{f(S,k)}$ is an l.u.b. of S.

The families learnable conservatively have been characterized by Kapur & Bilardi [5, 6]. Motoki [11] has also characterized a special version of conservative learning in an analogous way.

Theorem 3 *Let $\mathcal{F} = W_{I_1}, W_{I_2}, \ldots$ where $I_1, I_2, \ldots$ is a recursive enumeration of total indices. $\mathcal{F} \in$ CONSERVATIVE if and only if there exists a full-range l.u.b. function of type 0 for $\mathcal{F}$.*

It can easily be verified that the set *FIN* has the following characterization:

Proposition 4 *An indexed family $\mathcal{F}$ is in FIN if and only if there is a partial recursive function f that is good, full-range and, for any S and k, if $S \subseteq W_{f(S,k)}$, then, for any $L \in \mathcal{F}$, $S \subseteq L \Rightarrow L = W_{f(S,k)}$.*

FIN has also been characterized by Mukouchi [12] and Lange & Zeugmann [9] in somewhat different terms. Their characterizations are in terms of the existence of a recursive enumeration of a special kind of a tell-tale subset for each language in the family.

Due to the relationships in the previous section, we automatically have characterizations for the classes $SMON^d = SMON^\&$ (= *FIN*) and $WMON = WMON^\&$ (= *CONSERVATIVE*). We next characterize the class *SMON*.

Proposition 5 *An indexed family $\mathcal{F}$ is in SMON if and only if there is a partial recursive function f that is good, full-range and, for any S and k, if $S \subseteq W_{f(S,k)}$, then, for any $L \in \mathcal{F}$, $S \subseteq L \Rightarrow L \supseteq W_{f(S,k)}$.*

Proof: We define an IIM M that learns $\mathcal{F}$ strong-monotonically. Let $(D_1, k_1), (D_2, k_2), \ldots$ be a complete enumeration of pairs, where, for each $i \geq 1$, $D_i \in \Delta_{\mathcal{F}}$ and $k_i \in Z_+$. M is defined so as to change its guess only in case of inconsistency. Whenever M can change its guess (say, stage n), let the new guess be $f(D_i, k_i)$ for the least $i \leq n$ such that $D_i \subseteq content(\bar{t}_n) \subseteq W_{f(D_i,k_i)}$. If there is no $i \leq n$ that meets the requirement, output a 0. It is easy to see that this machine is strong-monotonic, since whenever the machine makes a guess, all the languages in the family consistent with the evidence have to be supersets of the guess. Further, it can be argued as follows that M learns $\mathcal{F}$. On a text t for $L \in \mathcal{F}$, clearly, the machine can never guess a superset of L. Since f is full-range, there must be a least i such that $W_{f(D_i,k_i)} = L$. Therefore, beyond some stage n, the machine must guess L since the condition $D_i \subseteq content(\bar{t}_n) \subseteq L$ will always be met.

Conversely, suppose M learns $\mathcal{F}$ strong-monotonically. The following procedure defines a function $f : \Delta \times Z_+ \mapsto Z_+ \cup \{0\}$. Let $S \in \Delta$ and $k \in Z_+$. Scan $\mathcal{F} = W_{I_1}, W_{I_2}, \ldots$ until an I_j is found such that $S \subseteq W_{I_j}$. (This scan terminates if and only if $S \in \Delta_{\mathcal{F}}$.) Then, let $f(S, k)$ be the output of M when the input is the lexicographic enumeration $s_1, s_2, \ldots, s_n$ of S ($n = |S|$) followed by $k - 1$ repetitions of s_n. Since $\mathcal{F}$ is learned, for any $S \in \Delta_{\mathcal{F}}$, this output must be defined. Further, f is good. Since M learns $\mathcal{F}$, for any $L \in \mathcal{F}$, there must be a sequence of the kind $s_1, s_2, \ldots, s_n$ followed by some number of repetitions of s_n on which M outputs an index for L. Hence, f is full-range. Finally, since M is strong-monotonic, whenever $f(S,k)$ is defined to be a non-zero value I, all the languages that include S must include W_I as well. ∎

SMON has also been characterized by Lange & Zeugmann [9] in terms of the existence of a special kind of a tell-tale subset for each language in a special enumeration of the family. We next characterize the class $MON^{\&}$. It is of interest to note that the strong-monotonicity constraint is a special case of the dual monotonicity constraint, where the first guess of the machine is assumed to be the empty set. Similarly, the dual strong-monotonicity constraint is a special case of the monotonicity constraint, where the first guess of the machine is assumed to be Σ^*. Our characterization of the class $MON^{\&}$ is similar in spirit to the following alternative characterization of the class of families (denoted as $CONS - CONSERVATIVE$) learnable by conservative learners that are consistent, i.e., whose non-zero guesses at any stage contain the evidence seen up to that stage. Lange & Zeugmann [9] have also obtained a similar characterization for conservative learning.

Proposition 6 *[5, 6] An indexed family $\mathcal{F}$ is in $CONS - CONSERVATIVE$ if and only if there is a recursive enumeration $(T_1, I_1), (T_2, I_2), \ldots$ of pairs such that, for each $h \in Z_+$, T_h is a finite set, I_h is a total index, W_{I_h} is an l.u.b. of T_h in $\mathcal{F}$ and $\mathcal{F} = \{W_{I_1}, W_{I_2}, \ldots\}$.*

It is easy to argue that for any family in the class $MON^{\&}$ there exists a learner that not only satisfies the required constraints but also conservativeness and consistency. We begin with some definitions.

Definition 7 Let $\mathcal{F}$ be an indexed family. We define a *trigger* relation $\mathcal{T}$ between three elements represented in a triple $< P, S, Q >$. The second component S is in $\Delta_{\mathcal{F}}$, while the other two components are total indices of languages in $\mathcal{F}$. Thus, for example, we may have $< I_5, \{1,3\}, I_1 >$ as one of the triples included in this relation. We will often use the symbol ϕ in the first component of the relation when we do not care about its value.

Definition 8 We say that the trigger relation $\mathcal{T}$ is *good* if the following conditions are met:

(1) For any triple $< \phi, S, Q >$ in $\mathcal{T}$, it is the case that S is an l.u.b. of W_Q in $\mathcal{F}$.

(2) For any triple $< P, S, Q >$ in $\mathcal{T}$ and any language $L \in \mathcal{F}$ such that $S \subseteq L$, there must be some triple $< P, S', R >$ in $\mathcal{T}$ such that $L = W_R$.

We say that a trigger relation $\mathcal{T}$ is *full-range* if, for every $L \in \mathcal{F}$, there is a triple of the form $< \phi, S, Q >$ in $\mathcal{T}$ such that $L = W_Q$.

Definition 9 A triple $< P, S, Q >$ is said to be *good* if, for any language $L \in \mathcal{F}$ such that $S \subseteq L$, the following conditions hold:

(1) $W_Q \setminus W_P \subseteq L$, and

(2) $(W_P \setminus W_Q) \cap L = \emptyset$.

Proposition 7 *If the triples $< P, S, Q >$ and $< P, S', R >$ are good, where $S \subseteq W_R$, then the triple $< Q, S', R >$ is also good.*

Proof: Since $< P, S, Q >$ is a good triple and $S \subseteq W_R$ we must have

(1) $W_Q \setminus W_P \subseteq W_R$ and

(2) $(W_P \setminus W_Q) \cap W_R = \emptyset$.

Similarly, since $< P, S', R >$ is a good triple we must have

(1) $W_R \setminus W_P \subseteq L$ and

(2) $(W_P \setminus W_R) \cap L = \emptyset$,

for any $L \in \mathcal{F}$ such that $S' \subseteq L$. Putting it together, we get

(1) $W_R \setminus W_Q \subseteq L$ and

(2) $(W_Q \setminus W_R) \cap L = \emptyset$,

for any $L \in \mathcal{F}$ such that $S' \subseteq L$. ∎

Theorem 4 *An indexed family $\mathcal{F}$ is in $MON^{\&}$ if and only if there is a good, full-range, enumerable trigger relation $\mathcal{T}$ such that every triple in $\mathcal{T}$ is good.*

Proof: Suppose first that an IIM M learns $\mathcal{F}$ while satisfying the monotonic and the dual monotonic constraint. We indicate how we can enumerate the relation $\mathcal{T}$. Let $\sigma_1, \sigma_2, \ldots$ be an enumeration of all finite sequences (possibly with repetition) of strings from Σ^*. For each σ_i, if $content(\sigma_i) \in \Delta_{\mathcal{F}}$, then run M on σ_i. If $M(\sigma_i) \neq 0$, then suppose P is the first non-zero guess made by the machine during its run. Output the triple $< P, content(\sigma_i), M(\sigma_i) >$. We claim that the relation $\mathcal{T}$, enumerated in this fashion, satisfies all the properties required of it.

Clearly, the enumerated relation $\mathcal{T}$ is a trigger relation. Let us first argue that it is good. Since any triple in the enumeration is of the form $< \phi, content(\sigma_i), M(\sigma_i) >$, and since a conservative machine must always guess an l.u.b. of the evidence, the first condition is met. Suppose there is a triple of the form $< P, S, Q >$ in $\mathcal{T}$ and a language L in the family such that $S \subsetneq L$. Such a triple must have been generated because there is a sequence σ_i such that M output P as its first non-zero guess on it and $M(\sigma_i) = Q$. Let t be a text for L that has σ_i as a prefix. Clearly, at some stage n, the machine when run on t must guess some index R such that $W_R = L$. By construction, the triple $< P, content(\bar{t}_n), R >$ must also be enumerated. Thus the second condition is also satisfied.

Clearly, since M learns each language $L \in \mathcal{F}$, $\mathcal{T}$ must be full-range. We still have to establish that any triple $< P, S, Q >$ in $\mathcal{T}$ is good. By construction, the triple $< P, S, Q >$ is in $\mathcal{T}$ if and only if there is a sequence σ_i ($content(\sigma_i) = S$) such that when M is run on σ_i, the machine outputs P as its first guess. Further, on seeing the entire sequence σ_i, it outputs Q. Since M is monotonic and dual monotonic, for any language $L \in \mathcal{F}$ such that $S \subseteq L$, the following conditions must hold:

(1) $W_Q \setminus W_P \subseteq L$, and

(2) $(W_P \setminus W_Q) \cap L = \emptyset$.

Thus, the triple $< P, S, Q >$ is good. This completes the demonstration that the relation $\mathcal{T}$ enumerated in this way has all the properties required of it.

Suppose we have a relation $\mathcal{T}$ satisfying all the properties in the statement of the theorem. We construct an IIM M to learn $\mathcal{F}$ as follows: At any stage $n \geq 1$, first scan the family $\mathcal{F}$ to determine whether $content(\bar{t}_n) \in \Delta_{\mathcal{F}}$. Let $< P_1, S_1, Q_1 >, < P_2, S_2, Q_2 >, \ldots$ be an enumeration of the trigger relation $\mathcal{T}$. M behaves differently depending on whether or not it has made a non-zero guess prior to this stage.

- If $(\forall m < n)(M(\bar{t}_m) = 0)$, scan the enumeration of $\mathcal{T}$ for the least $k \leq n$ such that $S_k \subseteq content(\bar{t}_n) \subseteq W_{Q_k}$. If such a triple is found, then output Q_k as the guess. Otherwise, output 0.
- If $(\exists m < n)(M(\bar{t}_m) \neq 0)$, maintain the previous guess if consistent with the previous input as well as with t_n. Otherwise, let I be the first non-zero guess that the machine made. Scan the enumeration of $\mathcal{T}$ for the least $k \leq n$ such that such that $P_k = I$ and $S_k \subseteq content(\bar{t}_n) \subseteq W_{Q_k}$. If such a triple is found, then output Q_k as the guess. Otherwise, output 0.

We first argue that M learns $\mathcal{F}$. Let $L \in \mathcal{F}$ and let $t = t_1, t_2, \ldots$ be a text for L. The machine when run on t must make a non-zero guess since the relation $\mathcal{T}$ is full-range and, at some stage n, there would be a triple $< P_k, S_k, Q_k >$ enumerated such that $S_k \subseteq content(\bar{t}_n) \subseteq W_{Q_k}$ and $L = W_{Q_k}$. Since the machine is conservative and consistent (by construction) and always guesses an l.u.b. of the evidence seen, the only way the machine could fail to converge to an index for L is by guessing an infinite number of guesses all different from L. Suppose I is the first non-zero guess that M makes on t. Clearly, there must be a triple $< I, S, Q >$ in $\mathcal{T}$ such that $L = W_Q$. Consider a stage at which all of S has been witnessed in the evidence and a triple of this form has already been enumerated. Suppose further that at some subsequent stage, the machine changes its guess to R. We claim that $W_R = W_Q$. Suppose the guess R is prompted by a triple $< I, S', R >$ in $\mathcal{T}$. Clearly, $(S \cup S') \subseteq (W_Q \cap W_R)$. Since $< I, S, Q >$ is in $\mathcal{T}$, it must be good. Hence, we have $W_Q \setminus W_I \subseteq W_R$. Likewise, since $< I, S', R >$ is in $\mathcal{T}$, we have $W_R \setminus W_I \subseteq W_Q$. For the same reason, we have $(W_I \setminus W_Q) \cap W_R = \emptyset$ and $(W_I \setminus W_R) \cap W_Q = \emptyset$. Thus, $W_R = W_Q$. We have shown that M converges to an index for L on t.

Suppose on some text for a language $L = W_Q$, the machine makes the guesses $I, \ldots, P, \ldots, R, \ldots, Q$, and it is the case that the machine exhibits a violation of monotonicity or dual monotonicity while proceeding from the guess P to R. Therefore,

(1) $W_R \setminus W_P \not\subseteq L$ (violation of dual monotonicity), or

(2) $(W_P \setminus W_R) \cap L \neq \emptyset$ (violation of monotonicity).

From the behavior of M, we can infer that there must be triples of the form $< I, S, P >$ and $< I, S', R >$ in $\mathcal{T}$, where $S \subseteq W_R$ and $S \cup S' \subseteq L$. By Proposition 7, the triple $< P, S', R >$

must also be good. Clearly, this leads to a contradiction. Hence, the machine M cannot violate monotonicity or dual monotonicity. ∎

This characterization of $MON^{\&}$ is useful to show that certain families are not in this class since a trigger relation of the type stipulated in the theorem cannot exist.

Proposition 8 *The following families are not in $MON^{\&}$:*

(1) (From Proposition 1(2).) Let $\Sigma = \{a,b,c\}$. Let $W_{I_1} = \{a\}^*$. For $k > 1$, let $W_{I_k} = \{a, a^2, \ldots a^{k-1}\} \cup \{c\}^*$. Further, for $k, m \geq 1$, let

$$W_{I_{k,m}} = \{a, a^2, \ldots a^k\} \cup \{c, c^2, \ldots c^m\} \cup \{b\}.$$

(2) (From [8].) Let $\Sigma = \{a,b\}$. Let $W_{I_1} = \{a\}^*$. For $k > 1$, let $W_{I_k} = \{a, a^2, \ldots a^{k-1}, b^k, b^{k+1}, \ldots\}$. Further, for $k, m \geq 1$, let

$$W_{I_{k,m}} = \{a, a^2, \ldots a^k, b^{k+1}, \ldots, b^m, a^{m+1}, a^{m+2}, \ldots\}.$$

Proof: (The same proof works for both the families.) Suppose $\mathcal{T}$ is claimed to exist and satisfy all the necessary requirements. Then there must be a triple $< P, S, I_1 >$ in $\mathcal{T}$. Let j^* be the largest integer j such that $a^j \in S$. There must also be a triple $< P, S', I_{j^*+1} >$ in $\mathcal{T}$. By Proposition 7, the triple $< I_1, S', I_{j^*+1} >$ must be good, but this is impossible. ∎

The characterization can also be used to show that certain families are in $MON^{\&}$.

Proposition 9 *The following families are in $MON^{\&}$:*

(1) The family of all finite sets over $\Sigma = \{a\}$.

(2) (From [8].) Let $\Sigma = \{a,b\}$. For any $m \geq 1$ and $k_1, k_2, \ldots, k_m \geq 1$, let

$$W_{I_{k_1,k_2,\ldots k_m}} = (\{a\}^* \setminus \{a^{k_1}, a^{k_2}, \ldots a^{k_m}\}) \cup \{b^{k_1}, b^{k_2}, \ldots b^{k_m}\}.$$

Proof:

(1) For any $j \geq 1$ and any S that contains a^j, let $< \{a^j\}, S, S >$ be in $\mathcal{T}$. It is easy to see that $\mathcal{T}$ satisfies all the requirements.

(2) For any $j \geq 1$, let $< I_j, \{b^j\}, I_j >$ be in $\mathcal{T}$. For any $\{k_1, k_2, \ldots, k_m\}$ such that j is in that set of numbers, let $< I_j, \{b^{k_1}, b^{k_2}, \ldots, b^{k_m}\}, I_{k_1,k_2,\ldots k_m} >$ be in $\mathcal{T}$. It is easy to check that $\mathcal{T}$ satisfies all the requirements. ∎

As regards other monotonicity classes, we note that MON has been characterized by Lange & Zeugmann [9]. The classes MON^d and $WMON^d$ have recently been characterized by Zeugmann, Lange and Kapur [16].

References

[1] Dana Angluin. Inductive inference of formal languages from positive data. *Information and Control*, 45:117–135, 1980.

[2] Robert Berwick. *The Acquisition of Syntactic Knowledge.* MIT press, Cambridge, MA, 1985.

[3] E. M. Gold. Language identification in the limit. *Information and Control*, 10:447–474, 1967.

[4] Klaus P. Jantke. Monotonic and non-monotonic inductive inference. *New Generation Computing*, 8:349–360, 1991.

[5] Shyam Kapur. *Computational Learning of Languages.* PhD thesis, Cornell University, September 1991. Technical Report 91-1234.

[6] Shyam Kapur and Gianfranco Bilardi. Language learning without overgeneralization. In *Proceedings of the 9th Symposium on Theoretical Aspects of Computer Science (Lecture Notes in Computer Science 577)*, pages 245–256. Springer-Verlag, 1992.

[7] Shyam Kapur and Gianfranco Bilardi. On uniform learnability of language families. *Information Processing Letters*, 1992. To appear.

[8] Steffen Lange and Thomas Zeugmann. Monotonic versus non-monotonic language learning. In *Proceedings of the 2nd International Workshop on Nonmonotonic and Inductive Logic (Lecture Notes in Artificial Intelligence Series)*, 1991.

[9] Steffen Lange and Thomas Zeugmann. Types of monotonic language learning and their characterization. In *Proceedings of the 5th Conference on Computational Learning Theory.* Morgan-Kaufman, 1992.

[10] Steffen Lange, Thomas Zeugmann, and Shyam Kapur. Class preserving monotonic language learning. In preparation, 1992.

[11] Tatsuya Motoki. Consistent, responsive and conservative inference from positive data. In *Proceedings of the LA Symposium*, pages 55–60, 1990.

[12] Yasuhito Mukouchi. Characterization of finite identification. 1992. To appear in AII'92.

[13] Masako Sato and Kazutaka Umayahara. Inductive inferability for formal languages from positive data. In *Proceedings of the Workshop on Algorithmic Learning Theory.* JSAI, 1991.

[14] E. Y. Shapiro. Inductive inference of theories from facts. Technical Report 192, Yale University, 1981.

[15] Rolf Wiehagen. A thesis in inductive inference. In *Proceedings of the 1st International Workshop on Nonmonotonic and Inductive Logic.* Springer-Verlag, 1991. Lecture Notes in Artificial Intelligence Vol. 543.

[16] Thomas Zeugmann, Steffen Lange, and Shyam Kapur. Characterizations of class preserving monotonic language learning. In preparation, 1992.

Prudence in Vacillatory Language Identification (Extended Abstract)

SANJAY JAIN
Department of Computer and Information Sciences
University of Delaware
Newark, DE 19716, USA
Email: sjain@cis.udel.edu

ARUN SHARMA
School of Computer Science and Engineering
The University of New South Wales
Sydney, NSW 2033, Australia
Email: arun@spectrum.cs.unsw.oz.au

Abstract

The present paper settles an open question about 'prudent' vacillatory identification of grammars from positive data only.

Consider a scenario in which a learner M is learning a language L from positive data. Three different criteria for success of M on L have been investigated in formal language learning theory. If M converges to a single correct grammar for L, then the criterion of success is Gold's seminal notion of **TxtEx**-identification. If M converges to a finite number of correct grammars for L, then the criterion of success is called **TxtFex**-identification. And, if M, after a finite number of incorrect guesses, outputs only correct grammars for L (possibly infinitely many distinct grammars), then the criterion of success is known as **TxtBc**-identification.

A learner is said to be *prudent* according to a particular criterion of success just in case the only grammars it ever conjectures are for languages that it can learn according to that criterion. This notion was introduced by Osherson, Stob, and Weinstein with a view to investigate certain proposals for characterizing natural languages in linguistic theory. Fulk showed that prudence does not restrict **TxtEx**-identification, and later Kurtz and Royer were able to show that prudence does not restrict **TxtBc**-identification. The present paper settles this question by showing that prudence does not restrict **TxtFex**-identification.

1 Introduction

A child (modeled as a machine) receives (in arbitrary order) all the well-defined strings of a language (a *text* for the language) L, and simultaneously conjectures a succession of grammars. A criterion of success is for the child to eventually conjecture a correct grammar for L and to never change its conjecture thereafter. If, in this scenario for success, the child machine is replaced by an algorithmic machine M, then we say that M **TxtEx**-*identifies* L.

TxtEx-identification is essentially Gold's [Gol67] seminal notion of identification in the limit of grammars for recursively enumerable languages from positive data only. The reader is directed to [Pin79, WC80, Wex82, OSW86] for a discussion of the influence of this paradigm on contemporary theories of natural language.

A major goal of linguistic theory is to characterize the class of natural languages. Any such characterization must account for the fact that children master natural languages in a few years time on the basis of rather casual and unsystematic exposure to it. Formal language learning theory provides a tool to evaluate proposals for characterizing the collection of natural languages by modeling the salient features of a proposal in the above paradigm (see Osherson, Stob, Weinstein [OSW84] for discussion of these issues.)

A collection of such proposals are known as "prestorage models" of linguistic development. A prestorage model assumes that an internal list of candidate grammars that coincides exactly with the collection of natural languages is available to a child. Language acquisition is thus a process of selecting a grammar from this list in response to linguistic input. Motivated by such models and with a view to investigate the effect of such a restriction, Osherson, Stob, and Weinstein [OSW82b] introduced the notion of "prudent" learning machines. According to their definition, prudent learners only conjecture grammars for languages they are prepared to learn. In other words, every incorrect grammar emitted by a prudent learner in response to any linguistic input is for some language that can be learned by the learner. Osherson, Stob, and Weinstein raised the natural question: "Does prudence restrict **TxtEx**-identification?" Fulk [Ful85, Ful90] provided the answer by establishing a surprising result that prudence does not restrict **TxtEx**-identification. He showed that given any learning machine M, a prudent learning machine M′ can be constructed which **TxtEx**-identifies every language **TxtEx**-identified by M.

However, the investigation of prudence for more general learning criteria, notably **TxtBc**-identification and **TxtFex**-identification, was left open by Fulk. Below, we informally describe these criteria.

A learning machine M is said to **TxtBc**-identify L just in case M, fed any text for L, outputs an infinite sequence of grammars such that after a finite number of incorrect guesses, M outputs only grammars for L. This criterion was first studied by Osherson and Weinstein [OW82a] and Case and Lynes [CL82], and is also referred to as "extensional" identification. A machine M is said to be **TxtBc**-prudent just in case any grammar ever output by M is for a language which M can **TxtBc**-identify.

Let b be a positive integer. A learning machine M is said to $\mathbf{TxtFex}_b$-identify L just in case M, fed any text for L, converges in the limit to a finite set, with cardinality

$\leq b$, of grammars for L. In other words, for any text T for L there exists a set D of grammars of L, cardinality of $D \leq b$, such that $\mathbf{M}$, fed T, outputs, after a finite number of incorrect guesses, only grammars from D. This notion was studied by Osherson and Weinstein [OW82a] and by Case [Cas88]. A machine $\mathbf{M}$ is said to be $\mathbf{TxtFex}_b$-prudent just in case any grammar ever output by $\mathbf{M}$ is for a language which $\mathbf{M}$ can $\mathbf{TxtFex}_b$-identify.

Fulk [Ful85] had conjectured that prudence was not likely to restrict **TxtBc**-identification. Kurtz and Royer [KR88] established that this conjecture was indeed true as they were able to show that for any learning machine $\mathbf{M}$, there exists a machine $\mathbf{M}'$ such that $\mathbf{M}'$ is **TxtBc** prudent and $\mathbf{M}'$ **TxtBc**-identifies every language which $\mathbf{M}$ **TxtBc**-identifies. However, they left the problem open for $\mathbf{TxtFex}_b$-identification. In the present paper, we settle this question by showing that prudence does not restrict $\mathbf{TxtFex}_b$-identification.

We now proceed formally. Section 2 introduces the notation, defines language learning machines, and describes various criteria of language identification. The main result of the paper is contained in Section 3.

2 Preliminaries

2.1 Notation

Any unexplained recursion theoretic notation is from [Rog67]. N denotes the set of natural numbers, $\{0,1,2,3,\ldots\}$. N^+ denotes the set of positive natural numbers, $\{1,2,3,\ldots\}$. Unless otherwise specified, i,j,m,n,s,t,x,y, with or without decorations[1], range over N. $\emptyset$ denotes the empty set. $\subseteq$ denotes subset. $\subset$ denotes proper subset. $\supseteq$ denotes superset. $\supset$ denotes proper superset. P and S, with or without decorations, range over finite sets. card(S) denotes the cardinality of S. Let $*$ denote a number such that $(\forall i)[i < * < \infty]$. Thus, card($A$) $< *$ denotes the fact that A is a finite set. We let b range over $N^+ \cup \{*\}$. D_x denotes the finite set with canonical index x [Rog67]. We sometimes identify finite sets with their canonical indices. We do this when we consider functions or machines which operate on complete knowledge of a finite set (equivalently, an argument which is a canonical index of the finite set), but when we want to display the argument simply as the set itself.

$\uparrow$ denotes undefined. $\max(\cdot), \min(\cdot)$ denote the maximum and minimum of a set, respectively, where $\max(\emptyset) = 0$ and $\min(\emptyset) = \uparrow$.

f, g and h, with or without decorations, range over *total* functions with arguments and values from N.

$\langle i,j \rangle$ stands for an arbitrary, computable, one-to-one encoding of all pairs of natural numbers onto N [Rog67]. Similarly, we can define $\langle \cdot, \ldots, \cdot \rangle$ for encoding multiple natural numbers onto N.

φ denotes a fixed *acceptable* programming system for the partial computable functions: $N \rightarrow N$ [Rog58, Rog67, MY78]. φ_i denotes the partial computable function

[1] Decorations are subscripts, superscripts and the like.

computed by program i in the φ-system. Φ denotes an arbitrary fixed Blum complexity measure [Blu67, HU79] for the φ-system.

W_i denotes domain(φ_i). W_i is, then, the r.e. set/language ($\subseteq N$) accepted (or equivalently, generated) by the φ-program i. $\mathcal{E}$ will denote the set of all r.e. languages. L, with or without decorations, ranges over $\mathcal{E}$. $\mathcal{L}$, with or without decorations, ranges over subsets of $\mathcal{E}$. $W_i^s \stackrel{\text{def}}{=} \{x \leq s : \Phi_i(x) \leq s\}$. $\mathcal{FIN} \stackrel{\text{def}}{=} \{L : \text{card}(L) < \infty\}$.

We sometimes consider partial computable functions with multiple arguments in the φ system. In such cases we implicitly assume that a $\langle\cdot, \ldots, \cdot\rangle$ is used to code the arguments, so, for example, $\varphi_i(x, y)$ stands for $\varphi_i(\langle x, y\rangle)$.

The quantifiers '$\overset{\infty}{\forall}$' and '$\overset{\infty}{\exists}$' mean 'for all but finitely many' and 'there exist infinitely many', respectively.

2.2 Language Learning Machines and Texts

We now consider language learning machines. Definition 1 below introduces a notion that facilitates discussion about elements of a language being fed to a machine.

Definition 1 A *sequence* σ is a mapping from an initial segment of N into $(N \cup \{\#\})$. The *content* of a sequence σ, denoted content(σ), is the set of natural numbers in the range of σ. The *length* of σ, denoted by $|\sigma|$, is the number of elements in σ.

Intuitively, #'s represent pauses in the presentation of data. We let σ and τ, with or without decorations, range over finite sequences. For $n \leq |\sigma|$, $\sigma[n]$ denotes the finite initial sequence of σ with length n. We say that $\sigma \subseteq \tau$ just in case σ is an initial segment of τ, that is, $|\sigma| \leq |\tau|$ and $\sigma = \tau[|\sigma|]$. SEQ denotes the set of all finite sequences. The set of all finite sequences of natural numbers and #'s, SEQ, can be coded onto N. This latter fact will be used implicitly in our proof of the main theorem.

Definition 2 A *language learning machine* is an algorithmic device which computes a mapping from SEQ into N.

We let $\mathbf{M}$, with or without decorations, range over learning machines.

Definition 3 A *text* T for a language L is a mapping from N into $(N \cup \{\#\})$ such that L is the set of natural numbers in the range of T. The *content* of a text T, denoted content(T), is the set of natural numbers in the range of T.

Intuitively, a text for a language is an enumeration or sequential presentation of all the objects in the language with the #'s representing pauses in the listing or presentation of such objects. For example, the only text for the empty language is just an infinite sequence of #'s.

We let T, with or without decorations, range over texts. $T[n]$ denotes the finite initial sequence of T with length n. Hence, domain($T[n]$) $= \{x : x < n\}$. We say that $\sigma \subset T$ just in case σ is an initial segment of T, that is, $\sigma = T[|\sigma|]$.

We next present three criteria for successful learning of languages by learning machines.

2.3 Language Identification Criteria

2.3.1 Explanatory Learning (TxtEx-identification)

In Definition 4 below we spell out what it means for a learning machine on a text to converge in the limit.

Definition 4 Suppose M is a learning machine and T is a text. $\mathbf{M}(T)\downarrow$ (read: $\mathbf{M}(T)$ *converges*) $\iff$ $(\exists i)(\overset{\infty}{\forall} n)\ [\mathbf{M}(T[n]) = i]$. If $\mathbf{M}(T)\downarrow$, then $\mathbf{M}(T)$ is defined = the unique i such that $(\overset{\infty}{\forall} n)[\mathbf{M}(T[n]) = i]$; otherwise, we say that $\mathbf{M}(T)$ *diverges* (written: $\mathbf{M}(T)\uparrow$).

The next definition describes the first criteria of success and is essentially Gold's paradigm of identification in the limit.

Definition 5 [Gol67]

(a) M **TxtEx**-*identifies* L (written: $L \in \mathbf{TxtEx}(\mathbf{M})$) $\iff$ ($\forall$ texts T for L)$(\exists i : W_i = L)[\mathbf{M}(T)\downarrow = i]$.

(b) $\mathbf{TxtEx} = \{\mathcal{L} : (\exists \mathbf{M})[\mathcal{L} \subseteq \mathbf{TxtEx}(\mathbf{M})]\}$.

The notation in the above definition is from [CL82]. The influence of Gold's paradigm [Gol67] to human language learning is discussed by Pinker [Pin79], Wexler and Culicover [WC80], Wexler [Wex82], and Osherson, Stob, and Weinstein [OSW82a, OSW84, OSW86].

2.3.2 Vacillatory Learning (TxtFex-identification)

In Definition 6 below we spell out what it means for a learning machine on a text to converge in the limit to a finite set of grammars.

Definition 6 [Cas88] $\mathbf{M}(T)$ *finitely-converges* (written: $\mathbf{M}(T)\Downarrow$) $\iff$ $\{\mathbf{M}(\sigma) : \sigma \subset T\}$ is finite, otherwise we say that $\mathbf{M}(T)$ *finitely-diverges* (written: $\mathbf{M}(T)\Uparrow$). If $\mathbf{M}(T)\Downarrow$, then we say that $\mathbf{M}(T)\Downarrow = P \iff P = \{i \mid (\overset{\infty}{\exists}\ \sigma \subset T)[\mathbf{M}(\sigma) = i]\}$.

Definition 7 [Cas88] Let $b \in N^+ \cup \{*\}$.

(a) M **TxtFex**$_b$-*identifies* L (written: $L \in \mathbf{TxtFex}_b(\mathbf{M})$) $\iff$ ($\forall$ texts T for L)$(\exists P \mid \mathrm{card}(P) \leq b \wedge (\forall i \in P)[W_i = L])[\mathbf{M}(T)\Downarrow = P]$.

(b) $\mathbf{TxtFex}_b = \{\mathcal{L} : (\exists \mathbf{M})[\mathcal{L} \subseteq \mathbf{TxtFex}_b(\mathbf{M})]\}$.

2.3.3 Behaviorally Correct Learning (TxtBc-identification)

Definition 8 [CL82, OW82b, OW82a]

(a) M **TxtBc**-*identifies* L (written: $L \in \mathbf{TxtBc}(\mathbf{M})$) $\iff$ ($\forall$ texts T for L)$(\overset{\infty}{\forall} n)[W_{\mathbf{M}(T[n])} = L]$.

(b) $\mathbf{TxtBc} = \{\mathcal{L} : (\exists \mathbf{M})[\mathcal{L} \subseteq \mathbf{TxtBc}(\mathbf{M})]\}$.

3 Prudence and Language Learning

The following definition describes the notion of prudence for each of the three criteria described in the previous section. The notion of prudence was first introduced by Osherson, Stob, and Weinstein [OSW82b].

Definition 9

(a) A machine $\mathbf{M}$ is **TxtEx**-*prudent* just in case $\{W_{\mathbf{M}(\sigma)} : \sigma \in \text{SEQ}\} = \mathbf{TxtEx}(\mathbf{M})$.

(b) A machine $\mathbf{M}$ is **TxtBc**-*prudent* just in case $\{W_{\mathbf{M}(\sigma)} : \sigma \in \text{SEQ}\} = \mathbf{TxtBc}(\mathbf{M})$.

(c) A machine $\mathbf{M}$ is $\mathbf{TxtFex}_b$-*prudent* just in case $\{W_{\mathbf{M}(\sigma)} : \sigma \in \text{SEQ}\} = \mathbf{TxtFex}_b(\mathbf{M})$.

Fulk showed the following result which says that prudence is not a restriction on **TxtEx**-identification.

Theorem 1 [Ful85, Ful90] *For each machine* $\mathbf{M}$ *there exists a* **TxtEx**-*prudent machine* $\mathbf{M}'$, *such that* $\mathbf{TxtEx}(\mathbf{M}) \subseteq \mathbf{TxtEx}(\mathbf{M}')$.

Fulk left the question of prudence open for **TxtBc**-identification and $\mathbf{TxtFex}_b$-identification. But, he conjectured that a counterpart of the above theorem was likely to hold for **TxtBc**-identification. Kurtz and Royer [KR88] showed that Fulk's conjecture was indeed true, as they established the following result.

Theorem 2 [KR88] *For each machine* $\mathbf{M}$ *there exists a* **TxtBc**-*prudent machine* $\mathbf{M}'$, *such that* $\mathbf{TxtBc}(\mathbf{M}) \subseteq \mathbf{TxtBc}(\mathbf{M}')$.

However, prudence for vacillatory identification remained open. We settle this question in the following main theorem of the present paper.

Theorem 3 *Let* $b \in N^+ \cup \{*\}$. *For each* $\mathbf{M}$ *there exists a* $\mathbf{TxtFex}_b$-*prudent machine* $\mathbf{M}''$, *such that* $\mathbf{TxtFex}_b(\mathbf{M}) \subseteq \mathbf{TxtFex}_b(\mathbf{M}'')$.

Our proof of the above theorem builds on machinery about vacillatory language identification. We first present these tools. First we define the notions of stabilizing and locking sequences for $\mathbf{M}$ on L.

Definition 10 (Based on [Cas88, Ful85]) Let $b \in N^+ \cup \{*\}$. Then $\langle \sigma, D \rangle$ is a $\mathbf{TxtFex}_b$-*stabilizing sequence* for $\mathbf{M}$ on L just in case the following hold:

(a) $\text{content}(\sigma) \subseteq L$,

(b) $\text{card}(D) \leq b$, and

(c) $(\forall \tau : \sigma \subseteq \tau \wedge \text{content}(\tau) \subseteq L)[\mathbf{M}(\tau) \in D]$.

Definition 11 (Based on [BB75, Cas88, Ful85]) Let $b \in N^+ \cup \{*\}$. Then $\langle \sigma, D \rangle$ is a $\mathbf{TxtFex}_b$-*locking sequence* for $\mathbf{M}$ on L just in case the following hold:

(a) $\langle \sigma, D \rangle$ is a $\mathbf{TxtFex}_b$-stabilizing sequence for $\mathbf{M}$ on L, and

(b) $(\forall j \in D)[W_j = L]$.

Lemma 1 (Based on [BB75, Cas88, Ful85]) *Suppose* $b \in N^+ \cup \{*\}$ *and* $\mathbf{M}$ $\mathbf{TxtFex}_b$-*identifies* L. *Then there exists a* $\mathbf{TxtFex}_b$-*locking sequence for* $\mathbf{M}$ *on* L.

We now sketch a proof of our main theorem.

PROOF OF THEOREM 3. Let M, b be given. Let MidentN be true iff M $\mathbf{TxtFex}_b$-identifies N. Let g be a recursive function such that, for all n, $W_{g(n)} = \{x : x < n\}$. Define M′ as follows.

$$\mathbf{M}'(\sigma) = \begin{cases} g(0), & \text{if content}(\sigma) = \emptyset; \\ g(n), & \text{if } \neg\text{MidentN} \wedge \text{content}(\sigma) = \{x : x < n\}; \\ \mathbf{M}(\sigma), & \text{otherwise.} \end{cases}$$

Let i_N be such that $W_{i_N} = N$. Let $C_N = \{N\}$, if MidentN; otherwise, let $C_N = \{L : (\exists n)[L = \{x : x < n\}]\}$.

It is easy to see that $\mathbf{TxtFex}_b(\mathbf{M}) \cup \{\emptyset\} \cup C_N \subseteq \mathbf{TxtFex}_b(\mathbf{M}')$.

Intuitively, M′ behaves just like M except with minor changes to allow it to identify $\emptyset$ and C_N.

Before constructing M″ as claimed in the theorem, we will define, using s-m-n theorem, a recursive function h, from $N \times \text{SEQ} \times \mathcal{FIN} \times \mathcal{FIN} \to N$. For defining h, we need a few functions. Let $\text{Good}(j, \sigma, P, S) \equiv$

the following three conditions hold
(a) $j \in P$,
(b) $\langle \sigma, P \rangle$ is the least $\mathbf{TxtFex}_b$-stabilizing sequence for M′ on W_j, and
(c) $S = W_j \cap \{x : x < \max(\{\max(P') : \langle \sigma', P' \rangle \leq \langle \sigma, P \rangle\})\}$,

Intuitively, h will be such that, if $\text{Good}(j, \sigma, P, S)$, then $W_{h(j,\sigma,P,S)} = W_j$; otherwise $W_{h(j,\sigma,P,S)}$ is a member of $C_N \cup \{\emptyset\}$. M″ will use this h to achieve its goal as claimed in the theorem.

We now define two predicates, plausible and impossible. Intuitively, $\text{plausible}(j, \sigma, P, S, t)$ is true just in case it can be verified, in at most t steps, that condition (a) and parts of (b) and (c) of $\text{Good}(j, \sigma, P, S)$ hold. And, $\text{impossible}(j, \sigma, P, S, t)$ is true just in case it can be verified, in at most t steps, that $\neg\text{Good}(j, \sigma, P, S)$.

Let

$\text{plausible}(j, \sigma, P, S, t) \equiv$

$[t > 0] \wedge [\text{card}(P) \leq b] \wedge [j \in P] \wedge$
$[S \subseteq \bigcup_{\langle \sigma', P' \rangle \leq \langle \sigma, P \rangle} \text{content}(\sigma')] \wedge [\text{content}(\sigma) \subseteq S \subseteq W_j^t] \wedge$
$(\forall \langle \sigma', P' \rangle < \langle \sigma, P \rangle)(\exists \tau : |\tau| \leq t)[[\text{content}(\sigma') \not\subseteq S] \vee [\text{card}(P') > b] \vee$
$[[\text{content}(\sigma') \subseteq \text{content}(\tau) \subseteq W_j^t] \wedge [\mathbf{M}'(\tau) \notin P']]]$;

$\text{impossible}(j, \sigma, P, S, t) \equiv$

$[\overline{S} \cap W_j^t \cap \bigcup_{\langle \sigma', P' \rangle \leq \langle \sigma, P \rangle} \text{content}(\sigma') \neq \emptyset] \vee$
$(\exists \tau \supseteq \sigma : \text{content}(\tau) \subseteq W_j^t \wedge |\tau| \leq t)[\mathbf{M}'(\tau) \notin P]$.

We now let h be a recursive function such that $W_{h(j,\sigma,P,S)} = \bigcup_t W_{h(j,\sigma,P,S),t}$, where

$$W_{h(j,\sigma,P,S),t} = \begin{cases} \emptyset, & \text{if } \neg\text{plausible}(j, \sigma, P, S, t); \\ W_j^t, & \text{if plausible}(j, \sigma, P, S, t) \\ & \quad \wedge \neg\text{impossible}(j, \sigma, P, S, t); \\ \{x : x \leq t\}, & \text{if MidentN} \wedge \text{plausible}(j, \sigma, P, S, t) \\ & \quad \wedge \text{impossible}(j, \sigma, P, S, t); \\ \{x : x \leq \max(W_{h(j,\sigma,P,S),t-1})\}, & \text{if } \neg\text{MidentN} \wedge \text{plausible}(j, \sigma, P, S, t) \\ & \quad \wedge \text{impossible}(j, \sigma, P, S, t); \end{cases}$$

Let
$C = \{W_j : (\exists\sigma, P)[[\langle\sigma, P\rangle$ is the least $\mathbf{TxtFex}_b$-stabilizing sequence for $\mathbf{M}'$ on $W_j]$
$\wedge\ j \in P]\}$
and
$C' = \{W_{h(j,\sigma,P,S)} : j \in N \wedge \sigma \in \mathrm{SEQ} \wedge P, S \in \mathcal{FIN}\}$.

Lemma 2 $C = C'$.

PROOF. It is easy to see that,

$$(\forall j, \sigma, P, S, t)[\text{plausible}(j, \sigma, P, S, t) \Rightarrow \text{plausible}(j, \sigma, P, S, t+1)]$$

and

$$(\forall j, \sigma, P, S, t)[\text{impossible}(j, \sigma, P, S, t) \Rightarrow \text{impossible}(j, \sigma, P, S, t+1)].$$

Thus, for each j, σ and finite sets P, S: $W_{h(j,\sigma,P,S)} \in \{W_j, \emptyset\} \cup C_N$. Moreover, if $W_{h(j,\sigma,P,S)} \notin \{\emptyset\} \cup C_N$, then there exists $\langle\sigma, P\rangle$ such that $\langle\sigma, P\rangle$ is the least $\mathbf{TxtFex}_b$-stabilizing sequence for $\mathbf{M}'$ on W_j and $j \in P$. It follows that $C' \subseteq C$.

For each $L \in C$, let $\langle\sigma, P\rangle$ be the least $\mathbf{TxtFex}_b$-stabilizing sequence for $\mathbf{M}'$ on L. Let $j \in P$ be such that $W_j = L$. Let $S = W_j \cap \bigcup_{\langle\sigma',P'\rangle \leq \langle\sigma,P\rangle} \text{content}(\sigma')$. It is easy to see that $(\exists t)[\text{plausible}(j, \sigma, P, S)]$ and $(\forall t)[\neg\text{impossible}(j, \sigma, P, S, t)]$. It follows that $W_{h(j,\sigma,P,S)} = W_j \in C'$. Thus $C \subseteq C'$. ∎ (Lemma 2)

We now construct $\mathbf{M}''$ as claimed in theorem. Intuitively, $\mathbf{M}''$ tries to find a stabilizing sequence $\langle\sigma, P\rangle$ for $\mathbf{M}'$ on the input language, and then outputs $h(j_0, \sigma, P, S)$, for a seemingly *best* $j_0 \in P$ and S. Let T be a text. Define $\text{match}(j, T[n]) = \max(\{s \leq n : T[s] \subseteq W_j^n \wedge W_j^s \subseteq T[n]\})$.

$\mathbf{M}''$ on input $T[n]$:
(1) Let $\langle\sigma, P\rangle$, be the least pair such that the following three conditions are satisfied.
(a) $\text{content}(\sigma) \subseteq T[n]$,
(b) $\text{card}(P) \leq b$,
(c) $(\forall\tau : \sigma \subseteq \tau \wedge |\tau| \leq n \wedge \text{content}(\tau) \subseteq \text{content}(T[n]))[\mathbf{M}'(\tau) \in P]$.
(2) Let $m = \max(\{\text{match}(j, T[n]) : j \in P\})$.
(3) Let $j_0 = \min(\{j \in P : \text{match}(j, T[n]) = m\})$.
(4) Output $h(j_0, \sigma, P, \text{content}(T[n]) \cap \bigcup_{\langle\sigma',P'\rangle \leq \langle\sigma,P\rangle} \text{content}(\sigma'))$.
End $\mathbf{M}''$

Clearly, $\mathbf{M}''$ outputs grammars only for languages in C'. Suppose $L \in C' = C$. Let σ, P, be such that $\langle\sigma, P\rangle$ is the least $\mathbf{TxtFex}$-stabilizing sequence for $\mathbf{M}'$ on L. Let $P_L = \{j \in P : W_j = L\}$. Let T be a text for L. It is easy to see that, for all but finitely many n, $\mathbf{M}''(T[n]) \in \{h(j, \sigma, P, S) : j \in P_L\}$. It follows that $L \in \mathbf{TxtFex}_b(\mathbf{M}'')$. ∎ (Theorem 3)

4 Conclusion

The problem of prudence for successful learning of languages from positive data only was described. It was shown that requiring vacillatory language learners to be prudent does not result in any loss of learning power. This result, together with previous results of Fulk and of Kurtz and Royer, settles the question of prudence for the three popularly investigated criteria of learning from positive data only.

References

[BB75] L. Blum and M. Blum. Toward a mathematical theory of inductive inference. *Information and Control*, 28:125–155, 1975.

[Blu67] M. Blum. A machine independent theory of the complexity of recursive functions. *Journal of the ACM*, 14:322–336, 1967.

[Cas88] J. Case. The power of vacillation. In D. Haussler and L. Pitt, editors, *Proceedings of the Workshop on Computational Learning Theory*, pages 133–142. Morgan Kaufmann Publishers, Inc., 1988.

[CL82] J. Case and C. Lynes. Machine inductive inference and language identification. *Lecture Notes in Computer Science*, 140:107–115, 1982.

[Ful85] M. Fulk. *A Study of Inductive Inference machines.* PhD thesis, SUNY at Buffalo, 1985.

[Ful90] M. Fulk. Prudence and other conditions on formal language learning. *Information and Computation*, 85:1–11, 1990.

[Gol67] E. M. Gold. Language identification in the limit. *Information and Control*, 10:447–474, 1967.

[HU79] J. Hopcroft and J. Ullman. *Introduction to Automata Theory Languages and Computation.* Addison-Wesley Publishing Company, 1979.

[KR88] S.A. Kurtz and J.S. Royer. Prudence in language learning. In D. Haussler and L. Pitt, editors, *Proceedings of the Workshop on Computational Learning Theory*, pages 143–156. Morgan Kaufmann Publishers, Inc., 1988.

[MY78] M. Machtey and P. Young. *An Introduction to the General Theory of Algorithms.* North Holland, New York, 1978.

[OSW82a] D. Osherson, M. Stob, and S. Weinstein. Ideal learning machines. *Cognitive Science*, 6:277–290, 1982.

[OSW82b] D. Osherson, M. Stob, and S. Weinstein. Learning strategies. *Information and Control*, 53:32–51, 1982.

[OSW84] D. Osherson, M. Stob, and S. Weinstein. Learning theory and natural language. *Cognition*, 17:1–28, 1984.

[OSW86] D. Osherson, M. Stob, and S. Weinstein. *Systems that Learn, An Introduction to Learning Theory for Cognitive and Computer Scientists.* MIT Press, Cambridge, Mass., 1986.

[OW82a] D. Osherson and S. Weinstein. Criteria of language learning. *Information and Control*, 52:123–138, 1982.

[OW82b] D. Osherson and S. Weinstein. A note on formal learning theory. *Cognition*, 11:77–88, 1982.

[Pin79] S. Pinker. Formal models of language learning. *Cognition*, 7:217–283, 1979.

[Rog58] H. Rogers. Gödel numberings of partial recursive functions. *Journal of Symbolic Logic*, 23:331–341, 1958.

[Rog67] H. Rogers. *Theory of Recursive Functions and Effective Computability.* McGraw Hill, New York, 1967. Reprinted, MIT Press 1987.

[WC80] K. Wexler and P. Culicover. *Formal Principles of Language Acquisition.* MIT Press, Cambridge, Mass, 1980.

[Wex82] K. Wexler. On extensional learnability. *Cognition*, 11:89–95, 1982.

Analogical Reasoning

IMPLEMENTATION OF HEURISTIC PROBLEM SOLVING PROCESS INCLUDING ANALOGICAL REASONING

Kazuhiro Ueda* Saburo Nagano
General Systems Studies
Graduate Division of International and Interdisciplinary Studies
University of Tokyo
3-8-1, Komaba, Meguro-ku, Tokyo, 153, JAPAN

July 27th, 1992

Abstract

This paper discribes a heuristic problem solver named HPSA. HPSA is constructed to explore human problem solving process with hypotheses formation including analogical inference. HPSA simulates two general types of human problem solving process; the first is deductive problem solving using domain-specific knowledge of the target domain and common knowledge, and the second is analogical reasoning executed between the target and source domains which are selected on the basis of some similarities. This system has the following advantages which most of precedent studies lack; that is, (1)HPSA enables simulation of a whole process of heuristic problem solving, besides either deductive problem solving or analogical reasoning, (2)problem solving with analogical reasoning can be executed from pragmatic aspects, *i.e.* goal-oriented problem solving and modification of pragmatic aspects can be simulated, (3)all phases of analogical reasoning are realized, and (4)multiple analogy is also realized. This problem solver is partly based on the observations on actual human problem solving processes with hypotheses formation. Hence, HPSA also tries to provide a cognitive simulation tool.

1 INTRODUCTION

Recently considerable numbers of studies have been made on problem solving or hypotheses formation by analogical reasoning. Most of these studies, however, can be said to have the following issues;

1. Many of precedent studies consider analogical reasoning process separately from whole process of problem solving [1] (the only exception is [3]). And some of them cannot explain a whole mechanism of analogical reasoning (for example, [2, 4, 7, 8]). As a result, these studies require quite complicated and unreasonable mechanisms of analogy.

*E-mail: ueda@kiso2.c.u-tokyo.ac.jp

[1] Problem solving processes which will be considered in this paper include hypotheses formation.

2. Many of these studies do not take pragmatic aspects such as goals or focuses of problem solving into account, though problem solving is, in general, goal-oriented (the exceptions are [1, 7]). Especially, problem solving systems should consider modification of these pragmatic aspects in the middle of process.

These matters should be considered as a whole. This paper discribes a heuristic problem solver named HPSA, which is implemented in order to overcome these issues. It is considered that the best way for this purpose is to design a system so as to simulate actual human problem solving. Hence, this system is built on the basis of a cognitive model of heuristic problem solving process, which is based on the analyses of protocol data of the interviews made by the authors.

In the next section, a cognitive experiment on heuristic problem solving will be discussed. In the third section, details of implementation of HPSA will be discussed and, in the fourth section, two examples will be shown to explain how HPSA works. And in the final section, the results of this study will be discussed.

2 COGNITIVE EXPERIMENT

In this section, we will introduce several terms necessary for the following discussions and discuss a cognitive experiment on heuristic problem solving process.

2.1 TERMS AND NOTATIONS

We will introduce the following terms and its notations.

target domain : target domain means the current problem denoted by t
K_t, R_t : indicate factual knowledge and rules specific to the target domain respectively

source domain : source domain means that domain selected as an analogous one to the target in the process of analogical reasoning, which is denoted by s
K_s, R_s : indicate factual knowledge and rules specific to the source respectively

Γ : indicates non domain-specific knowledge (*i.e.* common knowledge)
Λ : indicates general rules

goals : goals mean the goals of the problem solving, which are denoted by G
focuses : focuses mean the focused concepts which attract attention of a problem solver, which are denoted by F

problem solving type : means meta-type of the problem solving denoted by τ, which can be one of the following three types;

1. **why** — pursuit of the reasons why goals G arise
2. **how** — pursuit of the methods which can realize goals G
3. **what** — pursuit of what will be directly or indirectly caused in relation to goals G

2.2 OUTLINES OF A COGNITIVE EXPERIMENT

Actual human problem solving process has been studied by the authors; we have held interviews with researchers in the field of natural science and technology. This interview experiment is designed to explore how researchers solve problems and form hypotheses, especailly how they execute analogical reasoning in these processes. In this subsection, this cognitive experiment will be discussed. A cognitive model of heuristic problem solving process including analogy extracted from the present analyses of our interview protocols provides the basis of HPSA.

2.2.1 Method

The interviewees[2] were asked to dictate the outlines of their recent researches and to explain their mental processes with various reasonings in detail. Their verbalizations were recorded in tapes and these tapes were put in writing. These verbal protocols were arranged from the viewpoint of context-analysis; they were classified into respective cases first, then each case was classified into respective problem solving processes, each of which was consistent in contents, and finally these processes were recomposed in order of time.

Each problem solving process in each case was arranged in respect of the following five items; (1)setting up a problem, (2)focusing and working out a strategy, (3)inferences, (4)observations, and (5)experiments. In addition, each inference process was arranged in respect of the following seven items; (3-1)deduction, (3-2)screening, (3-3)problem solving by schemata, (3-4)analogical reasoning, (3-5)modeling, (3-6)induction or generalization, and (3-7)others.

All processes were analysed in more detail, some of which were modeled.

2.2.2 Results

In this subsection, part of the results will be discussed, which are considered as closely relevant to modeling a heuristic problem solving process.

1. 15 of a total of 36 inference processes involved analogical reasoning.
2. Deductive problem solving precedes analogical reasoning process even when the latter process is executed.
3. Problem solving process is, in general, goal-oriented. And goals or focuses of problem solving were often modified in the middle of a process.
4. Two domains were judged to be similar when they shared the goals or the knowledge relevant to the goals.
5. The knowledge and rules which were mapped from a source to the target were the components of a source that were causally relevant to the goals. The way to retrieve mapped knowledge and rules is slightly different between three types of problem solving process.
6. Interdomain analogy[6] has not been found yet so far.

Some of these analyses coincide with other cognitive works[5, 6]. Figure 1 illustrates a whole flow of a heuristic problem solving process which the present analyses may suggest. As is shown in Figure 1, the whole process includes following three subprocesses;

1. deductive problem solving process
2. modification of goals or focuses
3. analogical reasoning process

And the third subprocess – analogical reasoning process – can be said to include the following four phases;

- 3.1 selection of source analogues by calculation of similarities
- 3.2 search for mapped knowledge with rules in a source domain
- 3.3 mapping of knowledge with rules from a source to the target
- 3.4 generating novel knowledge and rules in the target

[2]The interviewees were selected from the leaders of some national projects or the researchers who were awarded from societies recently.

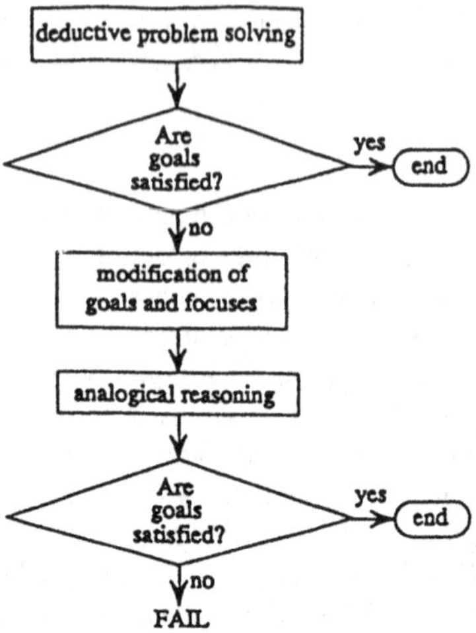

Figure 1: Whole Heuristic Problem Solving Process

3 DETAILS OF IMPLEMENTATION

In this section, the details of implementation will be discussed. Firstly the data structure will be explained. It will be followed by the details of implementation of heuristic problem solving process, in respect of the above mentioned subprocesses; **deductive problem solving process, modification of goals or focuses,** and **analogical reasoning process.**

3.1 DATA STRUCTURE

Each domain-specific knowledge with rules is expressed in the form of **frame representation** on this program. For example, such a domain named *insecticede-in-fiber* as follows can be expressed as shown in Figure 2.

> A single fiber consists of crystal domain and amorphous one. In general, an insecticide can penetrate not into crystal domain but into amorphous one. This fact is caused by another fact that an insecticide is soluble. And at a time, this fact implies that an insecticide can penetrate into the inside of a fiber.

In the slot of **explanation**, "(is-a $obj1 insecticide s-id1)" indicates that an object named $obj1 is an insecticide. This list form corresponds to the first-order predicate "is-a($obj1 insecticide)", which means that the object named $obj1 is an insecticide. Hence we will call the first element of a list **concept** and the rest elements except for the first and the last **arguments**. The last elements of these predicates indicate their identification numbers in each domain, which are introduced for the following three reasons; (1)they can simplify calculation of matching predicates in order to judge similarities, (2)they can simplify expressions of higher-order predicates which have lower-order predicates as their arugments, and (3)they enable deductive inference with higher-order predicates, which can be reduced to deduction with only first-order predicates. Higher order predicates are shown in Figure 12.

In the slot of **rule** in Figure 2, the first rule indicates a rule that a fact that $obj1 penetrates into $obj3 is caused by another fact that $obj1 is soluble. The first element of this rule indicates its identification number, the last but one does a conclusion of the rule, the last does type of

```
(make-frame insecticide-in-fiber
  (explanation
    (value (is-a $obj1 insecticide s-id1)
           (is-a $obj2 fiber s-id2)
           (is-a $obj3 amorphous-domain s-id3)
           (is-a $obj4 crystal-domain s-id4)
           (consists-of $obj2 $obj3 $obj4 s-id5)
           (inside $obj1 $obj2 s-id6)))
  (rule
    (value (s-rule-id1
             (soluble $obj1)
             (penetrate $obj1 $obj3)
            causality)
           (s-rule-id2
             (penetrate $obj1 $obj3)
             (consists-of $obj2 $obj3 $obj4)
             (inside $obj1 $obj2)
            implication))))
```

Figure 2: The Data of a Source *insecticide-in-fiber*

the rule, and the rest does an assertion of the rule[3].

3.2 DEDUCTIVE PROBLEM SOLVING PROCESS

This process is considered to be realized by a production system including bidirectical search method; forward and backward search. The details of production systems will be omitted, but only the whole strategy of this process will be shown as follows;

Procedure 1

if G exists
then execute a backward search to confirm whether G is realized or not
else execute a forward search starting from current K_t and Γ

3.3 MODIFICATION OF GOALS OR FOCUSES

This process modifies goals or focuses. Hence, this process is actually executed only if goals or focuses are specified in advance. All the mechanisms of this process have not been clarified yet, but the current implementation is as follows;

Procedure 2

for each $f \in F$
if $\exists f' : f, K_t, \Gamma \vdash^{M.A.R_t} f'$
then substitute f' for f in F and
for each $g \in G$
if the concept of g is a member of F

[3] In general, a rule have plural assertions.

then substitute f' for the concept of g

with a proviso that M indicates meta rules which allow a problem solver to modify goals or focuses

3.4 ANALOGICAL REASONING PROCESS

This process consists of four phases as discussed; that is, (1)selection of source analogues by calculation of similarities, (2)search for mapped knowledge with rules in a source domain, (3)mapping of knowledge from a source to the target, (4)generating novel knowledge and rules in the target.

In the first place, we will give a definition of three types of similarities.

Definition 1 (Similarities) *Three types of similarities can be defined as follows;*

1. **semantic similarities** : *two predicates are similar to each other if they have the concepts which are literally the same or in the same category.*
2. **pragmatic similarities** : *two predicates are similar to each other if they share semantically similar concepts, and either of the concepts is included in the focuses F or either of the predicates is included in the goals G.*
3. **structural similarities** : *two higher-ordered predicates are similar to each other if they are similar in the sense of other similarities.*

Definition 2 *Given two predicates p_1 and p_2,*
$p_1 \leftrightarrow_{sim} p_2 \stackrel{def}{=}$ *"p_1 and p_2 are similar in the sense of Definition 1"*

Next, the procedure of this process will be shown below. And in this procedure, *Sim* indicates similar predicates and M_k and M_r represent mapped predicates and mapped rules respectively.

Procedure 3

1. select a domain as a source which has Sim_s, s.t.

   (1)if goals or focuses exist,

   if $\exists k_s \in K_s, \exists k_t \in F_t \vee G : k_s \leftrightarrow_{sim} k_t$

   then append k_s to Sim_s and execute (2)

   with a proviso that $F_t = \{k_t \in K_t \mid k_t$ has an element of F as its concept $\}$

   (2)if $\exists k_s \in K_s, \exists k_t \in K_t \setminus F_t : k_s \leftrightarrow_{sim} k_t$

   then append k_s to Sim_s

2. for each selected source domain,

   (1)if $\exists k_s \in K_s, \exists r_s \in R_s : k_s, \Gamma \vdash^{r_s, \Lambda} Sim_s$

   then append k_s, r_s to M_k, M_r respectively

   This search is designed to include abduction

   (2)in the case of a problem solving whose type τ is what

   if $\exists k_s \in K_s, \exists r_s \in R_s : Sim_s, M_k, \Gamma \vdash^{r_s, M_r, \Lambda} k_s$

append k_s, r_s to M_k, M_r respectively

if G exists and $G_s \notin M_k \vee Sim_t$

then remove the domain S from the source analogues

with a proviso that G_s indicates G which is substituted on the basis of the correspondences between the source and the target

3. for each selected source domain,

   calculate the score of the domain by the following formula;

$$Score = \sum_{i \in M_k} order(i)$$

   with a proviso that $order(i)$ indicates the order of the predicate i

   In order of score, mapped knowledge and rules are substituted according to the correspondences between two domains.

4. if $\exists m_k \in M_K, \exists k_t \in K_t, \exists \gamma \in \Gamma : \gamma$ indicates that m_k contradicts k_t

   then remove m_k from M_k

In the first phase, if goals or focuses are specified, pragmatic similarities are calculated first, and then other similarities are judged. These calculations are based on matching $k_s \in K_s$ in a source with $k_t \in K_t$ in the target and G. That is to say, a potential source analogue will be selected in this phase. Besides, correspondences of objects, concepts, and predicates between two domains are simultaneously calculated by the system in this process. Even if the numbers of arguments of two similar predicates do not correspond to each other, the system generates new arguments in one predicate which would correspond to extra arguments in the other one.

The procedure of search for mapped knowledge and rules in the second phase is derived directly from the analyses of our interview protocols, which turns out to be a natural extension of Arima's **illustrative criterion** [1]. If goals exist and none of the goals are included in calculated mapped knowledge of the domain, a source domain will be excluded from the source analogues.

This search will be followed by calculation of the score of each domain of source analogues indicating validity as a source, which method is based on Gentner's **systematicity principle** [4]. The following two phases are to be executed in order of score, but, in example 4.2, the following two phases are executed in the reverse order.

In the third phase, mapped knowledge and rules, which are substituted on the basis of the correspondences calculated in the first phase, will be embedded in the target. If a proper correspondence is not calculated in the first phase, the system itself generates new objects or identification numbers which would correspond to non matched objects or identification numbers in the source respectively.

Finally, in the fourth phase, mapped knowledge and rules will be elaborated in order to acheive a possible solution or form a valid hypothesis in the target domain.

This program is written in Common Lisp and implemented on Sparc Station.

```
(make-frame sebum-in-fiber
  (explanation
    (value (is-a $obj1 sebum t-id1)
           (is-a $obj2 fiber t-id2)
           (is-a $obj3 amorphous-domain t-id3)
           (is-a $obj4 crystal-domain t-id4)
           (consists-of $obj2 $obj3 $obj4 t-id5)
           (soluble $obj1 t-id6)
           (not (observable-by-electron-microscope $obj1) t-id7)))
   (rule
     (value (t-rule-id1
              (not (observable-by-electron-microscope $obj1))
              (not (attached-outside $obj1 $obj2))
             implication))))
(set-goal '((attached-outside $obj1 $obj2 g-id1)))
(set-focus '(attached-outside))
```

Figure 3: The Data of the Target Domain *sebum-in-fiber*

```
1 -> (NOT (ATTACHED-OUTSIDE $OBJ1 $OBJ2))
```

Figure 4: The Infered Knowledge in the First Step

4 EXAMPLES

In this section, two examples will be shown to explain how HPSA works. Of the two following examples, the former is an example for explaining the whole process of heuristic problem solving, and the latter is one especially for explaining analogical reasoning process.

4.1 EXAMPLE 1

The target problem is as follows;

> To locate sebums which cause stain on cotton fabrics. It was already known that a sebum is soluble and that a fiber in cotton fabrics consists of crystal domain and amorphous one. Because, almost all experts believed that sebums were attached outside of fibers, the researchers tried to find sebums ouside of fibers by microscope, but in vain.

The system starts problem solving from this situation; that is, the goals, focuses, and domain-specific knowledge with rules are shown in Figure 3. And problem solving type τ of this simulation is why.

Firstly, this system executes deductive problem solving process. The result is as follows in Figure 4. Because this result is quite opposite to the given goal in Figure 3, goals and focuses are to be modified in the second subprocess. This process generates a modified focus "inside" and a modified goal "(inside $obj1 $obj2 g-id1)", using a meta rule in Figure 5.

This modification process is followed by analogical reasoning. As a result of the first phase, only one domain named *insecticide-in-fiber* is selected as an analogous domain *i.e.* source analog (the data of this domain is shown in Figure 2). In this phase, the following pairs of predicates are judged to be similar by the system; the first is a pair of (inside $obj1

```
(meta-rule-id1
  (not (attached-outside (? x) (? y)))
  (inside (? x) (? y))
 implication)
```

Figure 5: A Used Meta Rule in the Second Step

```
RULE ->(NEW-RULE-ID740
         (PENETRATE $OBJ1 $OBJ3) (CONSISTS-OF $OBJ2 $OBJ3 $OBJ4)
         (INSIDE $OBJ1 $OBJ2)
        IMPLICATION)
       (NEW-RULE-ID739
         (SOLUBLE $OBJ1)
         (PENETRATE $OBJ1 $OBJ3)
        CAUSALITY)
EXPLANATION ->(PENETRATE $OBJ1 $OBJ3 NEW-ID739)
              (INSIDE $OBJ1 $OBJ2 ADDED-T-ID734)
                (NOT (ATTACHED-OUTSIDE $OBJ1 $OBJ2) ADDED-T-ID733)
```

Figure 6: The Mapped Knowledge of Example 1

`$obj2 s-id6)` and `(inside $obj1 $obj2 g-id1)`, and the second is a pair of `(consists-of $obj2 $obj3 $obj4 s-id5)` and `(consists-of $obj2 $obj3 $obj4 t-id5)`.

Search for mapped knowledge with rules in the source domain is followed. In this case, all the searched knowledge with rules is embedded in the target domain without being removed. The mapped knowledge and rules are shown in Figure 6. CPU time was about 11.7 seconds.

4.2 EXAMPLE 2

The target of this example is as follows;

> To find correct relations which might explain the movements of two indicators. shown as Figure 7.

This problem situation can be expressed as Figure 8. In this case, problem solving type τ is what. HPSA starts heuristic problem solving without goals and focuses. It is needless to say that deductive problem solving is to fail. Accordingly the system executes analogical reasoning after inquiring of users whether analogy might be executed or not.

In analogical reasoning process, two domains are to be selected as source analogues, which correspond to Figure 9(a) and Figure 9(b) respectively.

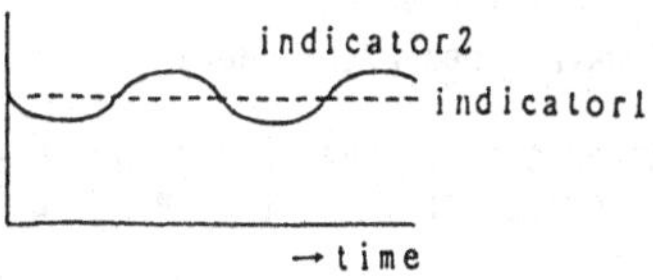

Figure 7: Movements of Certain Indicators

```
(make-frame indicator-chains
  (object (value indicator1 indicator2 number))
  (explanation
    (value (is-a $obj1 indicator1 t-id1)
           (is-a $obj2 indicator2 t-id2)
           (constance $obj1 number t-id3)
           (cyclic-fluctuation-curve1 $obj2 number t-id4)
           (synchronous t-id3 t-id4 t-id5))))
```

Figure 8: The Data of the Target Domain *indicator-chains*

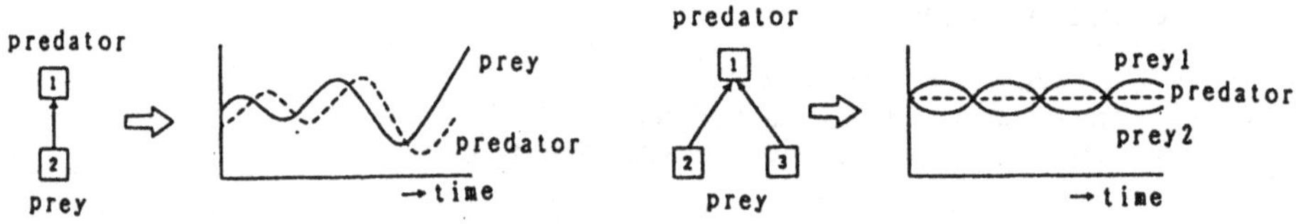

(a)in case of one predator and one prey (b)in case of one predator and two preys

Figure 9: Interaction between Predator and Preys

In the case of selecting the domain named *predator-chains1* which corresponds to Figure 9(a), the mapped knowledge calculated finally is shown in Figure 10, since the following two knowledges are given; the first knowledge indicates that the two concepts "**cyclic-growth-curve1**" and "**cyclic-fluctuation-curve1**" belong to the same catgory, and the second indicates that the two concepts "**synchronous**" and "**time-lag**" are contradictory to each other. The system also generates the concept "**cyclic-fluctuation-curve2**" as a correspondent one of the concept "**cyclic-growth-curve2**", based on the former of the above knowledge.

Because the above domain is inappropiate as a source, the other domain named *predator-chains2*, which corresponds to Figure 9(b), is to be selected as shown in Figure 11, which shows a knowledge state after executing deductive inference in the source. The final mapped knowledge is shown in Figure 12. In this case, a new object is predicted, which is named "$OBJ-P737" by the system. This example illustrates that HPSA can predict objects in the process of analogy. CPU time was about 21.4 seconds.

```
RULE ->(NEW-RULE-ID743
         (LOTKA-VOLTERRA-EQUATIONS1 $OBJ2 $OBJ-P740)
         (CYCLIC-FLUCTUATION-CURVE1 $OBJ2 NUMBER)
        IMPLICATION)
       (NEW-RULE-ID742
         (LOTKA-VOLTERRA-EQUATIONS1 $OBJ2 $OBJ-P740)
         (CYCLIC-FLUCTUATION-CURVE2 $OBJ-P740 NUMBER)
        IMPLICATION)
EXPLANATION ->(CYCLIC-FLUCTUATION-CURVE2 $OBJ-P740 NUMBER NEW-ID741)
              (LOTKA-VOLTERRA-EQUATIONS1 $OBJ2 $OBJ-P740 NEW-ID740)
```

Figure 10: The Mapped Knowledge in Selecting the Domain *predator-chains1*

```
(make-frame predator-chains2
  (explanation
    (value (is-a $obj1 predator s2-id1)
           (is-a $obj2 prey1 s2-id2)
           (is-a $obj3 prey2 s2-id3)
           (constance $obj1 number s2-id4)
           (cyclic-growth-curve1 $obj2 number s2-id5)
           (cyclic-growth-curve2 $obj3 number s2-id6)
           (Lotka-Volterra-equations2 $obj1 $obj2 $obj3 s2-id7)
           (switching-responce s2-id5 s2-id6 s2-id8)
           (synchronous s2-id4 s2-id5 s2-id6 s2-id9)))
  (rule
    (value (s1-rule-id1
             (Lotka-Volterra-equations2 $obj1 $obj2 $obj3)
             (constance $obj1 number)
            implication)
           (s1-rule-id2
             (Lotka-Volterra-equations2 $obj1 $obj2 $obj3)
             (cyclic-growth-curve1 $obj2 number)
            implication)
           (s1-rule-id3
             (Lotka-Volterra-equations2 $obj1 $obj2 $obj3)
             (cyclic-growth-curve2 $obj3 number)
            implication)
           (s1-rule-id4
             (Lotka-Volterra-equations2 $obj1 $obj2 $obj3)
             (switching-responce s2-id5 s2-id6)
            implication)
           (s1-rule-id5
             (Lotka-Volterra-equations2 $obj1 $obj2 $obj3)
             (synchronous s2-id4 s2-id5 s2-id6)
            implication))))
```

Figure 11: The Data of a Source Domain *predator-chains2*

5 CONCLUSIONS

It can be concluded that HPSA overcomes the issues which other studies have.

A method of problem solving by analogy is proposed by J.G.Carbonell[3]. His derivational analogy is similar to our goal-oriented problem solving method in that both methods consider pragmatic aspects of analogy. But, his study lacks sufficient consideration of a relationship between similarities and mapped knowledge with rules, and modification of pragmatic aspects. Purpose-directed analogy proposed by S.T.Kedar-Cabelli [7], is the same in the above points. HPSA, on the other hand, can consider causal relationship between similarities and mapped knowledge and modify some pragmatic aspects. Consideration of pragmatic aspects will reduce size of search space in retrieving analogous domains. In addition, multiple analogy can be executed.

This paper considers also all the four phases of analogy. Similarities are defined in detail except that analogical reasoning involves recategorization of concepts. Prediction of new objects is also realized. But modification of mapped knowledge and rules in the fourth phase has not been fully considered yet.

As a whole, HPSA can be said to simulate a heuristic problem solving process including analogy, and as a result, it can provide a cognitive simulation tool.

```
RULE ->(NEW-RULE-ID745
         (LOTKA-VOLTERRA-EQUATIONS2 $OBJ1 $OBJ2 $OBJ-P737)
         (CYCLIC-FLUCTUATION-CURVE1 $OBJ2 NUMBER)
        IMPLICATION)
       (NEW-RULE-ID744
         (LOTKA-VOLTERRA-EQUATIONS2 $OBJ1 $OBJ2 $OBJ-P737)
         (CONSTANCE $OBJ1 NUMBER)
        IMPLICATION)
       (NEW-RULE-ID743
         (LOTKA-VOLTERRA-EQUATIONS2 $OBJ1 $OBJ2 $OBJ-P737)
         (CYCLIC-FLUCTUATION-CURVE2 $OBJ-P737 NUMBER)
        IMPLICATION)
       (NEW-RULE-ID742
         (LOTKA-VOLTERRA-EQUATIONS2 $OBJ1 $OBJ2 $OBJ-P737)
         (SWITCHING-RESPONCE T-ID4 NEW-ID739)
        IMPLICATION)
       (NEW-RULE-ID740
         (LOTKA-VOLTERRA-EQUATIONS2 $OBJ1 $OBJ2 $OBJ-P737)
         (SYNCHRONOUS T-ID3 T-ID4 NEW-ID739)
        IMPLICATION)
EXPLANATION ->(SYNCHRONOUS T-ID3 T-ID4 NEW-ID739 NEW-ID740)
              (CYCLIC-FLUCTUATION-CURVE2 $OBJ-P737 NUMBER NEW-ID739)
              (LOTKA-VOLTERRA-EQUATIONS2 $OBJ1 $OBJ2 $OBJ-P737 NEW-ID738)
              (SWITCHING-RESPONCE T-ID4 NEW-ID739 NEW-ID737)
```

Figure 12: The Final Mapped Knowledge of Example 2

References

[1] Arima,J. A Logical Analysis of Relevance in Analogy, *Proc. of Workshop on Algorithmic Learning Theory ALT'91*, 1991

[2] Burstein,M.H. Concept Formation by Incremental Analogical Reasoning and Debugging, In R.S.Michalski, J.G.Carbonell, & T.M.Mitchell(ed.) *Machine Learning vol.2*, Tioga Publishing, Palo Alto, 1986

[3] Carbonell,J.G. Derivational Analogy : A Theoretical Reconstructive Problem Solving and Expertise Acquisition, In R.S.Michalski, J.G.Carbonell, & T.M.Mitchell(ed.) *Machine Learning vol.2*, Tioga Publishing, Palo Alto, 1986

[4] Gentner,D. Structure Mapping : A Theoretical Framework fo Analogy, *Cognitive Science, vol.7*, 1983

[5] Gick,M.L.,& Holyoak,K.J. Analogical Problem Solving, *Cognitive Psychology, vol.15*, 1980

[6] Holland,J.H.,Holyoak,K.J.,Nisbett,R.E.,& Thagard,P.R. Analogy In *Induction : Processes of inference, learning, and discovery*, MIT Press, 1986

[7] Kedar-Cabelli,S.T. Purpose-directed Analogy, *Proc.of the Cognitive Science Society Conference*, 1985

[8] Winston,P.H. Learning and Reasoning by Analogy, *Communications of ACM, vol.23*, 1980

PLANNING WITH ABSTRACTION BASED ON PARTIAL PREDICATE MAPPINGS

Yoshiaki OKUBO and Makoto HARAGUCHI
Department of Systems Science
Tokyo Institute of Technology
4259 Nagatsuta, Midori-ku, Yokohame 227, Japan

Abstract

Planning with abstraction has been studied to perform an efficient planning, and consists of processes for making abstraction, searching an abstract plan, and instantiating the abstract plan to obtain a final plan at a concrete level. If an abstract plan cannot be instantiated to any plan at the concrete level, it is no use obtaining the final plan. To avoid such an instantiation failure, each abstract plan must be instantiated to one or more concrete plans. This requirement is called Downward-Solution Property(DSP), and a system satisfying it has been already proposed by J. D. Tenenberg. However the very severe constraint in abstracting operators is assumed to satisfy DSP. As a result, only few abstract operators might be used for searching an abstract plan. This means that the planning ability at abstract level is decreased to satisfy DSP.

The purpose of this paper is to make the framework proposed by Tenenberg more flexible, still preserving DSP. For this purpose, the notion of a *partial predicate mapping* is introduced. A partial predicate mapping corresponds to a refinement of inheritance hierarchy used in abstracting operators. After partitioning a given concrete operator set, a partial predicate mapping is computed for each subset, and operators of the subset being instances under the partial predicate mapping are abstracted. Using abstract operators thus obtained, searching an abstract plan is carried out followed by an instantiation process. This paper contains some experimental results, and shows that the technique using partial predicate mappings is useful for improving planning process. Furthermore a theoretical result is also presented. It is shown that the proposed system satisfies DSP.

1 Introduction

Planning can be viewed as generating a plan, a sequence of operators, which changes a given initial state into a given goal state. STRIPS[1] is a familiar planning system consisting of a description language and a planning component. The former contains operators, background knowledge, a state space, and the latter performs planning using means-end analysis. Needless to say, STRIPS must find a plan in a huge search space. In order to reduce the size of the search space, planning using macro-operators [2] and planning with abstraction [3] [4] [5] [6] are very powerful. In this paper, the latter approach is focused on.

In planning with abstraction, it is expected that plans are found efficiently, since the search space is smaller than the concrete (original) one. Planning with abstraction is formed by the following :

1. **Abstracting a concrete problem :** Given an initial state and a goal state of a concrete problem, an abstract problem is constructed.

2. **Generating an abstract plan :** A plan for the abstracted problem, called an abstract plan, is generated by using the same strategy of the concrete level.

3. **Instantiating the abstract plan :** The abstract plan is instantiated in order to obtain a final plan for the original problem.

If an abstract plan cannot be instantiated to any concrete plan, it is no use obtaining the final plan. In order to avoid such a failure of instantiation process, it is required that any abstract plan must be instantiated to one or more concrete plans. This requirement is called *Downward-Solution Property* (DSP).

Tenenberg [3] [4] [5] has been already proposed a framework for constructing a planning system satisfying DSP. In the framework, abstraction is an act of changing a problem description language of a concrete planning system under a *predicate mapping* which maps concrete predicate symbols into abstract ones. However very severe constraint in abstracting operators is assumed to satisfy DSP. Because of the constraint, *few* abstract operators might be used to find abstract plans. It is undesirable to construct such a system, since the system seems to have poor planning ability at abstract level. In other words, to construct a system having more rich planning ability, an *appropriate* predicate mapping must be given to Tenenberg's framework, before planning. It is a very difficult task for system designer.

The purpose of this paper is to make the framework proposed by Tenenberg more flexible, still preserving DSP. For this purpose, the notion of a *partial predicate mapping* is newly introduced. Given a description language of a concrete system and a predicate mapping, the concrete operator set are partitioned using the predicate mapping. Then, a partial predicate mapping is computed for each subset. Intuitively, the computation of partial predicate mapping for each subset corresponds to providing a predicate mapping in order to abstract some operators of the subset, according to Tenenberg's framework. That is, based on the given predicate mapping, an appropriate predicate mapping can be generated. In our framework, the description language is abstracted by using *not only* the given predicate mapping *but also* the computed partial predicate mappings. Therefore, a system can be constructed such that the system satisfies DSP and has more rich planning ability at abstract level than Tenenberg's system.

2 Overview of Tenenberg's Abstraction Frameworks

In this section, we present frameworks for abstracting a clause set and for abstracting a description language of STRIPS, proposed by Tenenberg [4]. In the frameworks, abstraction is an act of changing representation based on a *predicate mapping*. By using the frameworks, we can obtain abstractions having very important properties.

2.1 Predicate Mappings

Given two first-order languages L and L' with the same set of constant and function symbols, a predicate mapping f is defined as an onto mapping $f : \Pi \overset{\text{onto}}{\mapsto} \Pi'$, where Π and Π' denote the sets of predicate symbols in L and L', respectively. f is clearly extensible to a mapping from L to L', $f : L \overset{\text{onto}}{\mapsto} L'$, so that any symbol except predicate is mapped into itself. An expression E in L is called an *instance* of an expression E' in L', if $f(E) = E'$. For example, a clause $C\ :\ p(x,a) \leftarrow p(x,b)$ is mapped into a clause $C'\ :\ f(p)(x,a) \leftarrow f(p)(x,b)$, and is therefore an instance of C'. Moreover the inverse mapping, $f^{-1}(\psi)$ is defined as the set of symbols mapped into ψ under f. $f^{-1}(\psi)$ is called the *instance set* of ψ under f.

2.2 Abstracting First-Order Theories

An abstraction of a clause set under a predicate mapping, is defined as follows. In the definition, for any clause C, $|C|$ denotes the number of literals in C, $neg(C)$ denotes the disjunction of all negative literals in C (or $\Box$, if there are none), and $pos(C)$ denotes the disjunction of all positive literals in C (or $\Box$, if there are none).

Definition 1 : *MembAbs* Clause Mapping [4]

Let $f : L \overset{onto}{\mapsto} L'$ be a predicate mapping and S be a concrete clause set in L. $MembAbs_f(S)$ defined as follows, is an abstract clause set of S under f.

$MembAbs_f(S) = \{ C' \mid$
for every $N \in f^{-1}(neg(C'))$, having $|neg(C')|$ distinct literals,
there exists $P \in f^{-1}(pos(C'))$ such that $N \vee P \in S \}$. ■

For instance, a clause $p' \leftarrow q'$ is not an abstract clause of a set $S = \{p \leftarrow q_1\}$, if q' and p' have two instances q_1, q_2 and a single instance p, respectively. In other words, in order for $p' \leftarrow q'$ to be an abstract clause of S, S should contain $p \leftarrow q_2$ in addition to $p \leftarrow q_1$. This definition results in the following theorem.

Theorem 1 : Proof-Theoretic Property [4]

Let $f : L \overset{onto}{\mapsto} L'$ be a predicate mapping, S be a Horn clause set in L, and G' be an atom in L'. If T' is a proof of G' from $MembAbs_f(S)$, then there exists a proof T of G from S such that G is an instance of G' under f, T is *isomorphic* to T', and for any node of T, the clause labelled at the node in T is an instance of a clause labelled at the corresponding node in T'. ■

By this theorem, it is insured that any proof at abstract level can be always instantiated to a proof at concrete level.

2.3 Abstracting Description Language of STRIPS

In here, we present a framework for abstracting a description language of STRIPS. At first, we define the description language.

2.3.1 Description Language

A description language of STRIPS is a quintuple $\Sigma = (L, E, O, K, \sigma)$, where L is a first-order language, E is a set of *primary predicates*, O is a set of *operators*, K, the *background knowledge*, is a non-empty set of Horn clauses of L, and σ, the *dynamic state space*, is a set whose elements are sets of ground atoms of E. Each primary predicate is a predicate in L. Non-primary predicate in L is called *secondary predicate*. Any primary predicate is assumed not to appear in a head of any clause in K. Each operator $o \in O$ has its name op, a list of variables $(arg_1, \ldots, arg_n)$, and a triple (Po, Do, Ao) of sets of atoms in L. For simplicity, o is written as $op(arg_1, \ldots, arg_n) : (Po, Do, Ao)$. Do and Ao must consist of atoms in E. All variables in (Po, Do, Ao) appear as arg_i, and others never. Alphabetic variants in operators are identified.

Each $T \in \sigma$ unioned with K is called a *state*. Applying an operator to a *dynamic state* $T \in \sigma$, T can be transformed into another dynamic state. Formally, o is applicable to a state S if there exists a ground substitution ϕ such that $S \vdash Po_\phi$, and S', the result of applying o to S, is defined as $(S \setminus Do_\phi) \cup Ao_\phi$. A problem is a pair $\rho = (T_0, G)$, where T_0 is an *initial dynamic state*, and G, a *goal*, is a set of ground atoms in L. A plan is a finite sequence of operators (fully instantiated), and the length of the plan is defined as the number of operators constructing the plan. Let $\omega = \langle o_1, \ldots, o_n \rangle$ be a length n plan, and S be a state. $result(\omega, S)$ denotes the state obtained by applying ω to S. ω solves $\rho = (T_0, G)$ if $result(\omega, T_0 \cup K) \vdash G$.

2.3.2 Abstract Description Language

In here, an abstract description language of STRIPS is represented. A planning system with the description language has very important property required in planning with abstraction. The property is called *Downward-Solution Property*(DSP). Intuitively, the property ensures that any generated abstract plan can be instantiated to a concrete plan. A precise definition of the abstract description language is as follows.

Let $\Sigma = (L, E, O, K, \sigma)$ be a description language of STRIPS and $f : L \overset{onto}{\mapsto} L'$ be a predicate mapping. An abstract description language of Σ, Σ', is defined as $(L', E', O', K', \sigma')$, where $E' = \{f(P) \mid P \in E\}, K' = MembAbs(K), \sigma' = \{f(T) \mid T \in \sigma\}$, under the restriction that any primary and secondary predicates never can be mapped into the same predicate under f. An abstract operator set O' is defined as follows. In the definition, $f(o) = op'(arg_1, \ldots, arg_n) : (f(Po), f(Do), f(Ao))$, if $o = op(arg_1, \ldots, arg_n) : (Po, Do, Ao)$ and $op' = f(op)$.

Definition 2 : Abstract Operators [4]

$O' = \{\ o' \mid o' \in f(O)$ and there exists $Q \subseteq O$ such that

1. for each $o \in Q$, $f(o) = o'$,
2. for each $P \in f^{-1}(Po')$ such that each element of P is atomic, there exists $o \in Q$ such that $P = Po$,
3. for each $o \in Q$, for each substitution ϕ, for each element $d_j \in Do_\phi$, for each atomic $d_k \in f^{-1}(f(d_j))$, either $d_k \in Do_\phi$ or $K \cup Po_\phi \vdash \neg d_k\ \}$. ■

A planning system with Σ' has the following important property, *Downward-Solution Property*.

- Any generated abstract plan can be instantiated to a concrete plan under the following restrictions : 1) the concrete plan is an *isomorphic* instance of the abstract plan under the predicate mapping, 2) each intermediate abstract state is *MembAbs* of the corresponding concrete state, 3) for each operator of the concrete plan, preconditions proofs of the operator and the corresponding abstract operator are related as in Theorem 1 .

3 Abstracting Description Language of STRIPS Based on Partial Predicate Mappings

3.1 Overview

A planning system proposed by Tenenberg provides an advantage such that useless abstract plans (such as cannot be instantiated) never can be generated, because of DSP. In order to satisfy DSP, however, very severe constraints in abstracting operators and background knowledge are assumed. In this paper, the constraint in abstracting operators, especially, the part of the constraint with respect to abstracting preconditions is focused on. The constraint requires that for each abstract operator, for each instance of preconditions of the abstract operator under a given predicate mapping, there must exist some concrete operator having the instance as its preconditions. Therefore, we cannot construct an abstract operator, even if there were all concrete operators required by the constraint, except *only one* operator. Even in such case, we would like to construct the abstract operator.

In this section, a process for abstracting a description language of STRIPS is proposed, realizing above desire. The process can be viewed as pre-process for performing planning with abstraction. The notion of a *partial predicate mapping* is introduced in the process. A partial predicate mapping is a function that maps a part of concrete predicates defined by a given predicate mapping into abstract ones. That is, the instance set of a predicate under the partial predicate mapping, is a subset of the instance set of a predicate under the predicate mapping. Therefore, by using the partial predicate mapping, it is possible to change concrete operators required by Tenenberg's constraint. The process consists of the following two subprocesses.

Computing Partial Predicate Mappings : Given a predicate mapping and a description language of STRIPS, partial predicate mappings of the predicate mapping are computed.

Abstracting Description Language : Under computed partial predicate mappings, the description language is abstracted.

3.2 Partial Predicate Mappings

The formal definition of partial predicate mappings is described as follows. In the definition, for any first-order languages L, *pred*(L) denotes the set of all predicate symbols in L, and *non-pred*(L) denotes the set of all non-predicate symbols in L.

Definition 3 : Partial Predicate Mappings

Let $f : L \overset{\text{onto}}{\mapsto} L'$ be a predicate mapping. $fp : Lp \overset{\text{onto}}{\mapsto} Lp'$ is a partial predicate mapping of f, if

1. $pred(Lp) \subseteq pred(L)$,
2. $non\text{-}pred(Lp) = non\text{-}pred(L)$,
3. $pred(Lp') \cap pred(L') = \phi$, and
4. for each $p \in pred(L')$, there exists $p_p \in pred(Lp')$ such that $fp^{-1}(p_p) \subseteq f^{-1}(p)$. ■

3.3 Computing Partial Predicate Mappings

Let $f : L \overset{\text{onto}}{\mapsto} L'$ be a predicate mapping, and $\Sigma = (L, E, O, K, \sigma)$ be a description language of STRIPS. Firstly, O is partitioned into several subsets, called *candidate sets.* Operators in a subset of each candidate set can be abstracted to the same operator. Each candidate set Q must satisfy the following two conditions.

1. All operators in Q can be mapped into the same abstract operator under f.
2. For each $o \in Q$, for each ground substitution ϕ, for each element $d_j \in Do_\phi$, for each atomic $d_k \in f^{-1}(f(d_j))$, either $d_k \in Do_\phi$ or $K \cup Po_\phi \vdash \neg d_k$.

For each candidate set, a partial predicate mapping of f is computed.

Let Q be a candidate set. Suppose that each operator in Q can be mapped into an abstract operator having $(p'_1, \ldots, p'_n)$ as its preconditions, under f. Assuming that for each p'_i,

$$P_i = \{\, p \mid p \in f^{-1}(p'_i) \text{ such that } p \text{ is an atom } \},$$

the product set $S = P_1 \times \cdots \times P_n$ can be considered as the instance set of $(p'_1, \ldots, p'_n)$ under f, consisting of only atoms. In addition, let P_Q be the set consisting of preconditions of all operators in Q. In order to compute a partial predicate mapping $fp : Lp \overset{\text{onto}}{\mapsto} Lp'$ of f for Q, a *partial product set* S_p of S such that $S_p \subseteq P_Q$, is exploited, where partial product sets are defined as follows:

Definition 4 : Partial Product Sets

Let A be a product set defined as $A_1 \times \cdots \times A_n$. $A_p = A'_1 \times \cdots \times A'_n$ is a partial product set of A, if each A'_i is a non-empty subset of A_i. ■

If $S_p = P'_1 \times \cdots \times P'_n$, then predicate symbols in P'_i are considered as predicate symbols in Lp. All predicate symbols in L mapped into others in $(p'_1, \ldots, p'_n)$ under f, are also in Lp. Lp consists of these predicate symbols and *non-pred*(L). For Lp', we can give quite new predicate symbols. Using fp to abstract operators in Q insures that for each instance of abstract preconditions (in the term of Lp'), there always exists some operator having the instance as its preconditions, that is, Tenenberg's constraint can be satisfied.

In general, there might exist two or more partial product sets being subset of P_Q. Then, the largest one is used as the basis of partial predicate mapping, since we would like to abstract as many operators in Q as possible. An non-deterministic algorithm for extracting a maximal (under subset relation) partial product set being a subset of P_Q, is shown in Figure 1. In fact, maximal partial product sets are extracted by this algorithm, and then the largest one is chosen. If there exist two or more such ones, some one is chosen from these, arbitrarily.

Example :

Figure 2 shows a predicate mapping $f : L \overset{\text{onto}}{\mapsto} L'$ and a candidate set Q. For simplicity, non-predicate symbols in L and L' are omitted. Preconditions of operators in Q can be mapped into (`filled, cookingVessele, inCooker, raw`) under f. The instance set of (`filled, cookingVessele, inCooker, raw`) under f, consisting of only atoms, S, is {(`filled, pot, inMicroWave, raw`), (`filled, bakingDish, inMicroWave, raw`), (`filled, pot, inOven, raw`), (`filled, bakingDish, inOven, raw`)}. The set consisting of preconditions of all operators in Q, P_Q, is {(`filled, pot, inMicroWave, raw`), (`filled, bakingDish, inMicroWave, raw`), (`filled, bakingDish, inOven, raw`)}. By the algorithm in Figure 1, the following two maximal partial product sets of S being subsets of P_Q can be extracted :

{(`filled, bakingDish, inMicroWave, raw`), (`filled, bakingDish, inOven, raw`)}

and

{(`filled, pot, inMicroWave, raw`), (`filled, bakingDish, inMicroWave, raw`)}.

The largest one, {(`filled, bakingDish, inMicroWave, raw`), (`filled, bakingDish, inOven, raw`)}, is used as the basis of partial predicate mapping $fp : Lp \overset{\text{onto}}{\mapsto} Lp'$ of f. `filled,`

```
INPUT(S, P_Q){ S is a product set, P_Q is a subset of S}
N ← S − P_Q
M ← S
select some Z from N
select some z from Z
let Del be a set of all elements of S containingz
N ← N − Del
M ← M − Del
DELs ← [Del ]
WHILE N ≠ φ BEGIN
   select some Z from N
   select some z from Z
   let Del be a set of all elements of S containing z,
         such that for each DEL ∈ DELs,
         there exists d ∈ DEL − Del such that d ∈ S − P_Q
   N ← N − Del
   M ← M − Del
   Insert Del into DELs
END
OUTPUT(M){M  is a maximal partial product set of  S  being a subset of  P_Q}
```

Figure 1: Algorithm for Extracting Maximal Partial Product Set

predicate mapping f :

```
f(filled) = filled,   f(raw)  = raw,
f(inMicroWave) = f(inOven)  = inCooker,
f(bakingDish) = f(pot)  = cookingVessele,
f(microWaved) = f(baked)  = cooked
```

candidate set Q :

```
microWaveInPot(A,B,C)
   P : filled(A), pot(B),
       inMicroWave(B,C), raw(A)
   D : raw(A)
   A : microWaved(A)

bakeInBakingDish(A,B,C)
   P : filled(A), bakingDish(B),
       inOven(B,C), raw(A)
   D : raw(A)
   A : baked(A)

microWaveInBakingDish(A,B,C)
   P : filled(A), bakingDish(B),
       inMicroWave(B,C), raw(A)
   D : raw(A)
   A : microWaved(A)
```

Figure 2: predicate mapping and candidate Set

partial predicate mapping fp :
  fp(filled) = pred1, fp(raw) = pred2,
  fp(inMicroWave) = fp(inOven) = pred3,
  fp(bakingDish) = pred4,
  fp(microWaved) = fp(baked) = pred5

Figure 3: partial predicate mapping

bakingDish, inMicroWave, inOven and raw are predicate symbols in Lp, and microWaved and baked mapped into others in (filled, cookingVessele, inCooker, raw) under f are also in Lp. The partial predicate mapping fp of f is shown in Figure 3. Each pred i is predicate symbol in Lp'.

3.4 Abstracting Description Language under Partial Predicate Mappings

Let $f : L \overset{onto}{\mapsto} L'$ be a predicate mapping, and $\Sigma = (L, E, O, K, \sigma)$ be a description language of STRIPS. Suppose that O is partitioned into n candidate sets, $Q_1, \ldots, Q_n$, and n partial predicate mappings $fp_1 : Lp_1 \overset{onto}{\mapsto} Lp'_1, \ldots, fp_n : Lp_n \overset{onto}{\mapsto} Lp'_n$, of f are computed for $Q_1, \ldots, Q_n$.

Abstract Operator Set :
Only operators in Q_i having atoms in Lp_i as preconditions, can be abstracted to the same abstract operator o'_i. The abstract operator set of O, O', consists of each o'_i.

Let $o_{i_1}, \ldots, o_{i_m}$ be operators in Q_i having atoms in Lp_i as preconditions and Fp be the set of computed partial predicate mappings. Preconditions of o'_i, Po'_i, are defined as

$$Po'_i = fp_i(Po_{i_1}) = \cdots = fp_i(Po_{i_m}),$$

and delete list of o'_i, Do'_i, as

$$Do'_i = f(Do_{i_1}) = \cdots = f(Do_{i_m}).$$

Add list of o'_i, Ao'_i, is defined as follows:

$$\begin{aligned} Ao_i' &= f(Ao_{i_1}) \cup \bigcup_{fp \in Fp_i} fp(Ao_{i_1}) \\ &= \quad \vdots \\ &= f(Ao_{i_m}) \cup \bigcup_{fp \in Fp_i} fp(Ao_{i_m}), \end{aligned}$$

where $Fp_i = \{ fp \mid fp : Lp \overset{onto}{\mapsto} Lp' \in Fp,$ for any o_{i_j}, Ao_{i_j} consists of atoms in $Lp \}$.

In Figure 2, Tenenberg's constraint in abstracting operators cannot be satisfied, since there exists no operator having (filled, pot, inOven, raw) as its preconditions. Although an abstract operator cannot be constructed by using Tenenberg's framework, the abstract operator shown in Figure 4 can be constructed by using our framework, where the set of computed partial predicate mappings, Fp, consists of the partial predicate mapping in Figure 3.

Abstract Background Knowledge :

```
cookInCookingVessele(A,B,C)
   P : pred1(A), pred4(B),
       pred3(B,C), pred2(A)
   D : raw(A)
   A : microWaved(A), pred5(A)
```

Figure 4: abstract operator

The abstract background knowledge of K, K', is defined as

$$K' = MembAbs_f(K) \cup MembAbs_{fp_1}((K)_{Lp_1}) \cup \cdots \cup MembAbs_{fp_n}((K)_{Lp_n}),$$

where $(K)_{Lp_j} = \{ \, C \mid C \in K$ such that C is a Horn clause in $Lp_j \, \}$.

Abstract Dynamic State Space :
The abstract dynamic state space of σ, σ', is defined as

$$\sigma' = \{ \, T' \mid \text{for any } T \in \sigma, T' = f(T) \cup fp_1((T)_{Lp_1}) \cup \cdots \cup fp_n((T)_{Lp_n}) \, \}.$$

Any subset T'_{sub} of $T' \in \sigma'$ such that $(T'_{sub})_L = (T')_L$, unioned with K', is called an abstract state.

4 Abstract Level Planning

In this section, planning using the abstract description language proposed in section 3, is explained.

Let $\rho = (T_0, G)$ be a concrete problem and $fp_1 : Lp_1 \overset{\text{onto}}{\mapsto} Lp'_1, \ldots, fp_n : Lp_n \overset{\text{onto}}{\mapsto} Lp'_n$ be partial predicate mappings of a predicate mapping $f : L \overset{\text{onto}}{\mapsto} L'$ that are computed to abstract a description language of STRIPS. The abstract problem ρ' of ρ is represented as a pair (T'_0, G'), where T'_0 is an *abstract initial state* and G', an *abstract goal*. These are defined as

$$T'_0 = f(T_0) \cup fp_1((T_0)_{Lp_1}) \cup \cdots \cup fp_n((T_0)_{Lp_n}),$$
$$G' = f(G).$$

Let Fp be the set of computed partial predicate mappings, o' be an abstract operator, and S'_i be an abstract state. o' is applicable to S'_i if there exists a ground substitution ϕ such that $S'_i \vdash Po'_\phi$. In order to generate S'_{i+1}, the result of applying o' to S'_i, Do'_ϕ must be removed from S'_i and then Ao'_ϕ added. However, it is not sufficient to remove *only* Do'_ϕ, since S'_i generally contains clauses in several abstract first-order languages. For example, if S'_i contains atoms d' and d'_p such that $f(d) = d', fp(d) = d'_p, fp \in Fp$, it is required to remove *not only* d' *but also* d'_p. Formally, the set of deleted abstract atoms, D'_ϕ, is defined as

$$D'_\phi = Do'_\phi \cup \{ \, d'_p \mid d' \in Do'_\phi, d \in f^{-1}(d'), fp(d) = d'_p, fp \in F_p \, \}.$$

Thus, $S'_{i+1} = (S'_i \setminus D'_\phi) \cup Ao'_\phi$. Definitions with respect to plans at abstract level (that is, the length, $\cdots$) are similar to ones at concrete level.

5 Instantiating Abstract Plan

Any plan generated at abstract level must be instantiated to a plan at concrete level in order to obtain a final plan for a original (concrete) problem. In this section, an instantiation process for abstract plan in section 4 is explained.

By using partial predicate mappings used to abstract a description language, any abstract plan can always be instantiated under following theorem.

Theorem 2 : Downward-Solution Property

Let $\Sigma = (L, E, O, K, \sigma)$ be a description language of STRIPS, $f : L \overset{\text{onto}}{\mapsto} L'$ be a predicate mapping, Fp be the set of partial predicate mappings of f used to abstract Σ, and $\omega' = \langle o'_1, \ldots, o'_n \rangle$ be an abstract plan for an abstract problem $\rho' = (T'_0, G')$. For each T_0 abstracted to T'_0 under computed partial predicate mappings, there exists a concrete plan $\omega = \langle o_1, \ldots, o_n \rangle$ such that

1. for each operator o_i in ω, o_i was abstracted to o'_i at abstracting stage,
2. $MembAbs_f(result(\langle o_1, \ldots, o_m \rangle, T_0 \cup K))$
$$= (result(\langle o'_1, \ldots, o'_m \rangle, T'_0 \cup K'))_{L'}, \quad 1 \leq m \leq n,$$
3. there exists $G \in f^{-1}(G')$ such that $result(\omega, T_0 \cup K) \vdash G$,
4. for each proof tree W' of preconditions Po'_{m+1} from
$$result(\langle o'_1, \ldots, o'_m \rangle, T'_0 \cup K'),$$
there exists proof tree W of preconditions Po_{m+1} from
$$result(\langle o_1, \ldots, o_m \rangle, T_0 \cup K),$$
such that W is isomorphic to W' and each node in W is an inverse image of corresponding node in W' under a partial predicate mapping in Fp used to generate $Po'_{m+1}, 1 \leq m < n$. ■

Proof :

Let $fp : Lp \overset{\text{onto}}{\mapsto} Lp'$ be a partial predicate mapping in Fp. Suppose that Po'_1 consists of k atoms $p'_1, \ldots, p'_k$ in Lp'. Let consider about a refutation proof tree W'_i of some p'_i from $T'_0 \cup K'$. The root node of W'_i is the empty clause, $\Box$, and each leaf node is either negated p'_i or clauses of $T'_0 \cup K'$ in Lp'. Let T_0 be a concrete state abstracted to T'_0. From the definitions of T'_0 and K', the following two properties are found. For each leaf node in W'_i except negated p'_i, 1) if the node is an atom, then there exists *some* atom in $T_0 \cup K$ being an instance of the atom under fp, 2) if the node is a rule, then for *each* instance of the precedent consisting of the same number literals as the precedent and *some* instance of the consequence of the rule, there exists a rule in K having these instances as its precedent and consequence, respectively. By traversing W'_i from the root node to negated p'_i using 1) and 2), it is shown that there exists a proof tree W_i of some p_i from $T_0 \cup K$, such that W_i is isomorphic to W'_i and each node in W_i is an instance of the corresponding node in W'_i under fp. Similarly, for any p'_i, this property can be shown. Thus, if there exists a proof of Po'_1 from $T'_0 \cup K'$, then there always exists a proof of some Po_1 from $T_0 \cup K$ such that $fp(Po_1) = Po'_1$. Further, from the definition of abstract operators, for any instance of Po'_1 under fp consisting of the same number atoms as Po'_1, there exists *some* operator o_1 in O having the instance as its precondition. That is, if o'_1 is applicable to $T'_0 \cup K'$,

Initial State :
beefSteak(beefSteak1), raw(beefSteak1), whole(beefSteak1),
countertop(table4), on(slicer3,table4), on(skillet2,table4),
slicer(slicer3), skillet(skillet2), refrigerator(refrige5),
empty(skillet2), in(beefSteak1,refrige5), closed(refrige5),
oven(oven6), handEmpty

Goal :
sliced(beefSteak1), baked(beefSteak1)

Figure 5: problem in kitchen-domain

Abstract Plan :	**Concrete Plan (Instantiated Plan) :**
OpenFoodSource(refrige5)	openRefrigerator(refrige5)
absGetFromIn(beefSteak1,refrige5)	getFromIn(beefSteak1,refrige5)
absPlaceOn(beefSteak1,table4)	placeOn(beefSteak1,table4)
absGetFromOn(slicer3,table4)	getFromOn(slicer3,table4)
putInPiecesFoodWithKnife (beefSteak1,slicer3,table4)	sliceBeefSteakWithSlicer (beefSteak1,slicer3,table4)
⋮	⋮
absGetFromOn(skillet2,table4)	getFromOn(skillet2,table4)
putInCooker(skillet2,oven6)	putInOven(skillet2,oven6)
CookInCookingVessele(beefSteak1,skillet2,oven6)	bakeInSkillet(beefSteak1,skillet2,oven6)

Figure 6: relation between abstract plan and concrete plan

then there always exists *some* o_1 that is applicable to $T_0 \cup K$ and was abstracted to o'_1 at the abstracting stage. It is not difficult to show

$$MembAbs_f(result(\langle o_1 \rangle, T_0 \cup K)) = (result(\langle o'_1 \rangle, T'_0 \cup K'))_{L'}.$$

For remaining operators in ω' and ω, it can be shown in similar way that these properties hold. Further, it is clear that there exists $G \in f^{-1}(G')$ such that $result(\omega, T_0 \cup K) \vdash G$, since it is equal to the case of preconditions proof.

Similarly, for any T_0 abstracted to T'_0, these properties hold. □

6 Experimental Results

A system consisting of *abstracting module* and *planning module*, has been implemented by SICStus Prolog on SparcStation. These modules are implementations of pre-process and planning with abstraction proposed in previous sections. Planning at abstract level is performed by using the same strategy of STRIPS, and instantiation of generated abstract plan is performed from the first operator to the last one.

A description language and a predicate mapping in kitchen-domain, similar to Tenenberg [3] [4] [5], are given to our system and the same description language is given to STRIPS. An example of problem in kitchen-domain is shown in Figure 5. Figure 6 shows the relation between abstract plan and concrete plan for the problem in Figure 5.

Figure 7 shows comparisons between planning times of our system and STRIPS. The lengths of the plans for Problem1 and Problem2 are both 11. The problem in Figure 5 is Problem1.

Given the same description language and predicate mapping to Tenenberg's framework, the constructed system by the framework cannot generate abstract plans for both problems. That is, our system has more rich planning ability than Tenenberg's. For each problem, the total time by our system is shorter than STRIPS's. This means that our abstraction technique is useful

	pre-process	problem 1		problem 2	
		our system	STRIPS	our system	STRIPS
planning at abstract level (sec.)		9.0		16.5	
instantiation (sec.)		0.18		0.19	
planning at concrete level (sec.)			15.6		25.4
total time (sec.)	103.1	9.18	15.6	16.69	25.4

Figure 7: Experimental results

for improving the planning process. Though one may claim that the total time must include the time for pre-process, the time at abstracting module, it would be canceled by solving more problems as long as the same problem description language is concerned, since the pre-process is never performed more than once.

7 Conclusion

In this paper, we proposed a framework for constructing a planning system satisfying DSP, which is more flexible than Tenenberg's framework. In our framework, a given predicate mapping is refined in order to abstract as many operator as possible. Therefore, our system have more rich planning ability than Tenenberg's. Furthermore, by some experimental results, it is shown that our technique using partial predicate mappings is useful for improving planning process.

Acknowledgment

We would like to thank anonymous referees of ALT'92 for their helpful comments.

References

[1] Fikes R. and Nilsson N., *STRIPS : a new approach to the application of theorem proving to problem solving*, Artificial Intelligence 2, 1971.

[2] S.Yamada, *Selective Learning of Macro-Operators with Perfect Causality* (in Japanese), Journal of JSAI, Vol,4, No.3, 1989,

[3] Josh D. Tenenberg, *Abstraction in Planning*, Technical Report ♯250, Univ.of Rochester, 1988.

[4] Josh D. Tenenberg, *Inheritance in Automated Planning*, In Proc. of The Second International Conference in Principles of Knowledge Representation and Reasoning, 1989.

[5] Josh D. Tenenberg, *Abstraction in Planning*, In Reasoning about Plans, James F. Allen et al., Morgan Kaufmann Publishers, 1991.

[6] Craig A. Knoblock, *A Theory of Abstraction for Hierarchical Planning*, In Pual Benjamin, editor, Proc. of the Workshop on Change of Representation and Inductive Bias, MA,Kluwer, 1989.

Approximate Learning

LEARNING k-TERM MONOTONE BOOLEAN FORMULAE

Yoshifumi Sakai Akira Maruoka
Faculty of Engineering
Tohoku University
Aza Aoba, Aramaki, Aoba, Sendai 980, Japan

Abstract

Valiant introduced a computational model of learning by examples, and gave a precise definition of learnability based on the model. Since then, much effort has been devoted to characterize learnable classes of concepts on this model. Among such learnable classes is the one, denoted k-term MDNF, consisting of monotone disjunctive normal form formulae with at most k terms. In literature, k-term MDNF is shown to be learnable under the assumption that examples are drawn according to the uniform distribution. In this paper we generalize the result to obtain the statement that k-term MDNF is learnable even if positive examples are drawn according to such distribution that the maximum of the ratio of the probabilities of two positive examples is bounded from above by some polynomial.

1 Introduction

L.G.Valiant introduced a computational model of learning by examples, and gave the definition of learnability for a class of target concepts[5]. Since then, various classes of concepts represented by some restricted Boolean formulae have been shown to be learnable in literature, and some classes not to be learnable. For example, in the distribution free learning model, where examples of a target concept are generated according to some fixed but unknown arbitrary probability distribution, the class of disjunctive normal form formulae with at most k literals in each term, denoted k-DNF, is shown to be learnable by examples[5], whereas the class of disjunctive normal form monotone formulae with at most k terms, denoted k-term MDNF, is not learnable unless RP = NP[2]. By a kind of duality, the class of conjunctive normal form formulae with at most k literals in each clause, denoted k-CNF, is learnable, while the class of conjunctive normal form formulae with at most k clauses, denoted k-clause CNF, is not learnable unless RP = NP. On the other hand, in the distribution specific setting, where examples are generated according to the uniform distribution, k-term MDNF is learnable, and learning algorithms for this class are proposed in [3][1][4]. In particular, it is shown in [4] that this class is also learnable even in the case where positive examples are generated according to the uniform distribution, while negative examples are generated according to some unknown arbitrary distribution. The problem whether the class of general disjunctive normal form formulae is learnable remains open in both cases mentioned above. In particular, in the distribution free setting, the problem of deciding the learnability of this class seems to be important in this field.

In this paper, we give a learning algorithm for k-term MDNF in the uniform distribution setting, which is much simpler than the ones previously given in literature in the sense that the behavior of the algorithm is simple, and hence the correctness of the algorithm is easily verified. Because of the simplicity of the algorithm we can establish somewhat stronger statement, namely, k-term MDNF is learnable when positive examples are drawn according to such distribution that the maximum of the ratio of the probabilities of two positive examples is bounded from above by some polynomial.

In section 2, we illustrate Valiant's learning model, and give the definition of learnability. In section 3, on the assumption that positive examples are drawn according to the distribution deviated from the uniform distribution in the way mentioned above, we give an algorithm that learns k-term MDNF together with the proof of its correctness.

2 Preliminaries

Let f be a Boolean function of n variables $x_1, \ldots, x_{n-1}$ and x_n. Let the set of n variables be denoted X. Given f, a vector $v \in \{0,1\}^n$ is called a positive example (resp. negative example) of f if and only if $f(v) = 1$ (resp. $f(v) = 0$). Let D_f^+ (resp. D_f^-) denote a probability distribution on $f^{-1}(1)$ (resp. $f^{-1}(0)$). D_f^+ (resp. D_f^-) is simply written as D^+ (resp. D^-) when no confusion arises. Positive examples (resp. negative examples) are assumed to be generated independently according to D^+ (resp. D^-). In this paper we restrict ourselves to learning algorithms that take as input only positive examples. In order to get positive examples, a learning algorithm calls an oracle, denoted POS(), which produces positive examples independently according to D^+. So we only need the probability distribution D^+ on $f^{-1}(1)$.

In the following, we often identify a Boolean formula with the Boolean function that it represents. Thus, we regard the class F of Boolean formulae as the corresponding class of Boolean functions. For a given Boolean formula f, let $\mathcal{D}_f$ denote the set of probability distribution on $f^{-1}(1)$. Let F be a class of Boolean formulae (or the corresponding class of Boolean functions), and let $\mathcal{D}_F$ denote $\bigcup_{f \in F} \mathcal{D}_f$. In PAC learning model, probability distribution according to which examples are drawn is usually assumed to be either arbitrary, or to be uniform. The former is called the distribution free setting while the latter is called the distribution specific setting. In the following definition we consider more general setting where examples are assumed to be generated according to a probability distribution which belongs to an arbitrarily given class $\mathcal{D} \subseteq \mathcal{D}_F$. Given a class of Boolean functions F, F_n denotes the set of Boolean functions of n variables in F. In the following, f also represents $\{v | f(v) = 1\}$, so we write $D^{-1}(f)$ to represent $\sum_{f(v)=1} D^+(v)$.

Definition 1 *Let $\mathcal{D} \subseteq \mathcal{D}_F$. A class of formulae F is called learnable under the class of distributions $\mathcal{D}$ if and only if there exists a polynomial $poly(\cdot,\cdot,\cdot)$ and a learning algorithm $L_{F,\mathcal{D}}$ with oracle POS() such that for any n, $f \in F_n$, $0 < \varepsilon, \delta < 1$ and $D_f^+ \in \mathcal{D} \cap \mathcal{D}_f$, the algorithm $L_{F,\mathcal{D}}$ halts in time $poly(n, \frac{1}{\varepsilon}, \frac{1}{\delta})$ and outputs a formula $g \in F_n$ that with probability at least $1-\delta$ satisfies $D_f^+(f\Delta g) < \varepsilon$ and $g \subseteq f$, where $f\Delta g$ denotes the formula $(f\wedge\bar{g})\vee(\bar{f}\wedge g)$.*

In general a learning algorithm is assumed to use negative examples, which are generated according to D^-, as well as positive examples, and accordingly the hypothesis g it produces is required to satisfy $D_f^-(f\Delta g) < \varepsilon$ as well as $D_f^+(f\Delta g) < \varepsilon$. Since we deal with learning algorithms which use only positive examples, we simply replaced the condition $D_f^-(f\Delta g) < \varepsilon$ with the somewhat stronger condition $g \subseteq f$ in the definition of learnability.

f and g in the definition above are called a target function and a hypothesis, respectively. A literal is either a Boolean variable or its negation, and a conjunction of finite literals is called a monomial (or a term). For constant k, k-term MDNF denotes the class of monotone disjunctive normal form formulae with up to k terms. Let Term(f) denote the set of terms of a disjunctive normal form formula f. Let Var(t) denote the set of variables that appear in term t.

In general there exist a number of formulae in k-term MDNF that represent the same function. In this paper we assume, without loss of generality, that the Boolean formula representing a target function is the one that has the fewest number of terms among such formulae so that, if f is such a monotone Boolean formula, then for any distinct terms t and t' of f, $\mathrm{Var}(t) \not\subseteq \mathrm{Var}(t')$ holds.

For $v \in \{0,1\}^n$ and $1 \le i \le n$, let v_i denote the ith component of v. Likewise, for $v \in \{0,1\}^n$ and $a = \{x_{a_1}, \ldots, x_{a_j}\} \subseteq X$, let v_a denote $(v_{a_1}, \ldots, v_{a_j})$ and $v_a = 0$ means that $v_{a_1} = 0, \ldots, v_{a_j} = 0$. Natural logarithm is written as "ln". $|S|$ denotes the cardinality of a set S.

For any $f \in k$-term MDNF and any $a \subseteq X$, let f_a denote the disjunction of terms of f with no variable in a, that is,

$$\bigvee_{t \in \{t' \in \mathrm{Term}(f) | \mathrm{Var}(t') \cap a = \phi\}} t.$$

For constant k, let A_{k-1} denote the set $\{a \in 2^X | |a| \le k-1\}$. Before closing this section, we give a series of lemmas which will be used in the following section.

Lemma 2 *Let $f \in k$-term MDNF, and $a \subseteq X$. For any Boolean function $h : \{0,1\}^n \to \{0,1\}$ that does not depend on any variable in a,*

$$|\{v \in f_a | h(v) = 1\}| = 2^{|a|} |\{v \in f | h(v) = 1, v_a = 0\}|.$$

Proof Note that $\{v \in f_a | h(v) = 1\}$ denotes the set of v's in f that satisfy both $f_a(v) = 1$ and $h(v) = 1$, and that $\{v \in f | h(v) = 1, v_a = 0\}$ denotes the set of v's in f that satisfy $f_a(v) = 1$, $h(v) = 1$ and $v_a = 0$. Thus, since both f_a and h are functions that only depend on variables in $X \backslash a$, the lemma follows. □

Lemma 3 ([3]) *Let $f \in k$-term MDNF. For any term $t \in \mathrm{Term}(f)$, there exists a set $a \in A_{k-1}$ such that $f_a = t$.*

Proof For every $t' \in \mathrm{Term}(f)\backslash\{t\}$, choose a variable $x' \in \mathrm{Var}(t')\backslash\mathrm{Var}(t)$, and let a be the set of all such variables. Note that by the assumption that the target formula f has the fewest terms, $\mathrm{Var}(t')\backslash\mathrm{Var}(t) \neq \phi$ holds. Since $|\mathrm{Term}(f)\backslash\{t\}| \leq k-1$, it is clear that $a \in A_{k-1}$. It is easy to see that $f_a = t$. □

Definition 4 *Let* $0 \leq \zeta \leq 1$ *and* $\eta \geq 1$. $\mathcal{U}_{\zeta,\eta}$ *is the class of* $D_f^+ \in \mathcal{D}_F$ *such that for any* $v \in f^{-1}(1)$,

$$\frac{\zeta}{|f|} \leq D_f^+(v) \leq \frac{\eta}{|f|}.$$

In particular, $\mathcal{U}_{1,1}$ *denotes the class of uniform distributions with respect to* D^+.

3 Learning k-term MDNF under $\mathcal{U}_{\zeta,\eta}$

The main result of this paper is stated as the next theorem.

Theorem 5 *For any* ζ *and* η *such that* $\eta/\zeta \leq q$ *for some polynomial* q *in* n, $\frac{1}{\epsilon}$ *and* $\frac{1}{\delta}$, k-term MDNF *is learnable under* $\mathcal{U}_{\zeta,\eta}$.

Definition 6 *Let* f *be a disjunctive normal form formula.* $t \in \mathrm{Term}(f)$ *is the largest term of* f *under* D_f^+ *if and only if for any term* $t' \in \mathrm{Term}(f)$, $D_f^+(t) \geq D_f^+(t')$.

Note that in general the largest term is not necessarily unique. In the following, when D_f^+ is clear from the context, the largest term of f under D_f^+ is simply referred to as the largest term of f. Throughout the present paper, let v denote the random variable that takes vectors in $f^{-1}(1)$ according to D_f^+.

Definition 7 *For* $0 \leq \beta \leq 1$ *and* $0 \leq \gamma \leq 1$, $\mathcal{U}'_{\beta,\gamma}$ *is the class of* $D_f^+ \in \mathcal{D}_{k\text{-term MDNF}}$ *that satisfies the following two conditions:*
1. *for any* $a \in A_{k-1}$ *such that* $f_a \neq 0$,

$$\Pr[v_a = 0 | f_a(v) = 1] \geq \beta,$$

and
2. *for any* $a \in A_{k-1}$ *with* $f_a \neq 0$, *any* $x_i \in X$, *and any of the largest term* t *of* f_a *under* D_f^+ *if* $x_i \notin \mathrm{Var}(t)$ *then*

$$\Pr[v_i = 1 | v_a = 0] \leq \gamma.$$

Note that when D_f^+ is the uniform distribution the right hand side of the inequality in the Condition 1 of the definition above is simply written as $1/2^{|a|}$, and that the Condition 2 is restated as follows: Let $\bigcap \mathrm{Var}(t)$ denote the intersection of all the sets of variables in the largest terms of f_a under D_f^+; For any $a \in A_{k-1}$ with $f_a \neq 0$ and any $x_i \in X\backslash(a \cup (\bigcap \mathrm{Var}(t))$

$$\Pr[v_i = 1 | v_a = 0] \leq \gamma.$$

Lemma 8 ([4]) *Let* $0 < p \leq 1$, $0 < \delta < 1$, *and* $m \geq \frac{32}{p} \ln \frac{1}{\delta}$. *If* $\Pr[E] \geq p$, *then* E *occurs at most* $\frac{3}{4}pm$ *in* m *independent Bernoulli trials with probability at most* δ. *Similarly, if* $\Pr[E] \leq \frac{1}{2}p$, *then* E *occurs at least* $\frac{3}{4}pm$ *in* m *independent Bernoulli trials with probability at most* δ. □

Lemma 9 *For any* β *and* γ *such that* $\frac{1}{q_1} \leq \beta \leq 1$ *and* $0 \leq \gamma \leq e^{-\frac{1}{q_2}}$ *for some polynomials* q_1 *and* q_2 *in* n, $\frac{1}{\epsilon}$ *and* $\frac{1}{\delta}$, k-term MDNF *is learnable under* $U'_{\beta,\gamma}$.

Proof To prove this lemma, we show that L given in Figure 1 is the algorithm that learns k-term MDNF. L takes as input only positive examples of target function $f \in$ k-term MDNF and with high probability returns as output hypothesis g consisting of every term t of f with $D^+(t) \geq \frac{\epsilon}{k}$ so that g approximates f accurately enough. As is shown in Figure 1 algorithm L calls procedure DOMINANT_VARS given in Figure 2.

First we briefly explain how procedure DOMINANT_VARS works. DOMINANT_VARS takes as input a real number $0 \leq p \leq 1$ and a multiset S of vectors in $f^{-1}(1)$, and returns as output a class M of sets of variables in X. In what follows, a subset w of X is also thought of as the term composed of variables in the set w. So the value that the procedure returns is regarded as a disjunctive normal form formula consisting of terms corresponding to sets in M. Taking a to be a set in A_{k-1}, the procedure makes subset S_a of S consisting of vectors v's in S such that $v_a = 0$. Note that S_a turns out to be a set of positive examples for f_a. If S_a contains sufficiently many positive examples, namely, $|S_a| \geq \frac{3}{4}\beta p|S|$, then it produces set $w_a = \{x_i \in X |$ for any $v \in S_a$, $v_i = 1\}$. In fact, if a is such that f_a becomes a long term, condition $|S_a| \geq \frac{3}{4}\beta p|S|$ is not satisfied (For ease of explanation, we assume that the probabilistic distribution D^+ is such that for a short term t $D^+(t)$ is large, and for a long term t $D^+(t)$ is small). If a is such that $f_a = t$ for some short term t (hence, $\{v|t(v) = 1\}$ being large) in target function f, then w_a hopefully corresponds to term t, while if a is such that f_a consists of terms more than two, then w_a may corresponds to no term in target function f. In the later case, w_a possibly corresponds to the set of variables that belong to any of somewhat larger ones among these terms. The procedure produces the set W of such subsets w_a, by repeating what is mentioned above for each a in A_{k-1}, and then gets the set M by discarding subsets in W that are not maximal in W. It will be shown later that, if S consists of sufficiently many positive examples, then the set M, with sufficiently high probability, satisfies the following properties: (i) M contains any set Var(t) such that $D^+(t) \geq p$, and (ii) For any other set w in M there exists t in Term(f) such that $w \subseteq$ Var(t) and $D^+(t) \geq \frac{\beta p}{2k}$. These properties will be referred to as $E_1^{p,S}$ and $E_2^{p,S}$, respectively. The proof of the correctness of the algorithm is based on these properties.

We are now ready to explain an outline of algorithm L. L takes as input n, ϵ, and δ, and calls POS() sufficiently many time to obtain the multiset S of positive examples. Then L calls DOMINANT_VARS twice with parameters $\frac{\epsilon}{k}$, S, and then with $\frac{\beta\epsilon}{2k^2}$, S, obtaining M_1 and M_2, respectively. Finally L produces as output $M_1 \cap M_2$. Putting $p = \frac{\epsilon}{k}$, the property $E_1^{\frac{\epsilon}{k},S}$ mentioned above guarantees that M_1 contains all Var(t)'s (good subsets) such that $D^+(t) \geq \frac{\epsilon}{k}$. Unfortunately M_1 may also contains subsets (bad subsets) that corresponds to no term in f. But by $E_2^{\frac{\epsilon}{k},S}$ such a subset w satisfies the condition that $w \subseteq$ Var(t)

input: n, ε, δ
output: g

begin
    call POS() $\max\{\frac{64k^2}{\beta^2\varepsilon}\ln\frac{4n^{k+1}}{\delta}, \frac{8k^2}{3\beta^2\ln(1/\gamma)\varepsilon}\ln\frac{4n^{k+1}}{\delta}\}$ times,
        and let S be the multiset of positive examples that POS() returns ;
    M_1 :=DOMINANT_VARS($\frac{\varepsilon}{k}, S$) ;
    M_2 :=DOMINANT_VARS($\frac{\beta\varepsilon}{2k^2}, S$) ;
    $g := \bigvee_{w \in M_1 \cap M_2}(\bigwedge_{x_i \in w} x_i)$
end.

Figure 1: Learning algorithm L

and $D^+(t) \geq \frac{\beta\varepsilon}{2k^2}$ for some t in Term(f). So we need to get rid of the bad subsets without discarding the good ones. To do so, L produces M_2 with p being $\frac{\beta\varepsilon}{2k^2}$, and makes $M_1 \cap M_2$. Again noting the properties $E_1^{\frac{\beta\varepsilon}{2k^2}}$ and $E_2^{\frac{\beta\varepsilon}{2k^2}}$, we can easily see that L does what we wanted: Subset $w = \mathrm{Var}(t)$ in M_1 such that $D^+(t) \geq \frac{\varepsilon}{k}$ is not discarded by taking the intersection, while $w \subsetneq \mathrm{Var}(t)$ in M_1 such that $D^+(t) \geq \frac{\beta\varepsilon}{2k^2}$ is discarded unless $w = \mathrm{Var}(t)$. In what follows, we present the argument more precisely.

Let $f \in k$-term MDNF be the target function, and D_f^+ be in $\mathcal{U}'_{\beta,\gamma}$. D_f^+ is simply written as D^+. The following fact states the property that the set M that DOMINANT_VARS returns satisfies.

For any $0 \leq p \leq 1$, we denote the set $\{\mathrm{Var}(t) | t \in \mathrm{Term}(f) \text{ and } D^+(t) \geq p\}$ by T_p. For any $0 \leq p \leq 1$ and multiset S of vectors, let events $E_1^{p,S}$ and $E_2^{p,S}$ concerning M that DOMINANT_VARS produces with parameters p and S be defined as follows:
$E_1^{p,S}$: $T_p \subseteq M$,
$E_2^{p,S}$: for any $w \in M$, there exists $w' \in T_{\frac{\beta p}{2k}}$ such that $w \subseteq w'$,
and let $E^{p,S}$ be the product event of $E_1^{p,S}$ and $E_2^{p,S}$, that is ,
$E^{p,S}$: $E_1^{p,S}$ and $E_2^{p,S}$.

Fact 10 *Let* $0 < p \leq 1$, $m \geq \max\{\frac{32}{\beta p}\ln\frac{4n^{k+1}}{\delta}, \frac{4}{3\beta\ln(1/\gamma)p}\ln\frac{4n^{k+1}}{\delta}\}$, *and let* $\boldsymbol{S}$ *be the random variable according to the probability distribution* $(D^+)^m$. *Then,* $\Pr[E^{p,\boldsymbol{S}}] \geq 1 - \frac{\delta}{2}$.

Proof of Fact 10 The random variable $\boldsymbol{S}$ corresponds S obtained in L. Let $\boldsymbol{S}_a$ and $\boldsymbol{w}_a$ denote the corresponding random variables of S_a and w_a that appear in DOMINANT_VARS, respectively, where $a \in A_{k-1}$. Let A denote the set $\{a \in A_{k-1} || S_a| \geq \frac{3}{4}\beta pm\}$, and $\boldsymbol{A}$ the corresponding random variable. A' denotes the set $\{a \in A_{k-1} | \Pr[v_a = 0] \geq \beta p\}$, and A'' denotes the set $\{a \in A_{k-1} | \Pr[v_a = 0] \leq \frac{1}{2}\beta p\}$. Let events E_1' with respect to $\boldsymbol{A}$ and E_2' with respect to $\boldsymbol{w}_a$ for every $a \in \boldsymbol{A}$ be defined as follows:
E_1' : $A' \subseteq \boldsymbol{A}$ and $A'' \cap \boldsymbol{A} = \phi$,

```
procedure DOMINANT_VARS(p, S)
begin
    W := φ ;
    for a ∈ A_{k-1} do
    begin
        S_a := {v ∈ S | v_a = 0} ; (* multiset *)
        if |S_a| ≥ (3/4)βp|S| then
        begin
            w_a := {x_i ∈ X | for any v ∈ S_a, v_i = 1} ;
            W := W ∪ {w_a}
        end
    end ;
    M := {w ∈ W | for any w' ∈ W\{w}, w ⊈ w'} ;
    return M
end.
```

Figure 2: Procedure DOMINANT_VARS

E_2' : for any $a \in A$ and the largest term t of f_a, $w_a \subseteq \mathrm{Var}(t)$.
Let E' be the product event of E_1' and E_2'. Moreover, let events E_1'' and E_2'' with respect to W be defined as follows:
$E_1'' : T_p \subseteq W$,
E_2'' : for any $w \in W$, there exists $w' \in T_{\frac{\beta p}{2k}}$ such that $w \subseteq w'$,
and let E'' be the product event of E_1'' and E_2''.

To establish $\Pr[E^{p,S}] \geq 1-\frac{\delta}{2}$, it suffices to verify that E' implies E'' which in turn implies $E^{p,S}$, and that $\Pr[E'] \geq 1-\frac{\delta}{2}$.

We start with showing that E' implies E''. To do so, assume that E' holds. Let t be any term in T_p, namely, a term of f such that $D^+(t) \geq p$. By Lemma 3, there exists $a \in A_{k-1}$ such that $t = f_a$. Then, because of the Condition 1 of the definition of $\mathcal{U}'_{\beta,\gamma}$, $\Pr[v_a = 0] \geq \Pr[f_a(v) = 1]\beta = \Pr[t(v) = 1]\beta = D^+(t)\beta \geq \beta p$. We therefore have $a \in A$ by E_1'. So, since t is the largest term of $f_a(= t)$, we have $w_a \subseteq \mathrm{Var}(t)$ by E_2'. On the other hand, if $x_i \in \mathrm{Var}(t)$, then $v_i = 1$ holds for any $v \in f^{-1}(1)$ such that $v_a = 0$. This is because the fact that $f_a = t$ implies that $t(v) = 1$ for such v. We therefore have $x_i \in w_a$ by the definition of w_a, hence obtaining $w_a \supseteq \mathrm{Var}(t)$. Thus we have $w_a = \mathrm{Var}(t)$ which together with $a \in A$ implies $T_p \subseteq W$, hence E_1'' holds.

Next we show that E_2'' holds. Since w is written as w_a for some $a \in A$, it follows from E_2' that there exists the largest term t of f_a such that $w_a \subseteq \mathrm{Var}(t)$. Since $a \notin A''$ by E_1', $\Pr[v_a = 0] > \frac{1}{2}\beta p$ holds. So, since $v_a = 0$ implies $f_a(v) = 1$, we have $\Pr[f_a(v) = 1] \geq \Pr[v_a = 0] > \frac{1}{2}\beta p$. On the other hand, since $|\mathrm{Term}(f_a)| \leq k$ and t is the largest term of f_a, we have $\Pr[t(v) = 1 | f_a(v) = 1] \geq \frac{1}{k}$. Thus, noting that $t \subseteq f_a$, we have

$$D^+(t) \;=\; \Pr[t(v) = 1]$$

$$\begin{aligned} &= \Pr[t(v) = 1, f_a(v) = 1] \\ &= \Pr[f_a(v) = 1] \cdot \Pr[t(v) = 1 | f_a(v) = 1] \\ &> \frac{\beta p}{2k}, \end{aligned}$$

establishing E_2'', and hence E'' together with E_1''.

We now show that E'' implies $E^{p,S}$. To do so, assume that E'' holds. Notice that, since any distinct sets in $\{\mathrm{Var}(t) | t \in \mathrm{Term}(f)\}$ are incomparable (with respect to the inclusion relation of sets), so are those in $T_{\frac{\beta p}{2k}}$. Hence it is easily seen that $w \not\subseteq w'$ holds for any $w \in W \cap T_{\frac{\beta p}{2k}}$ and $w' \in W \setminus \{w\}$ by E_2''. So any set in $W \cap T_{\frac{\beta p}{2k}}$ is maximal among all sets in W, hence, by the definition of M, $W \cap T_{\frac{\beta p}{2k}} \subseteq M$ holds. On the other hand, by E_1'' and the fact that $T_p \subseteq T_{\frac{\beta p}{2k}}$, we have $T_p \subseteq W \cap T_{\frac{\beta p}{2k}}$. Thus we have $T_p \subseteq W \cap T_{\frac{\beta p}{2k}} \subseteq M$, establishing $E_1^{p,S}$. Furthermore, since $M \subseteq W$ and E_2'' hold, we have $E_2^{p,S}$ which, together with $E_1^{p,S}$, establishes $E^{p,S}$.

Finally we shall establish $\Pr[E'] \geq 1 - \frac{\delta}{2}$ by showing $\Pr[\bar{E}'] \leq \frac{\delta}{2}$, where $\bar{E}'$ denotes the complementary event of E'. Similarly, let $\bar{E}_1'$ and $\bar{E}_2'$ be the complementary events of E_1' and E_2', respectively, that is,

$\bar{E}_1'$: $A' \setminus \boldsymbol{A} \neq \phi$ or $A'' \cap \boldsymbol{A} \neq \phi$,

$\bar{E}_2'$: there exist $a \in \boldsymbol{A}$ and the largest term t of f_a such that $w_a \setminus \mathrm{Var}(t) \neq \phi$.

Put $\delta' = \frac{\delta}{4n^{k+1}}$. In order to show $\Pr[\bar{E}'] \leq \frac{\delta}{2}$, it suffices to verify both $\Pr[\bar{E}_1'] \leq n^k \delta'$ and $\Pr[\bar{E}_2'] \leq n^{k+1}\delta'$. This is because

$$\begin{aligned} \Pr[\bar{E}'] &= \Pr[\overline{E_1' \wedge E_2'}] \\ &= \Pr[\bar{E}_1' \vee \bar{E}_2'] \\ &\leq \Pr[\bar{E}_1'] + \Pr[\bar{E}_2'] \\ &\leq n^k\delta' + n^{k+1}\delta' \\ &\leq \frac{\delta}{2}. \end{aligned}$$

First we shall show that the first inequality holds. Let a be such that $a \in A'$, or equivalently $\Pr[v_a = 0] \geq \beta p$. If event $v_a = 0$ is thought of as success, event $|\boldsymbol{S}_a| < \frac{3}{4}\beta p m$ (or equivalently $a \notin \boldsymbol{A}$) corresponds to the event that less than $\frac{3}{4}\beta p m$ successes occur in m independent Bernoulli trials with probability of success at least βp. From Lemma 8 and $m \geq \frac{32}{\beta p} \ln \frac{1}{\delta'}$, it follows that $\Pr[a \notin \boldsymbol{A}] = \Pr[|\boldsymbol{S}_a| < \frac{3}{4}\beta p m] \leq \delta'$. Similarly, for $a \in A''$ we have $\Pr[a \in \boldsymbol{A}] = \Pr[|\boldsymbol{S}_a| \geq \frac{3}{4}\beta p m] \leq \delta'$. Furthermore, we have $A' \cap A'' = \phi$ and $A' \cup A'' \subseteq A_{k-1}$. Therefore, noting that $|A_{k-1}| = \sum_{i=0}^{k-1} \binom{n}{i} \leq n^k$, we have

$$\begin{aligned} \Pr[\bar{E}_1'] &\leq \Pr[A' \setminus \boldsymbol{A} \neq \phi] + \Pr[A'' \cap \boldsymbol{A} \neq \phi] \\ &\leq \sum_{a \in A'} \Pr[a \notin \boldsymbol{A}] + \sum_{a \in A''} \Pr[a \in \boldsymbol{A}] \\ &\leq \sum_{a \in A'} \delta' + \sum_{a \in A''} \delta' \\ &\leq \sum_{a \in A_{k-1}} \delta' \end{aligned}$$

$$\leq n^k\delta'.$$

Next we shall show that the second inequality holds. Since $S_a \neq \phi$ for $a \in A$, it is easy to see that if $a \in A$ then $f_a \neq 0$. Clearly if $\bar{E}'_2$ holds, then there exist $a \in \boldsymbol{A}$, the largest term t of f_a, and x_i with $x_i \notin \mathrm{Var}(t)$ such that $v_i = 1$ and $v_a = 0$ holds $|\boldsymbol{S}_a|$ times, where $|\boldsymbol{S}_a| \geq \frac{3}{4}\beta pm$. On the other hand, by the Condition 2 of the definition of $\mathcal{U}'_{\beta,\gamma}$, $\Pr[v_i = 1 | v_a = 0] \leq \gamma$. So, since $|A_{k-1}| \leq n^k$, $|X| = n$ and $\frac{3}{4}\beta pm \geq \log_\gamma \delta'$ hold, we have

$$\begin{aligned}
\Pr[\bar{E}'_2] &\leq \sum_{a\in A_{k-1}} \sum_{x_i \in X} \Pr[v_i = 1 \text{ and } v_a = 0 \text{ occur } |\boldsymbol{S}_a| \text{ times, and } |\boldsymbol{S}_a| \geq \frac{3}{4}\beta pm] \\
&\leq \sum_{a\in A_{k-1}} \sum_{x_i \in X} (\Pr[v_i = 1 \text{ and } v_a = 0])^{\frac{3}{4}\beta pm} \\
&\leq \sum_{a\in A_{k-1}} \sum_{x_i \in X} (\Pr[v_i = 1 | v_a = 0])^{\frac{3}{4}\beta pm} \\
&\leq \sum_{a\in A_{k-1}} \sum_{x_i \in X} \gamma^{\frac{3}{4}\beta pm} \\
&\leq n^{k+1}\gamma^{\log_\gamma \delta'} \\
&= n^{k+1}\delta'.
\end{aligned}$$

Thus, $\Pr[E'] \geq 1 - \frac{\delta}{2}$ is established. □

In the following, we shall show that with probability at least $1-\delta$ the hypothesis g that L outputs satisfies $D^+(f\Delta g) < \varepsilon$ and $\mathrm{Term}(g) \subseteq \mathrm{Term}(f)$, and that L halts in time polynomial in n, $\frac{1}{\varepsilon}$ and $\frac{1}{\delta}$, and hence completing the proof of the lemma.

The fact above implies that $\Pr[E^{\frac{\varepsilon}{k},S}] \geq 1 - \frac{\delta}{2}$ and $\Pr[E^{\frac{\beta\varepsilon}{2k^2},S}] \geq 1 - \frac{\delta}{2}$. We therefore have

$$\begin{aligned}
\Pr[E^{\frac{\varepsilon}{k},S} \text{ and } E^{\frac{\beta\varepsilon}{2k^2},S}] &\geq 1 - \left((1 - \Pr[E^{\frac{\varepsilon}{k},S}]) + (1 - \Pr[E^{\frac{\beta\varepsilon}{2k^2},S}])\right) \\
&\geq 1 - \delta.
\end{aligned}$$

So it suffices to verify the conditions about error of hypothesis assuming that $E^{\frac{\varepsilon}{k},S}$ and $E^{\frac{\beta\varepsilon}{2k^2},S}$. Note that $E^{\frac{\varepsilon}{k},S}$ and $E^{\frac{\beta\varepsilon}{2k^2},S}$ state the conditions concerning M_1 and M_2 in algorithm L, respectively. It is shown that, if $T_{\frac{\varepsilon}{k}} \subseteq M_1 \cap M_2$ and $M_1 \cap M_2 \subseteq T_0$ ($= \mathrm{Term}(f)$), then the conditions about error of hypothesis g are satisfied. This is because $\mathrm{Term}(g) \subseteq \mathrm{Term}(f)$, hence $g \subseteq f$. Furthermore, since $\mathrm{Term}(g) = M_1 \cap M_2 \subseteq T_0 = \mathrm{Term}(f)$ and since $D^+(t) < \frac{\varepsilon}{k}$ for any $t \in \mathrm{Term}(f)\backslash\mathrm{Term}(g)$ by assumption $T_{\frac{\varepsilon}{k}} \subseteq M_1 \cap M_2$, we have

$$\begin{aligned}
D^+(f\Delta g) &= D^+(g^{-1}(0)) \\
&= 1 - D^+(g^{-1}(1)) \\
&\leq 1 - \left(1 - \sum_{t\in\mathrm{Term}(f)\backslash\mathrm{Term}(g)} D^+(t)\right) \\
&< 1 - (1 - k \cdot \frac{\varepsilon}{k}) \\
&= \varepsilon.
\end{aligned}$$

Assuming that $E^{\frac{\epsilon}{k},S}$ and $E^{\frac{\beta\epsilon}{2k^2},S}$, we shall show that $T_{\frac{\epsilon}{k}} \subseteq M_1 \cap M_2$ and $M_1 \cap M_2 \subseteq T_0$. Since $E_1^{\frac{\epsilon}{k},S}$ and $E_1^{\frac{\beta\epsilon}{2k^2},S}$ imply that $T_{\frac{\epsilon}{k}} \subseteq M_1$ and $T_{\frac{\beta\epsilon}{2k^2}} \subseteq M_2$, and since $T_{\frac{\epsilon}{k}} \subseteq T_{\frac{\beta\epsilon}{2k^2}}$ holds by the definition, we have $T_{\frac{\epsilon}{k}} \subseteq M_1 \cap M_2$. We now assume in contradiction that there exists $w \in (M_1 \cap M_2)\backslash T_0 \subseteq (M_1 \cap M_2)\backslash T_{\frac{\beta\epsilon}{2k^2}}$. Then, by $E_2^{\frac{\epsilon}{k},S}$, there exists $w' \in T_{\frac{\beta\epsilon}{2k^2}}$ such that $w \subset w'$. But this contradicts the minimality of M_2, because $w \in M_2$, $w' \in T_{\frac{\beta\epsilon}{2k^2}} \subseteq M_2$. Thus we have $M_1 \cap M_2 \subseteq T_0$.

It only remains to estimate the time complexity of L. Because $|A_{k-1}| \leq n^k$, DOMINANT_VARS produces W in time $O(n^k|S|)$, and because $|W| \leq n^k$, it is easily verified that M is produced from W in time $O(n^{2k})$. Therefore, DOMINANT_VARS returns M in time $O(n^k(n^k + |S|))$. Because DOMINANT_VARS returns M such that $|M| \leq n^k$, L produces the hypothesis g in time $O(n^k \log_2 n)$, by sorting M_1 and M_2, and then obtaining $M_1 \cap M_2$. The cardinality of S obtained in L is $O(\frac{q_1{}^2 q_2}{\epsilon} \ln \frac{n}{\delta})$ for fixed k, hence L halts in time $O(n^k(n^k + \frac{q_1{}^2 q_2}{\epsilon} \ln \frac{n}{\delta}))$. □

Proof of Theorem 5 It suffices to show that $\mathcal{U}_{\zeta,\eta} \subseteq \mathcal{U}'_{\frac{1}{2^{k-1}q}, e^{-\frac{1}{2kq^3}}}$, because by Lemma 9 k-term MDNF is learnable under $\mathcal{U}'_{\frac{1}{2^{k-1}q}, e^{-\frac{1}{2kq^3}}}$, and hence under $\mathcal{U}_{\zeta,\eta}$. Let $f \in$ k-term MDNF and $D_f^+ \in \mathcal{U}_{\zeta,\eta}$. We shall show D_f^+ satisfies both Condition 1 and 2 in Definition 7 for $\beta = \frac{\zeta}{2^{k-1}\eta}$ $(\geq \frac{1}{2^{k-1}q})$ and $\gamma = 1 - \frac{\zeta^3}{2k\eta^3}$ $(\leq e^{-\frac{\zeta^3}{2k\eta^3}} \leq e^{-\frac{1}{2kq^3}})$.

Suppose that $a \in A_{k-1}$ such that $f_a \not\equiv 0$. Then, noting that $|a| \leq k-1$, we have by Lemma 2

$$
\begin{aligned}
\frac{\Pr[v_a = 0]}{\Pr[f_a(v) = 1]} &\geq \frac{|\{v \in f | v_a = 0\}|\zeta/|f|}{|f_a|\eta/|f|} \\
&= \frac{|\{v \in f | v_a = 0\}|}{|f_a|} \cdot \frac{\zeta}{\eta} \\
&= \frac{\zeta}{2^{|a|}\eta} \\
&\geq \frac{\zeta}{2^{k-1}\eta}.
\end{aligned}
$$

To show Condition 2, let t be the largest term of f_a and let $x_i \in X$. Since the condition holds trivially in the case that $x_i \in a$, we shall consider the case that $x_i \notin a$. Since $|\mathrm{Term}(f_a)| \leq k$, $\Pr[t(v) = 1 | f_a(v) = 1] \geq \frac{1}{k}$ holds. We therefore have by Lemma 2

$$
\begin{aligned}
& \Pr[v_i = 1 | v_a = 0] \\
= \; & 1 - \Pr[v_i = 0 | v_a = 0] \\
\leq \; & 1 - \Pr[v_i = 0, t(v) = 1 | v_a = 0] \\
= \; & 1 - \Pr[v_i = 0 | t(v) = 1, v_a = 0] \cdot \Pr[t(v) = 1 | v_a = 0] \\
= \; & 1 - \frac{\Pr[v_i = 0, t(v) = 1, v_a = 0]}{\Pr[t(v) = 1, v_a = 0]} \cdot \frac{\Pr[t(v) = 1, v_a = 0]}{\Pr[v_a = 0]} \\
\leq \; & 1 - \frac{|\{v \in f | v_i = 0, t(v) = 1, v_a = 0\}|\zeta/|f|}{|\{v \in f | t(v) = 1, v_a = 0\}|\eta/|f|} \cdot \frac{|\{v \in f | t(v) = 1, v_a = 0\}|\zeta/|f|}{|\{v \in f | v_a = 0\}|\eta/|f|}
\end{aligned}
$$

$$
\begin{aligned}
&= 1-\frac{\zeta^2}{\eta^2}\cdot\frac{|\{v\in f_a|v_i=0,t(v)=1\}|/2^{|a|}}{|\{v\in f_a|t(v)=1\}|/2^{|a|}}\cdot\frac{|\{v\in f_a|t(v)=1\}|/2^{|a|}}{|f_a|/2^{|a|}}\\
&= 1-\frac{\zeta^2}{\eta^2}\cdot\frac{|\{v\in t|v_i=0\}|}{|t|}\cdot\frac{|\{v\in f_a|t(v)=1\}|\eta/(|f|\eta)}{|f_a|\zeta/(|f|\zeta)}\\
&\leq 1-\frac{\zeta^3}{\eta^3}\cdot\frac{1}{2}\cdot\frac{\Pr[t(v)=1,f_a(v)=1]}{\Pr[f_a(v)=1]}\\
&= 1-\frac{\zeta^3}{2\eta^3}\cdot\Pr[t(v)=1|f_a(v)=1]\\
&\leq 1-\frac{\zeta^3}{2\eta^3}\cdot\frac{1}{k}\\
&= 1-\frac{\zeta^3}{2k\eta^3}.
\end{aligned}
$$

Thus, $\mathcal{U}_{\zeta,\eta}\subseteq\mathcal{U}'_{\frac{\zeta}{2^{k-1}\eta},1-\frac{\zeta^3}{2k\eta^3}}$ holds and we have established the theorem. □

Acknowledgments

We would like to thank Eiji Takimoto at faculty of engineering, Tohoku University for his helpful comments and suggestions.

References

[1] Gu, Q. P. and Maruoka, A., Learning Monotone Boolean Functions by Uniformly Distributed Examples, *SIAM Journal on Computing*, to appear.

[2] Kearns, M., Li, M., Pitt, L. and Valiant, L. G., On the Learnability of Boolean Formulae, *In proceedings of the 19th Annual ACM Symposium on Theory of Computing*, 1987, pp.285–295.

[3] Kucera, L., Marchetti-Spaccamela, A. and Protasi, M., On the learnability of DNF formulae, *Lecture Notes in Computer Science*, 1988, 317:347–361.

[4] Ohguro, T. and Maruoka, A., A learning algorithm for monotone k-term DNF, *Proceeding of FUJITSU IIAS-SIS Workshop on Computational Learning Theory*, 1989.

[5] Valiant, L. G., A theory of the learnable, *Communications of the ACM*, 27(11), 1984, pp.1134–1142

Some Improved Sample Complexity Bounds in the Probabilistic PAC Learning Model

Jun-ichi Takeuchi

C&C Information Technology Research Labs., NEC Corp.,
4-1-1 Miyazaki, Miyamaeku, Kawasaki, Kanagawa 216, Japan
e-mail tak@ibl.cl.nec.co.jp

Abstract

Various authors have proposed probabilistic extensions of Valiant's PAC learning model in which the target to be learned is a conditional (or unconditional) probability distribution. In this paper, we improve upon the best known upper bounds on the sample complexity of learning an important class of stochastic rules called 'stochastic rules with finite partitioning' with respect to the classic notion of distance between distributions, the Kullback-Leibler divergence (KL-divergence). In particular, we improve the upper bound of order $O(\frac{1}{\epsilon^2})$ due to Abe, Takeuchi, and Warmuth [2] to a bound of order $O(\frac{1}{\epsilon})$. Our proof technique is interesting for at least two reasons: First, previously known upper bounds with respect to the KL-divergence were obtained using the uniform convergence technique, while our improved upper bound is obtained by taking advantage of the properties of the maximum likelihood estimator. Second, our proof relies on the fact that only a linear number of examples are required in order to distinguish a true parametric model from a bad parametric model. The latter notion is apparently related to the notion of discrimination proposed and studied by Yamanishi, but the exact relationship is yet to be determined.

1 Introduction

Since Valiant's introduction of the PAC learning model [6] for boolean functions, several extensions of the model to the learning of probability distributions were made. Yamanishi [7] and Kearns and Schapire [3] considered the problem of learning stochastic rules (or probabilistic concepts), which is the problem of learning conditional distributions. Abe and Warmuth [1] looked at the sample as well as computational complexity of learning probability distributions using probabilistic automata as hypotheses. Yamanishi called his learning model the 'stochastic PAC learning model'. Here we use the term the 'probabilistic PAC learning model' to refer to these models collectively.

In this paper, we concentrate on the sample complexity aspect of the learning problem for stochastic rules. In particular, we improve upon the best known upper bounds on the sample complexity for learning the important class of stochastic rules called 'the stochastic rules with finite partitioning' (c.f. Yamanishi [7]), *with respect to* the classic notion of distance among distributions, the *Kullback-Leibler divergence* (KL-divergence, for short). Yamanishi proved an upper bound on the sample complexity of learning stochastic rules with finite partitioning, which was linear in $\frac{1}{\epsilon}$ using the property of the MDL estimator. But this result is with respect to a relatively loose notion of distance called the Hellinger distance.[1] With respect to the Kullback-Leibler divergence, a similar upper bound had not been obtained. The best known bound to date was of order $O(\frac{1}{\epsilon^2})$ due to Abe, Takeuchi, and Warmuth [2]. The quadratic dependence on $\frac{1}{\epsilon}$ was due to the fact that Hoeffding's inequality was used to bound the estimation error for

[1] It is well-known that the Kullback-Leibler divergence between two distributions (or stochastic rules) bounds from above the Hellinger distance between the two but the converse does not necessarily hold, even allowing polynomial blow-ups.

frequency estimates, together with a uniform convergence argument. Thus, it was clear that some other proof technique was needed to improve this result.

The main result we obtain is an upper bound on the sample complexity of learning stochastic rules with finite partitioning of order $O(\frac{1}{\epsilon})$. The proof technique used to show this result combines the uniform convergence method and the properties of the maximum likelihood estimator. This is done in a number of steps. First, we show that when given a single parametric model (with finite partitioning), the sample complexity for the parameter estimation with respect to KL-divergence is linear. Here we use a property of KL-divergence and the maximum likelihood estimator (MLE). For example, in the case of binomial distributions[2], the Taylor expansion of KL-divergence around the true parameter(p) is $O(\frac{\Delta^2}{p(1-p)})$. We therefore need only to show that with high probability the MLE is $O((\epsilon p(1-p))^{\frac{1}{2}})$ close to the true parameters in order to show that its KL-divergence with respect to the true distribution is within ϵ. This is possible with a linear sample size in $\frac{1}{\epsilon}$, since the variance of MLE is proportional to $\frac{p(1-p)}{N}$. Secondly, we show that when given a pair of parametric models (with finite partitioning), the true model and one additional model, the sample size required to be able to select the true model is again linear. Here we make use of an evaluation function (mapping each of the two models to a real number), which resembles the (log) likelihood function, but is interestingly different. We then use a uniform convergence argument, which also takes advantage of the properties of the maximum likelihood estimator. In other words, for each of a relatively small, finite set of parameter vector pairs (for the two models), we upper bound the probability that our evaluation function gives a higher score to the worse model, *given* that those two particular vectors were obtained as *maximum likelihood estimates.* This is how we use a uniform convergence method as in [2], and still manage to obtain a sample complexity bound which is linear. Finally, we extend the foregoing argument to the general case in which we have a finite number of models from among which we must select an approximately best model. We apply a general sample complexity upper bound to the specific case of the stochastic decision lists (c.f. [7, 3]). We improve the bound of [2] on this class with respect to ϵ, although the bound is worse with respect to the number of attributes in the domain.

The tight sample complexity bound (linear in $\frac{1}{\epsilon}$ with respect to the KL-divergence) we obtained was made possible by restricting our attention to the class of *stochastic rules with finite partitioning.* This restriction was needed in the second step in the above description of the proof, the process of selecting a true model from a pair of candidate models. The evaluation measure used is the maximum likelihood (possible with a given model), with a modification which resembles the *Laplace estimator.* The Laplace estimator is a modification of the maximum likelihood estimator used often in practice for estimating the parameter of a multinomial distribution. It is obtained by, in effect, adding a single example to the sample belonging to each class[3]. In comparing two distinct parametric models with finite partitioning (which are nothing but the partitions themselves), we add into the input sample, a small number of examples (proportional to the sample size) to each cell of the *refinement* of the two partitions. Using the likelihood function accordingly defined, we are able to select the true model with high probability with a linear number of examples. We then extend this method to the case with any finite number of candidate models. In doing so, we compare each pair of models separately, using each time an evaluation function defined with respect to the refinement of the partitions of that particular pair of stochastic rules. Note that if we insisted on using the same evaluation function on all candidate models by taking the refinement of all the candidate models, it would result in a combinatorial explosion. In this sense, it seems that our proof confirms the importance of the notion of discrimination between *two* models studied in detail by Yamanishi in [8], although the discrimination problem that appeared in our proof represents

[2]In this paper, we let 'binomial distribution' denote a probability distribution over $\{0,1\}$. Also, 'multinomial distribution' denotes a probability distribution over $\{1,\cdots,s\}$ $(s \geq 2)$.

[3]It is also similar to the notion of $\epsilon-$Bayesian shifts in [2] and [1].

only the simplest case.

The rest of the paper is organized as follows. Section 2 gives some necessary definitions for our discussion. In Section 3, we state and prove the sample complexity result for the parameter estimation problem for a *single* parametric model with finite partitioning. In Section 4, we prove the result on the problem of hypothesis testing in a single parametric model. In section 5, we present the upper bounds on learning stochastic rules with finite partitioning, and for the specific class of stochastic decision rules. In Section 6, we mention some open problems.

2 Preliminaries

2.1 The Probabilistic PAC Learning Model

Let X and Y be finite sets. We call X the domain and Y the range. A stochastic rule over the domain X and the range Y is a conditional probability distribution over Y given $x \in X$.

A learning algorithm for stochastic rules is given a training sample drawn from an unknown probability distribution over $X \times Y$. We write the distribution as $p^*(x, y)$. We can rewrite $p^*(x,y)$ as

$$p^*(x,y) = f(x)h^*(y|x). \tag{1}$$

Here $h^*(y|x)$ is the target stochastic rule. Let $\xi = \{(x_1, y_1), (x_2, y_2), \cdots, (x_N, y_N)\}$ denote the input sample, where each $(x_i, y_i) \in X \times Y$ is independently drawn from p^*. We assume that h^* can be represented by a member of some class C of representations of stochastic rules (We call C the target class.). In this paper, we restrict our attention to the case in which $X = \{0,1\}^n, (n = 1, 2, \cdots)$ and $Y = \{0,1\}$, i.e. C is a class of stochastic rule representations over the boolean domain. We let $C^{n,s}$ denote the subclass of C of size at most s, defined over $\{0,1\}^n$. The learning algorithm is assumed to output as hypotheses a member of another class H of stochastic rule representations. In this paper, we assume that $C = H$. We let c denote the "representation mapping", i.e. $c(h)$ is the stochastic rule represented by h.

A leaning algorithm is given as input an accuracy parameter ε, a confidence parameter δ, and a sample ξ, and is to output $h \in H$.

A learning algorithm is said to PAC learn a target class C in terms of a hypothesis space H with respect to a distance measure d, if for arbitrary distribution f over X, for arbitrary $\varepsilon > 0, \delta \in (0,1)$, s, and n, using the above protocol, the algorithm outputs a representation of a stochastic rule $h \in H$ satisfying $d(h^*|c(h)) \leq \varepsilon$ with probability at least $1 - \delta$ whenever $N \geq g(\frac{1}{\varepsilon}, \frac{1}{\delta}, s, n)$($g$ is some polynomial function.). Here, g is an upper bound on the sample complexity of learning C in terms of H. Normally, the learning algorithm's running time is required to be bounded by a polynomial function. But in this paper, we drop this requirement and concentrate on the sample complexity of PAC learning.

We also consider the learning problem for probability distributions. In this case, the target to be learn is an unknown probability distribution p^* over some domain D, where the learning sample $\xi = \{x_1, x_2, \cdots, x_N\}$ is independently drawn from p^*. The notion of PAC learning is defined analogously as in the case of stochastic rules.

We reviewed the definition of KL-divergence for stochastic rules as distance measure.

Definition 1 (KL-divergence of stochastic rules) *Let h^* and h be stochastic rules over domain X and range Y and $f(x)$ a distribution over X. The KL-divergence of h with respect to h^* and f is defined as*

$$d_f(h^*|h) = \sum_{x \in X} f(x) \sum_{y \in Y} h^*(y|x) \log \frac{h^*(y|x)}{h(y|x)}. \tag{2}$$

In this paper, we let log denote the natural logarithm.

We also define the KL-divergence for probability distributions.

Definition 2 (KL-divergence of probability distribution) *Let p^* and p be probability distributions over finite domain D. The KL-divergence of p with respect to p^* is defined as*

$$d(p^*|p) = \sum_{x \in D} p^*(x) \log \frac{p^*(x)}{p(x)}. \tag{3}$$

Note that $d(p^*|p) \geq 0$ and the equality holds if and only if $p^* = p$. Also, the KL-divergence plays a key role in the following fundamental relation characterizing how far one multinomial distribution, p is from another, q: The probability that the most likely sample for p is obtained when actually drawn from q is proportional to $\exp(-Nd(p|q))$.

We can calculate the KL-divergence of $g(x)h(y|x)$ with respect to $f(x)h^*(y|x)$ by considering them as probability distributions over $X \times Y$,

$$d(f \cdot h^*|g \cdot h) = d_f(h^*|h) + d(f|g). \tag{4}$$

Since $d(f|g) \geq 0$, we have

$$d(f \cdot h^*|g \cdot h) \geq d_f(h^*|h). \tag{5}$$

(equality holds if and only if $g = f$). Therefore, having $d(f \cdot h^*|g \cdot h) \leq \varepsilon$ is a sufficient condition for $d_f(h^*|h) \leq \varepsilon$.

We also use the notation $d_2(x|y) \equiv x \log \frac{x}{y} + (1-x) \log \frac{1-x}{1-y}$, which is a real-valued function over $[0,1]^2$.

2.2 Knowledge Representations

In this paper, we consider the class of 'stochastic rules with finite partitioning', defined and studied by Yamanishi [7]. These are stochastic rules representable by a (finite) partition of the domain X, and a (conditional) probability for each cell D^i of the partition (specifying $p^*(1|x)$ for any $x \in D^i$).

Definition 3 (Stochastic Rules with Finite Partitioning) *A finite set $\{D^i | i \in [s]\}$ is called a partition of X if $\bigcup_{i \in [s]} D^i = X$ and $D^i \bigcap D^j = \emptyset (i \neq j)$, where $[s]$ denotes the finite set $\{1, 2, \cdots, s\}$. We call each D^i a cell of the partition. Let $\{p^i | i \in [s]\}$ denote the parameters associated with $\{D^i\}$, where $p^i \in [0,1] (i \in [s])$. We let the pair $h = (\{D^i\}, \{p^i\})$ represent the stochastic rule (conditional probability distribution) $c(h)$ defined by*

$$\forall i \in [s] \ \forall x \in D^i \ \ c(h)(y|x) = \begin{cases} p^i, & \textit{if } y = 1, \\ 1 - p^i, & \textit{otherwise.} \end{cases} \tag{6}$$

Let M denote the class of all finite partitions of X. For arbitrary $m \in M$, define $\Theta(m) = [0,1]^{|m|}$. Let C_{FPX} denote the class of stochastic rules with finite partitioning, i.e. $C_{FPX} = \{c((m, \theta)) | m \in M, \theta \in \Theta(m)\}$.

We also consider the class of representations for probability distributions called histogram.

Definition 4 (Histograms) *Let $\{D^i | i \in [s]\}$ be a partition of a finite domain D. Let $\{p^i | i \in [s]\}$ denote the parameters associated with $\{D^i\}$, where $\sum_{i \in [s]} p^i = 1, \forall i \in [s] \ p^i \geq 0$. We let the pair $p = (\{D^i\}, \{p^i\})$ represent the probability distribution $c(p)$ defined by*

$$\forall i \in [s] \ \forall x \in D^i \ \ c(p)(x) = \frac{p^i}{|D^i|}. \tag{7}$$

Let W denote the class of all finite partitions of D. For arbitrary $w \in W$, define $\Phi(w) = \{(x^1, x^2, \cdots, x^{|w|}) | \sum_{i=1}^{|w|} x^i = 1, \forall i \in [|w|] \ x^i \geq 0\}$. Let C_{HD} denote the class of histograms, i.e. $C_{HD} = \{c((w, \phi)) | w \in W, \phi \in \Phi(w)\}$.

We abuse notation and let $\{D^i\}$ denote the class of histograms $\{(\{D^i\},\phi)|\phi \in \Phi(\{D^i\})\}$ having $\{D^i\}$ as the partition. In this case we refer to $\{D^i\}$ as a *histogram model*. Similarly, we use the term *finite partitioning model* for $\{(m,\theta)|\theta \in \Theta(m)\}$.

Note that if p_1 and p_2 belong to a histogram model $\{D_i\}$, then $d(p_1|p_2)$ can be written as follows,

$$d(p_1|p_2) \;=\; \sum_{x\in X} p_1(x)\log\frac{p_1(x)}{p_2(x)} = \sum_{i=1}^{s}\sum_{x\in D^i}\frac{p_1^i}{|D^i|}\log\frac{\frac{p_1^i}{|D^i|}}{\frac{p_2^i}{|D^i|}} = \sum_{i=1}^{s} p_1^i\log\frac{p_1^i}{p_2^i}. \tag{8}$$

Namely, we can look upon $d(p_1|p_2)$ as the divergence between two multinomial distributions.

3 The Parameter Estimation Problem

In this section, we consider the problem of learning an arbitrary distribution when given a histogram model as hypothesis space H(Note that we make no assumption about the target distribution.). From (8), this problem is the same as the parameter estimation problem for multinomial distribution. The following theorem gives a sample complexity bound for this problem.

Theorem 1 (Convergence Rate of an Estimator for Multinomial Distribution) *Let p^* be an unknown probability distribution over $D = \{1,2,\cdots,s\}$, $\xi = \{x_1,\cdots,x_N\}$ a learning sample independently drawn from p^*, N^i the number of occurrences of $i \in \xi$. For all $\varepsilon \in (0,\frac{1}{2})$,for arbitrary $\alpha \in (0,\frac{1}{4})$, if we define a distribution q by*

$$q(i) = \left(1-\frac{\varepsilon}{2}\right)\frac{N^i}{N} + \frac{\varepsilon}{2s} \quad i \in D, \tag{9}$$

then

$$\Pr\{d(p^*|q) > 4\varepsilon + 6\alpha\varepsilon\} < 2s\exp\left(-\frac{\alpha\varepsilon N}{3s}\right). \tag{10}$$

From this Theorem, if $N \geq O\left(\frac{s}{\varepsilon}\log\frac{s}{\delta}\right)$, it is guaranteed that we can choose $q \in C$ such that $d(p^*|q) \leq \varepsilon$ with probability at least $1-\delta$.

Theorem 1 is derived from the following two Lemmas.

Lemma 1 (Sufficient Condition for Parameter Estimation) *Let p and q be probability distributions over $D = \{0,1\}$ and r an arbitrary member of $(0,\frac{1}{2})$. Define a distribution $p_\varepsilon$over D as follows:*

$$p_\varepsilon(1) = \left(1-\frac{\varepsilon}{2}\right)p(1) + \frac{r\varepsilon}{2}. \tag{11}$$

If

$$d(p_\varepsilon|q) < \alpha\varepsilon r \quad \left(\alpha < \frac{1}{4}\right) \quad \text{and} \quad \frac{r\varepsilon}{2} < q(1) < 1 - \frac{(1-r)\varepsilon}{2}, \tag{12}$$

then

$$d(p|q) < 2\{p(1)(1-r) + (1-p(1))r\}\varepsilon + 6\alpha\varepsilon r. \tag{13}$$

We omit the proof from this abstract.

Lemma 2 (Convergence Rate of an Estimator for Binomial Distribution) *Let p^* be an unknown probability distribution over $D = \{0,1\}$, $\xi = \{x_1, \cdots, x_N\}$ a learning sample independently drawn from p^*, N^+ the number of occurrence of $1 \in \xi$, and r an arbitrary member of $(0, \frac{1}{2})$. Define a distribution p_ϵ over D by*

$$p_\epsilon(1) = \left(1 - \frac{\epsilon}{2}\right) p^*(1) + \frac{r\epsilon}{2}. \tag{14}$$

Define our estimator for p^*, q, as follows:

$$q(1) = \left(1 - \frac{\epsilon}{2}\right) \frac{N^+}{N} + \frac{r\epsilon}{2}. \tag{15}$$

then, for arbitrary $\epsilon \in (0, \frac{1}{2})$, for arbitrary $\alpha \in (0, \frac{1}{4})$,

$$\Pr\{d(p_\epsilon|q) > \alpha\epsilon r\} < 2\exp\left(-\frac{\alpha\epsilon r N}{3}\right), \tag{16}$$

$$\Pr\{d(q|p_\epsilon) > 3\alpha\epsilon r\} < 2\exp\left(-\frac{\alpha\epsilon r N}{3}\right). \tag{17}$$

Now we give the proof of Theorem 1, given Lemma 1 and Lemma 2.

Proof of Theorem 1: For two distributions p and q over D, for each $i \in D$, $d_2(p^i|q^i)$ is the KL-divergence between the binomial distributions over $\{i, D - \{i\}\}$ induced by p and q. Then, from (16) of Lemma 2,

$$\forall i \in D \;\; \Pr\left\{d_2(p_\epsilon(i)|q(i)) > \frac{\alpha\epsilon}{s}\right\} \leq 2\exp\left(-\frac{\alpha\epsilon N}{3s}\right). \tag{18}$$

Hence,

$$\Pr\left\{\exists i \; d_2(p_\epsilon(i)|q(i)) > \frac{\alpha\epsilon}{s}\right\} \leq 2s\exp\left(-\frac{\alpha\epsilon N}{3s}\right). \tag{19}$$

Now suppose for arbitrary $i \in D$ $d_2(p_\epsilon(i)|q(i)) \leq \frac{\alpha\epsilon}{s}$. Then substituting $\frac{1}{s}$ for r in (13) of Lemma 1,

$$\forall i \in D \;\; 2\left\{p(i)(1 - \frac{1}{s}) + (1 - p(i))\frac{1}{s}\right\}\epsilon + \frac{6\alpha\epsilon}{s} \geq d_2(p(i)|q(i)). \tag{20}$$

Summing up for all $i \in D$ we get

$$4\epsilon + 6\alpha\epsilon \geq \sum_{i=1}^{s} p(i)\log\frac{p(i)}{q(i)} + (s-1)\sum_{i=1}^{s}\frac{1-p(i)}{s-1}\log\frac{\frac{1-p(i)}{s-1}}{\frac{1-q(i)}{s-1}} \geq d(p|q). \tag{21}$$

Thus $d(p|q) > 4\epsilon + 6\alpha\epsilon$ implies that $\exists i \; d_2(p_\epsilon(i)|q(i)) > \frac{\alpha\epsilon}{s}$. By (19), this probability is bounded above by $2s\exp\left(-\frac{\alpha\epsilon N}{3s}\right)$. ∎

We omit the proof of Lemma 2 (see [5]) and just note that it can be proved using Bernstein's inequality(Lemma 3, see for example [4]) and other inequalities concerning KL-divergence (Lemma 4).

Lemma 3 (Bernstein's Inequality) *Let X_i's$(i = 1, 2, \cdots, N)$ are independent random variables with 0 means and bounded ranges:$|X_i| \leq M$. Write V_i for the variance of X_i. Suppose $\sum_{i=1}^N V_i \leq NV$. The following inequality holds.*

$$\Pr\left\{\left|\frac{\sum_{i=1}^N X_i}{N}\right| > \beta\right\} \leq 2\exp\left(-\frac{\beta^2 N}{2V + \frac{2}{3}M\beta}\right). \tag{22}$$

Lemma 4 (Inequalities for KL-divergence of binary distributions) *Let κ denote $p-q$. The following inequalities hold.*

$$d_2(p|q) \le \frac{\kappa^2}{p(1-p)}, \quad if\ \frac{1-p}{2} \ge \kappa \ge -\frac{p}{2}, \tag{23}$$
$$d_2(p|q) \le 2|\kappa|, \quad if\ \frac{1-p}{2} \ge \kappa \ge -\frac{p}{2}, \tag{24}$$
$$d_2(p|q) \ge \frac{\kappa^2}{4p(1-p)}, \quad if\ -1.6(1-p) \le \kappa \le 1.6p. \tag{25}$$

We omit the proof from this abstract.

4 The Hypothesis Testing Problem

For learning a histogram (or equivalently a multinomial distribution), with more than one models given as candidate, the following problem of 'hypothesis testing' becomes important. That is, given a multinomial distribution p, a testing sample ξ drawn from a multinomial distribution p^*, can we decide whether p is an ε-good approximation of p^*, i.e. $d(p^*|p) \le \varepsilon$. This problem is similar to the problem of 'discrimination' considered in [8].

We can obtain the following theorem concerning this problem.

Theorem 2 (Hypothesis Testing of Multinomial Distribution) *Let D be a finite set $\{1, 2, \cdots, s\}$, p^* a distribution over D, p a distribution over D satisfying $d(p^*|p) \ge \frac{11\varepsilon}{2}$ and $\frac{\varepsilon}{2s} \le p(i) \le \frac{(1-\frac{1}{s})\varepsilon}{2}$, and ξ a sample of size N independently drawn from p^*. Let N^i denote the number of occurrences of $i \in \xi$. Define a distribution q by,*

$$q(i) = \left(1 - \frac{\varepsilon}{2}\right)\frac{N^i}{N} + \frac{\varepsilon}{2s}. \tag{26}$$

Then, we have

$$\Pr\left\{d(q|p) < \frac{w\varepsilon}{s}\right\} < 2\exp\left(-\frac{b^2\varepsilon N}{3s}\right), \tag{27}$$

where $w < \frac{1}{16}$ and $b = \frac{1-4\sqrt{w}}{3}$.

By this Theorem 2, we can conclude that p is a good model if $d(q|p) < \frac{w\varepsilon}{s}$.

Proof: We can essentially reduce this to a similar statement about binomial distributions. From the premise $d(p^*|p) \ge \frac{11}{2}\varepsilon$,

$$\exists i \in [s]\ \ d_2(p^*(i)|p(i)) \ge 2\left\{p^*(i)(1-\frac{1}{s}) + (1-p^*(i))\frac{1}{s}\right\}\varepsilon + \frac{3\varepsilon}{2s}. \tag{28}$$

Let u be an i which satisfied the above inequality, from Lemma 5 to be stated later substituting $\frac{1}{s}$ for r,

$$\Pr\left\{d_2(q(u)|p(u)) < \frac{w\varepsilon}{s}\right\} < 2\exp\left(-\frac{b^2\varepsilon N}{3s}\right), \tag{29}$$

where $b = \frac{1-4\sqrt{w}}{3}$. Hence, from $d_2(q(u)|p(u)) < d(q|p)$,

$$\Pr\left\{d(q|p) < \frac{w\varepsilon}{s}\right\} < 2\exp\left(-\frac{b^2\varepsilon N}{3s}\right). \tag{30}$$

∎

Lemma 5 (Hypothesis Testing of Binomial Distribution) *Let D be a set $\{0,1\}$, p^* the distribution over D, p the distribution over D satisfying $d(p^*|p) \geq 2\{p^*(1)(1-r)+p^*(0)r\}\varepsilon + \frac{3r\varepsilon}{2}$ $(r \in (0,\frac{1}{2}))$ and $\frac{r\varepsilon}{2} \leq p(1) \leq \frac{(1-r)\varepsilon}{2}$ $(r \leq \frac{1}{2})$, and ξ a sample of size N independently drawn from p^*. Let N^+ denote the number of occurrences of $1 \in \xi$. Define a distribution q by,*

$$q(1) = \left(1 - \frac{\varepsilon}{2}\right)\frac{N^+}{N} + \frac{r\varepsilon}{2}. \tag{31}$$

Then we have

$$\Pr\{d(q|p) < w\varepsilon r\} < 2\exp\left(-\frac{b^2 r\varepsilon N}{3}\right), \tag{32}$$

where $0 < w < \frac{1}{16}$ and $b = \frac{1-4\sqrt{w}}{3}$.

We omit the proof(see [5]) and just note that it can be proved using Bernstein's inequality and Lemma 4.

5 Sample Complexity Bounds

First, we propose PAC learning strategies for learning histograms and stochastic rules with finite partitioning, when given two candidate models. We then generalize these methods to the case with any finite number of candidate models.

Learning Strategy 1 (histogram version)

Let $\{D_1^i | i \in [s_1]\}$ and $\{D_2^i | i \in [s_2]\}$ be histogram models over finite domain D, and ξ a sample. Let N_a^i denote the number of occurrences of $x \in \xi$ belonging to D_a^i $(a = 1, 2)$. Now, let $\{D_3^i | i \in [s_3]\}$ denote the refinement[4] of $\{D_1^i\}$ and $\{D_2^i\}$. We write N_3^i for the number of examples belonging to D_3^i, too. Let k_a^i be the number of cells in $\{D_3^i\}$ included in D_a^i (a=1,2). Define $r_a^i = \frac{k_a^i}{s_3}(a = 1,2)$, $r_3^i = r = \frac{1}{s_3}$. Now, define

$$q_a^i = \left(1 - \frac{\varepsilon}{2}\right)\frac{N_a^i}{N} + \frac{r_a^i\varepsilon}{2}, \quad \textit{and} \quad q_a = (\{D_a^i\}, \{q_a^i\}). \tag{33}$$

Note that

$$q_a^i = \sum_{j \in \{t | D_3^t \subseteq D_a^i\}} q_3^j.$$

Define an evaluation function l by,

$$l(\{D_1^i\}, \{D_2^i\}, \xi) = \sum_{i=1}^{s_1} q_1^i \log\frac{\bar{q}_1^i}{|D_1^i|} - \sum_{i=1}^{s_2} q_2^i \log\frac{\bar{q}_2^i}{|D_2^i|}, \tag{34}$$

where $\bar{q}_a^i$'s are rounded values[5] and let $\bar{q}_a$ denote $(\{D_a^i\}, \{\bar{q}_a^i\})$. Now we define our learning function L as follows,

$$L(\{D_1^i\}, \{D_2^i\}, \xi) = \begin{cases} (\{D_1^i\}, \{\bar{q}_1^i\}), & \text{if } l(\{D_1^i\}, \{D_2^i\}, \xi) \geq 0, \\ (\{D_2^i\}, \{\bar{q}_2^i\}), & \text{otherwise.} \end{cases} \tag{35}$$

[4] Let $\{D_1^i | i \in [s_1]\}$ and $\{D_2^i | i \in [s_2]\}$ be partitions of X. We call $\{D_1^i \bigcap D_2^j | i \in [s_1], j \in [s_2], D_1^i \bigcap D_2^j \neq \emptyset\}$ the refinement of $\{D_1^i\}$ and $\{D_2^i\}$. The size of the refinement, s_3 is not greater than $s_1 s_2$.

[5] In general, we let $\bar{q}$ denote the rounded vlue of q obatined as follows. For each q_a^i, define grid points $\{t^i \in [0,1] | i = \pm 1, \pm 2, \cdots\}$, by $t^i = \frac{1}{2}r_a^i\epsilon$, $t^{k+1} = t^k + (c\epsilon t^k(1-t^k))^{\frac{1}{3}}(s_1 s_2 s_a)^{-\frac{1}{2}}$ $(k = 1, 2, \cdots, u)$, where c is some small constant number, $u = \max_{i \in \mathbb{N}} t^i \leq \frac{1}{2}$, and $t^{-i} = 1 - t^i$. We then define $\bar{q}$ as the nearest grid point to q.

Learning Strategy 2 (stochastic rule version)

Let X be a finite domain and $Y = \{0,1\}$ the range. We write Y_p for $\{\{0\},\{1\}\}$. Let $\{D_1^i | i \in [s_1]\}$ and $\{D_2^i | i \in [s_2]\}$ be partitions over X, and ξ a sample. Let N_a^i denote the number of examples in ξ belonging to $D_a^i \times Y$ $(a = 1,2)$ and N_a^{i+} belonging to $D_a^i \times \{1\}$.

Then, let $\{D_3^i | i \in [s_3]\}$ denote the refinement of $\{D_1^i\}$ and $\{D_2^i\}$. We write N_3^i for the number of examples in ξ belonging to $D_3^i \times Y$, N_3^{i+} belonging to $D_3^i \times \{1\}$. Let k_a^i be the number of cells in $\{D_3^i\}$ included in D_a^i (a=1,2). Define $r_a^i = \frac{k_a^i}{2s_3} (a = 1,2)$, $r_3^i = r = \frac{1}{2s_3}$. Define

$$g_a^i = \left(1 - \frac{\varepsilon}{2}\right) \frac{N_a^i}{N} + r_a^i \varepsilon, \tag{36}$$

$$v_a^{i+} = \left(1 - \frac{\varepsilon}{2}\right) \frac{N_a^{i+}}{N} + \frac{r_a^i \varepsilon}{2}, \tag{37}$$

$$q_a^{i+} = \frac{v_a^{i+}}{g_a^i} = \frac{\left(1 - \frac{\varepsilon}{2}\right) N_a^{i+} + \frac{r_a^i \varepsilon}{2} N}{\left(1 - \frac{\varepsilon}{2}\right) N_a^i + r_a^i \varepsilon N}. \tag{38}$$

Now we define an evaluation function by,

$$\begin{aligned} l_{fp}(\{D_1^i\},\{D_2^i\},\xi) &= \sum_{i=1}^{s_1} g_1^i \left(q_1^{i+} \log \bar{q}_1^{i+} + (1 - q_1^{i+}) \log (1 - \bar{q}_1^{i+})\right) \\ &\quad - \sum_{i=1}^{s_2} g_2^i \left(q_2^{i+} \log \bar{q}_2^{i+} + (1 - q_2^{i+}) \log (1 - \bar{q}_2^{i+})\right), \end{aligned} \tag{39}$$

where $\bar{q}_a^{i+}$'s are rounded values defined as $\bar{q} = \frac{\bar{v}}{\bar{g}}$. Note that

$$l_{fp}(\{D_1^i\},\{D_2^i\},\xi) = l(\{D_1^i\} \times Y_p, \{D_2^i\} \times Y_p, \xi) - l(\{D_1^i \times Y\}, \{D_2^i \times Y\}, \xi). \tag{40}$$

Finally, we define our learning function L_{fp} as follows,

$$L_{fp}(\{D_1^i\},\{D_2^i\},\xi) = \begin{cases} (\{D_1^i\},\{\bar{q}_1^i\}), & \text{if } l_{fp}(\{D_1^i\},\{D_2^i\},\xi) \geq 0, \\ (\{D_2^i\},\{\bar{q}_2^i\}), & \text{otherwise.} \end{cases} \tag{41}$$

The following lemma verifies the validity of the two learning strategies just defined.

Lemma 6 (Comparing Two Models)

1. *Let $\{D_1^i | i \in [s_1]\}$ and $\{D_2^i | i \in [s_2]\}$ be histogram models over finite domain D and ξ a sample of size N drawn from a distribution $c(p^*)(p^* \in \{(\{D_1^i\},\phi) | \phi \in \Phi(\{D_1^i\})\})$. The following inequality holds.*

$$\Pr\left\{d(c(p^*)|c(L(\{D_1^i\},\{D_2^i\},\xi))) > \frac{11\varepsilon}{2}\right\} < 2^{\frac{7}{3}s_3+1} s_1 s_2 \exp\left(-\frac{\varepsilon N}{2^7 3^2 (s_1 s_2)^2}\right). \tag{42}$$

2. *Let $\{E_1^i | i \in [s_1]\}$ and $\{E_2^i | i \in [s_2]\}$ be finite partitioning models over the finite domain X and the range $Y = \{0,1\}$ and ξ a sample of size N drawn from $f(x) \cdot c(h^*)(y|x)$, where $h^* \in \{(\{E_1^i\},\theta) | \theta \in \Theta(\{E_1^i\})\}$ and f is an arbitrary probability distribution over X. The following inequality holds.*

$$\Pr\left\{d_f(c(h^*)|c(L(\{E_1^i\},\{E_2^i\},\xi))) > \frac{11\varepsilon}{2}\right\} < 2^{\frac{7}{3}s_3+3} s_1 s_2 \exp\left(-\frac{\varepsilon N}{2^{11} 3^2 (s_1 s_2)^2}\right). \tag{43}$$

We can easily generalize this result to the case with any finite number of models.

Theorem 3 (Sample Complexity Bound of Learning Histograms) *Let W be a finite class of histogram models over finite domain D. Let $s = \max_{w \in W} |w|$. There exists an algorithm which takes as input a parameter ε and a sample ξ of size N drawn from $c(p^*)$, where p^* is an arbitrary member of $\{(w,\phi)|w \in W, \phi \in \Phi(w)\}$ and output $p = (w,\phi),(w \in W, \phi \in \Phi(w))$ such that for arbitrary $\varepsilon \in [0,\frac{1}{2}]$, for arbitrary $\delta > 0$,*

$$d(c(p^*)|c(p)) < \frac{11\varepsilon}{2} \tag{44}$$

with probability at least $1-\delta$, whenever

$$N > \frac{2^7 3^2 s^4}{\varepsilon}\left(\log\frac{1}{\delta} + \log\left(2^{\frac{7}{3}s+1}s^2|W|\right)\right). \tag{45}$$

Theorem 4 (Sample Complexity Bound of Learning Stochastic Rules) *Let M be a finite class of finite partitioning models over the finite domain X and the range $Y = \{0,1\}$. Let $s = \max_{m \in M} |m|$. There exists an algorithm which takes as input a parameter ε and a sample ξ of size N drawn from $c(h^*)\cdot f(x)$, where h^* is an arbitrary member of $\{(m,\theta)|m \in M, \theta \in \Theta(m)\}$ and f is an arbitrary probability distribution over X, and output $h = (m,\theta),(m \in M, \theta \in \Theta(m))$ such that for arbitrary $\varepsilon \in [0,\frac{1}{2}]$, for arbitrary $\delta > 0$,*

$$d_f(c(h^*)|c(h)) < \frac{11\varepsilon}{2} \tag{46}$$

with probability at least $1-\delta$, whenever

$$N > \frac{2^{11} 3^2 s^4}{\varepsilon}\left(\log\frac{1}{\delta} + \log\left(2^{\frac{7}{3}s+3}s^2|M|\right)\right). \tag{47}$$

In this abstract, we only give a proof sketch of Lemma 6, and omit the proofs of theorems.

Proof sketch of Lemma 6: Here, we give the proof sketch only for the second case (stochastic rules). Let m_a denote $\{E_a^i\}$ $(a = 1,2,3)$, FP_a denote finite partition models $\{(m_a,\theta)|\theta \in \Theta(m_a)\}$, h denote $L_{fp}(\{E_1^i\},\{E_2^i\},\xi)$, and p^* denote $f \cdot c(h^*)$. We also define histogram models (over $X \times Y$) $D_a = \{(\mu_a,\phi)|\phi \in \Phi(\mu_a)\}(a = 1,2,3$ and $\mu_a = \{E_a^i \times Y\})$ and $H_a = \{(\nu_a,\phi)|\phi \in \Phi(\nu_a)\}(a = 1,2,3$ and $\nu_a = \{E_a^i\} \times Y_p)$. Define e to be the member of H_3 in which each cell (in $\{E_3^i\} \times Y_p$) receives the same probability $\frac{1}{2s_3}$. Similarly, define u to be the member of D_3 in which each cell receives the same probability $\frac{1}{s_3}$. Define $p_\varepsilon^* = (1-\frac{\varepsilon}{2})c(p^*) + \frac{\varepsilon}{2}c(e)$, $f_\varepsilon = (1-\frac{\varepsilon}{2})f + \frac{\varepsilon}{2}c(u)$, and $h_\varepsilon^*(y|x) = p_\varepsilon^*(x,y)(f_\varepsilon(x))^{-1}$.

We argue by cases: (1)$h \in FP_1$ and (2)$h \in FP_2$. The case (1) reduced to the parameter estimation problem because FP_1 contains h^*. Hence, $\Pr\{d_f(c(h^*)|c(h)) > \frac{11\varepsilon}{2}\}$ is bounded by $O\left(s_1 \exp\left(-\frac{\varepsilon N}{s_1 s_2}\right)\right)$ from Lemma 1.

So we assume the case (2). We define $H_{2\varepsilon} = \{(1-\frac{\varepsilon}{2})p + \frac{\varepsilon}{2}c(e)|p \in H_2\}$ and $D_{2\varepsilon} = \{(1-\frac{\varepsilon}{2})p + \frac{\varepsilon}{2}c(u)|p \in D_2\}$. Now, we define $g_{2opt} = \arg\min_{p \in D_{2\varepsilon}} d(p_\varepsilon^*|p)$, $v_{2opt} = \arg\min_{p \in H_{2\varepsilon}} d(p_\varepsilon^*|p)$, and $q_{2opt}(y|x) = v_{2opt}(x,y)(g_{2opt}(x))^{-1}$. We let $\{g_{2opt}^i\}$,$\{v_{2opt}^{i+}\}$, and $\{q_{2opt}^{i+}\}$ denote the parameters associated with g_{2opt}, v_{2opt}, and q_{2opt} respectively. There are two cases. (a) For each $i \in [s_2]$, the estimation error for g^i and v^i are less than $((r r_2^i g_{2opt}^i(1-g_{2opt}^i)\varepsilon)^{\frac{1}{2}}$ and $(r r_2^i v_{2opt}^{i+}(1-v_{2opt}^{i+})\varepsilon)^{\frac{1}{2}}$ respectively, and (b) otherwise. The probability that (b) happens can be bounded above by $O\left(s_2 \exp\left(-\frac{\varepsilon N}{s_1 s_2^3}\right)\right)$, similarly to the probability of a parameter estimation failure.

Now, we consider case (a). Note that

$$\begin{aligned}
l_{fp}(\xi) &= d_g(c(q_3)|c(\bar{q}_2)) - d_g(c(q_3)|c(h_\varepsilon^*)) + d_g(c(q_1)|c(h_\varepsilon^*)) - d_g(c(q_1)|c(\bar{q}_1)) \\
&= d(c(q_3 \cdot g_3)|c(\bar{q}_2)\cdot f_\varepsilon) - d(c(q_3 \cdot g_3)|c(h_\varepsilon^*)\cdot f_\varepsilon) + d_g(c(q_1)|c(h_\varepsilon^*)) - d_g(c(q_1)|c(\bar{q}_1)) \\
&= d(c(q_3 \cdot g_3)|c(\bar{q}_2)\cdot f_\varepsilon) - d(c(q_3 \cdot g_3)|p_\varepsilon^*) + d_g(c(q_1)|c(h_\varepsilon^*)) - d_g(c(q_1)|c(\bar{q}_1)).
\end{aligned}$$

Hence, it suffices to show that

$$\Pr\{d(c(q_3 \cdot g_3)|c(\bar{q}_2 \cdot f_\epsilon)) - d(c(q_3 \cdot g_3)|p_\epsilon^*) + d_g(c(q_1)|c(h_\epsilon^*)) - d_g(c(q_1)|c(\bar{q}_1)) < 0\} \tag{48}$$

is small for each grid point $\bar{q}_2^{i+}$ which satisfies $d_f(c(h^*)|c(\bar{q}_2)) \geq \frac{11\epsilon}{2}$ and the condition (a), since the summation of these probabilities bounds the probability that $L_{fp}(\xi)$ is $\frac{11\epsilon}{2}$-bad.

Since $d(c(p^*)|c(\bar{q}_2 \cdot f_\epsilon)) \geq d_f(c(h^*)|c(\bar{q}_2)) \geq \frac{11\epsilon}{2}$, by Theorem 2 (substituting $w = \frac{1}{64}$),

$$\Pr\left\{d(c(q_3 \cdot g_3)|c(\bar{q}_2 \cdot f_\epsilon)) < \frac{\varepsilon}{64 s_3}\right\} \leq O\left(\exp\left(-\frac{\varepsilon N}{s_3}\right)\right). \tag{49}$$

From Lemma 2,

$$\Pr\left\{d(c(q_3 \cdot g_3)|p_\epsilon^*) > \frac{\varepsilon}{128 s_3}\right\} \leq O\left(s_3 \exp\left(-\frac{\varepsilon N}{s_3^2}\right)\right), \tag{50}$$

and from the definition of grid points,

$$d_g(c(q_1)|c(\bar{q}_1)) \leq \frac{c\varepsilon}{s_1 s_2}. \tag{51}$$

Since having $d(c(q_3 \cdot g_3)|c(\bar{q}_2 \cdot f_\epsilon)) > \frac{\epsilon}{64 s_3}$, $d(c(q_3 \cdot g_3)|p^*) \leq \frac{\epsilon}{128 s_3}$, and $d_g(c(q_1)|c(\bar{q}_1)) \leq \frac{\epsilon}{128 s_3}$(now we let $c = 2^{-7}$) is sufficient for $l_{fp}(\xi) \geq 0$, we get for a particular grid point $\bar{q}_a^i$,

$$\Pr\{d(c(q_3 \cdot g_3)|c(\bar{q}_2 \cdot f_\epsilon)) - d(c(q_3 \cdot g_3)|p^*) + d_g(c(q_1)|c(h_\epsilon^*)) - d_g(c(q_1)|c(\bar{q}_1)) < 0\}$$
$$\leq O\left(s_3 \exp\left(-\frac{\varepsilon N}{(s_3)^2}\right)\right).$$

Since the number of grid point satisfying the condition (a) is less than $2(\sqrt{c})^{-s_3} = 2^{\frac{7}{2}s_3+1}$, we obtain the desired bound via a uniform convergence argument. ∎

As an application of Theorem 4, we derive an upper bound on the sample complexity of learning stochastic decision lists(c.f. [7, 3]) . We first review the definition of stochastic decision lists.

Definition 5 (Stochastic Decision Lists $DL^{n,k}$) *A stochastic decision list h representing a stochastic rule with finite partitioning $c(h)$ over the domain $\{0,1\}^n$ and the range $\{0,1\}$, is a list*

$$(t_1, p_1), ..., (t_s, p_s), \tag{52}$$

where each t_i is a conjunction of Boolean literals (except t_s =' true') and each p_i is a member of $[0,1]$. For an assignment x, $c(h)(1|x)$ is defined to be p_j, where j is the least index such that t_j is made true by x, and $h(0|x) = 1 - h(1|x)$. Let $DL^{n,k}$ be the class of all such lists of each t_i has at most k literals.

Note that $c(DL^{n,k})$ is a class of stochastic rules with finite partitioning. We let $M(DL^{n,k})$ denote the class of finite partitioning models such that

$$c(DL^{n,k}) = \left\{c\left((m, \theta(m))\right) | m \in M(DL^{n,k}), \theta \in \Theta(m)\right\}.$$

Each member of $M(DL^{n,k})$ has at most $O(n^k)$ parameters, and $|M(DL^{n,k})| = O(n^k!)$. Therefore we can obtain the following theorem.

Theorem 5 (Sample Complexity Bound of Learning Decision Lists) *The sample complexity of learning $DL^{n,k}$ is upper bounded by*

$$O\left(\frac{n^{4k}}{\varepsilon}\left(\log\frac{1}{\delta} + kn^k \log n\right)\right). \tag{53}$$

This bound is to be contrasted with Yamanishi's bound for the same class with respect to the Hellinger distance [7]: $O\left(\frac{1}{\epsilon}\log\frac{1}{\delta} + \frac{s^*}{\epsilon}\log\frac{n}{\epsilon}\right)$, where s^* is the size of the target rule ($s^* \leq O(n^k)$). Note that our bound compares favorably with Yamanishi's bound with respect to ε, although it is less favorable with respect to n.

6 Conclusions

The improved upper bound presented above was obtained by restricting our attention to the subclass of stochastic rules with *finite partitioning*, and the sample complexity upper bound we obtained was polynomial in the number of cells of the most complex model in the target class. Stochastic decision lists happen to fall into this class, and specifically always have a number of cells which is linear in the size of the rule. The analogous statement is not true of other classes of stochastic rules, such as probabilistic automata (as acceptors) or convex combinations of a finite set of stochastic rules. For example, probabilistic automata have a partition size which is exponential in n, the string length. The proof technique used in [2] applied in these cases as well. It would be interesting, therefore, to see if we can exploit a technique akin to ours presented in the present paper, and obtain similar sample complexity upper bounds for a more general class of stochastic rules, including for example, the other classes mentioned above.

Acknowledgements

The author thanks Prof. David Haussler of UCSC and Dr. Naoki Abe of NEC corporation for teaching him Bernstein's inequality without which he could not have obtained the results in this paper. He also thanks Dr. Abe and Dr. Kenji Yamanishi of NEC corporation for many helpful suggestions and advices. The author would like to thank the anonymous referees for useful comments.

References

[1] Abe, N. & Warmuth, M.(1992). On the computational complexity of approximating distributions by probabilistic automata. *Machine Learning, a special issue for COLT '90,* 9(2/3).

[2] Abe, N., Takeuchi, J., & Warmuth, M.(1991). Polynomial Learnability of Probabilistic Concepts with respect to the Kullback-Leibler Divergence. *Proceedings of the Forth Annual Workshop on Computational Learning Theory* (pp. 277-289), Rochester, NY: Morgan Kaufmann.

[3] Kearns, M. & Schapire, R.(1990). Efficient distribution-free learning of probabilistic concepts. *Proceedings of the 31st Symposium on Foundations of Computer Science*, (PP. 382-391), St. Louis, Missouri.

[4] Pollard, D.(1984).Convergence of Stochastic Processes. (pp. 191-193), Springer Verlag.

[5] Takeuchi, J.(1992).Some Improved Sample Complexity Bounds in the Probabilistic PAC Learning Model. *To appear.*

[6] Valiant, L.G.(1984). A theory of the learnable. *Communications of the ACM, 27,* 1134-1142.

[7] Yamanishi, K.(1992). A learning criterion for stochastic rules. *Machine Learning, a special issue for COLT '90,* 9(2/3).

[8] Yamanishi, K.(1992). Probably Almost Discriminative Learning. *Proceedings of the Fifth Annual Workshop on Computational Learning Theory*, Rochester, NY: Morgan Kaufmann.

AN APPLICATION OF BERNSTEIN POLYNOMIALS IN PAC MODEL

Masahiro Matsuoka
International Institute for Advanced Study of
Social Information Science (IIAS-SIS)
FUJITSU LABORATORIES LTD
140, Miyamoto, Numazu, Shizuoka 410-03, Japan

E-mail : mage%iias.flab.fujitsu.co.jp@uunet.uu.net

Abstract

We study a problem for learning the class of Lipschitz bounded, continuous and real valued functions in terms of Bernstein polynomials in the PAC model [2]. Let f be a Lipschitz bounded continuous function with constant L. We intend to approximate the function f with accuracy ε and confidence δ. By using Bernstein polynomials of degree $n = \lceil (\frac{3L}{\varepsilon})^2 \rceil$, we will construct a polynomial time algorithm which will learn an ε-approximation to the function in probability $1-\delta$ on the uniform distribution over [0, 1]. This algorithm requires a sample of size $\lceil (n+1) \ln (\frac{n+1}{\delta}) \rceil$. This approximate learning is assumed to ideal machine but in practice we have to do the task by using real machine with finite resources. We also consider the robustness of Bernstein polynomial for machine epsilons.

1 Introduction

We study a learning problem for continuous functions on [0, 1] in the PAC model [2] in terms of Bernstein polynomials. Bernstein polynomials [4][5] are well known as an instance of the Weierstrass approximation theorem to continuous functions. Every continuous function on [0, 1] can be uniformly approximated by the polynomials. We describe Bernstein polynomials in the following.

Bernstein polynomials. Let the set of all continuous functions on [0, 1] be denoted by $C[0,1]$. Let $f \in C[0,1]$. We define the (n-th) Bernstein polynomial associated with f by

$$\mathrm{B}_n(f)(x) = \sum_{k=0}^{n} f(\tfrac{k}{n})\binom{n}{k} x^k (1-x)^{n-k}.$$

Then, for any $\varepsilon > 0$, there is a positive integer n_0 such that $|f(x) - \mathrm{B}_n(f)(x)| \leq \varepsilon$ for all $n \geq n_0$ and any $x \in [0, 1]$.

In this paper we use Bernstein polynomials as the class of representations for Lispschitz continuous functions on [0, 1] in the PAC model. We will construct a polynomial time algorithm which will learn an ε-approximation to the function with probability $1-\delta$ on the uniform distribution over [0, 1]. For this our purpose, we modify the above definition in the next section, and obtain the notion of *Pseudo Bernstein polynomials.*

2 Pseudo Bernstein polynomials

We define *Pseudo Bernstein polynomials* and *ε-approximation* as the following.

Pseudo Bernstein polynomials. Let $f \in C[0,1]$. Let a sequence of $n+1$ distinct points of [0, 1] be denoted by $S = \{t_0, t_1, \dots, t_n\}$, where $t_0 < t_1 < \cdots < t_n$. We define a Pseudo Bernstein polynomial associated with f and S by

$$\mathrm{PB}_n(f;S)(x) = \sum_{k=0}^{n} f(t_k)\binom{n}{k}x^k(1-x)^{n-k}.$$

ε-approximation. For any $\varepsilon > 0$, if $\|f - \mathrm{PB}_n(f;S)\|_1 \le \varepsilon$, then $\mathrm{PB}_n(f;S)(x)$ is said to be *ε-approximate* to f with L1-norm which is defined by $\|f-g\|_1 = \int_0^1 |f(x) - g(x)|dx, f, g \in C[0,1]$.

Furthermore, we introduce terminology and notation to clarify our argument.

Lattice points and base points. We define $\ell(n) = \{0, \frac{1}{n}, \frac{2}{n}, \dots, \frac{n-1}{n}, 1\}$ as the sequence of *lattice points* on [0, 1]. We also define $b(n) = \{t_0, t_1, \dots, t_n\}$ as a sequence of *base points* for $\ell(n)$, where $t_k \in b(n)$ satisfies the inequality $\left|\frac{k}{n} - t_k\right| \le \frac{1}{n+1}$ $(k = 0, 1, \dots, n)$.

Lipschitz condition. Let f be a continuous function on [0, 1]. If there is a constant $\mathrm{L} \ge 0$ such that

$$|f(x_1) - f(x_2)| \le \mathrm{L}|x_1 - x_2| \ (x_1, x_2 \in [0,1]),$$

then we say that f satisfies the Lipschitz condition or f is a Lipschitz continuous function. L is said to the Lipschitz bound on f.

Theorem 1 gives a sufficient condition for existence of Pseudo Bernstein polynomials of ε-approximation.

Theorem 1. *Let f be a Lipschitz continuous function on [0, 1] and its Lipschitz bound $\mathrm{L} > 0$. Let $\mathrm{PB}_n(f;S)$ be a Pseudo Bernstein polynomial associated with f and S. For any $\varepsilon > 0$, if $n = \lceil (\frac{3\mathrm{L}}{\varepsilon})^2 \rceil$ and S is a sequence of base points of $\ell(n)$, then $\mathrm{PB}_n(f;S)$ is ε-approximate to f with L1-norm over $C[0,1]$.*

We use the following, Lemma 1 and Lemma 2, in the proof of Theorem 1.

Lemma 1. *Let f be a Lipschitz continuous function on [0, 1] and its Lipschitz bound $\mathrm{L} > 0$. Let $\mathrm{B}_n(f)$ be the (n-th) Bernstein polynomial associated with f. For any $\varepsilon > 0$, if $n \ge (\frac{3\mathrm{L}}{2\varepsilon})^2$, then $\|f - \mathrm{B}_n(f)\|_1 \le \varepsilon$.*

Proof. By Popoviciu's estimation [5], we can bound the convergent rate of Bernstein polynomials for $f \in C[0,1]$ such that $\|f - \mathrm{B}_n(f)\| \le \frac{3}{2}\omega(f, \frac{1}{\sqrt{n}})$. Here $\| \ \|$ denotes sup-norm and let $\omega(f, \frac{1}{\sqrt{n}}) = sup\{|f(x) - f(y)| : |x - y| \le \frac{1}{\sqrt{n}}\}$ $(x, y \in [0,1])$, that is called the modulus of continuity of f. Since L is the Lipschitz bound on f, the value of $\omega(f, \frac{1}{\sqrt{n}})$ is at most $\frac{L}{\sqrt{n}}$. Therefore, if degree of $B_n(f)$ is set to $n \geq \left(\frac{3L}{2\varepsilon}\right)^2$ for any $\varepsilon > 0$, $|f(x) - B_n(f)(x)| \leq \varepsilon$ for all $x \in [0, 1]$. In this case it is clear that

$$\begin{aligned} \|f - B_n(f)\|_1 &= \int_0^1 |f(x) - B_n(f)(x)| dx \\ &\leq \varepsilon \int_0^1 dx = \varepsilon. \end{aligned}$$

Lemma 2. *Let f be a Lipschitz continuous function on $[0, 1]$ and its Lipschitz bound $L > 0$. Let $B_n(f)$ be the (n-th) Bernstein polynomial associated with f. Let S be a sequence of $n+1$ distinct points of $[0, 1]$. Let $PB_n(f; S)$ be a Pseudo Bernstein polynomial associated with f and S. For any $\varepsilon > 0$, if each $t_k \in S$ satisfies the inequality $\left|\frac{k}{n} - t_k\right| \leq \frac{\varepsilon}{L}$ $(k = 0, 1, \ldots, n)$, then $\|B_n(f) - PB_n(f; S)\|_1 \leq \varepsilon$.*

Proof. Each $t_k \in S$ satisfies the inequality $\left|\frac{k}{n} - t_k\right| \leq \frac{\varepsilon}{L}$ $(k = 0, 1, \ldots, n)$. It is easy to see that

$$\begin{aligned} |B_n(f)(x) - PB_n(f; S)(x)| &= \left| \sum_{k=0}^{n} \{f(\frac{k}{n}) - f(t_k)\} \binom{n}{k} x^k (1-x)^{n-k} \right| \\ &\leq \left| \sum_{k=0}^{n} L \left| \frac{k}{n} - t_k \right| \binom{n}{k} x^k (1-x)^{n-k} \right| \\ &\leq \left| \varepsilon \sum_{k=0}^{n} \binom{n}{k} x^k (1-x)^{n-k} \right| = \varepsilon \end{aligned}$$

for all $x \in [0, 1]$. The metric between $B_n(f)$ and $PB_n(f)$ in L1-norm is that

$$\begin{aligned} \|B_n(f)(x) - PB_n(f; S)(x)\|_1 &= \int_0^1 \left| \sum_{k=0}^{n} \{f(\frac{k}{n}) - f(t_k)\} \binom{n}{k} x^k (1-x)^{n-k} \right| dx \\ &\leq \varepsilon \int_0^1 dx = \varepsilon. \end{aligned}$$

This completes the proof of Lemma 2.

Now we give the proof of the Theorem 1.

Proof of Theorem 1. By Lemma 1, we obtain $\|f - B_n(f)\|_1 \leq \frac{\varepsilon}{2}$ for any $\varepsilon > 0$ and degree of this $B_n(f)$ is $n = \lceil \left(\frac{3L}{\varepsilon}\right)^2 \rceil$. By assumption $S = \{t_0, t_1, \ldots, t_n\}$ is a sequence of base points of $\ell(n)$. Then each of them satisfies the inequality $\left|\frac{k}{n} - t_k\right| \leq \frac{3}{2\sqrt{n}}$ $(k = 0, 1, \ldots, n)$. Therefore, by lemma 2, we also get $\|B_n(f) - PB_n(f; S)\|_1 \leq \frac{\varepsilon}{2}$ for any $\varepsilon > 0$. Here it can be used the triangle inequality in L1-norm, we conclude that

$$\begin{aligned} \|f - PB_n(f; S)\|_1 &\leq \|f - B_n(f)\|_1 + \|B_n(f) - PB_n(f; S)\|_1 \\ &\leq \frac{\varepsilon}{2} + \frac{\varepsilon}{2} = \varepsilon. \end{aligned}$$

3 Algorithm PBA

Before we start to composite the approximation algorithm based on these Pseudo Bernstein polynomials, we give the definitions of the sampling oracle EX(f) and a sample of f as the following.

The sampling oracle EX(f) and a sample of f. The *sampling oracle* EX(f) takes a positive integer m and outputs a set $\sigma = \{(t_0, f(t_0)), (t_1, f(t_1)), \dots, (t_{m-1}, f(t_{m-1}))\}$ in time of $O(m)$. We call this set σ a sample of f. For each pair $(t_i, f(t_i))$ of σ, t_i is a random point according to the uniform distribution over [0, 1].

We consider the method of constructing a sequence of base points under the uniform distribution over [0, 1]. It can be given as a Corollary of the Coupon Collector Problem [3], and we will obtain base points with probability.

Coupon Collector Problem [3]. Let $A_1, A_2, \dots, A_\gamma$ be events. Each event is independent and has probability at least η. If the number of trials is $m = \lceil \frac{1}{\eta} \ln(\frac{\gamma}{\delta}) \rceil$, then every events occurred at least once in m independent trials with probability at least $1 - \delta$.

Proof. The probability, which the event A_i, $(1 \le i \le \gamma)$ never occurred in m independent trials, is at most $(1-\eta)^m$. Then the probability that there exist an event never occurred in m independent trials is at most $\gamma(1-\eta)^m$. We bound this probability $\gamma(1-\eta)^m \le \gamma \exp(-m\eta) \le \delta$. Hence, if the number of trials $m = \lceil \frac{1}{\eta} \ln(\frac{\gamma}{\delta}) \rceil$, then every events occurred at least once in m independent trials with probability at least $1 - \delta$.

Corollary 1. *Let $S = \{t_0, t_1, \dots, t_m\}$ be a set of m points generated by independent random selection according to the uniform distribution over* [0, 1]. *If $m = \lceil (n+1) \ln(\frac{n+1}{\delta}) \rceil$, then, for each k $(= 0, 1, \dots, n)$, there is some $t \in S$ such that $|\frac{k}{n} - t| \le \frac{1}{n+1}$ with probability at least $1 - \delta$.*

Proof. For $t \in [0, 1]$ and i $(= 0, 1, \dots, n)$, A_i is an event that $t \in [\frac{i}{n+1}, \frac{i+1}{n+1})$ with probability $\frac{1}{n+1}$. As can be seen from Coupon Collector Problem above, if the number of trials is $m = \lceil (n+1) \ln(\frac{n+1}{\delta}) \rceil$, then every events occurred at least once in m independent trials with probability at least $1 - \delta$.

We describe the main result of this paper as the following.

Theorem 2. *Assume $0 < \varepsilon < 1$ and $0 < \delta < 1$. Let f be a continuous function and satisfies the Lipschitz condition with the Lipschitz bound $L > 0$. If a sample of f is of size $\lceil (n+1) \ln(\frac{n+1}{\delta}) \rceil$, where $n = \lceil (\frac{3L}{\varepsilon})^2 \rceil$, then Algorithm PBA outputs an n-th Pseudo Bernstein polynomial which is ε-approximate to f with probability at least $1 - \delta$. In this case, the Algorithm PBA runs in time polynomial of $\frac{1}{\varepsilon}$, $\frac{1}{\delta}$, and L.*

Algorithm PBA.

Input :

- A positive integer L of the Lipschitz bound on f.
- Positive fractions ε and δ.
- The sampling oracle EX(f).

Output :

- A Pseudo Bernstein polynomial of degree $n = \lceil (\frac{3L}{\varepsilon})^2 \rceil$ or $\emptyset$.

Procedure :

1. $n \leftarrow \lceil (\frac{3L}{\varepsilon})^2 \rceil, \sigma_0 \leftarrow \emptyset, \sigma_1 \leftarrow \emptyset$.
2. Request a sample σ_0 of size m, where
$$m = \lceil (n+1) \ln(\tfrac{n+1}{\delta}) \rceil$$
and
$$\sigma_0 = \{(t_0, f(t_0)), \ldots, (t_{m-1}, f(t_{m-1}))\}.$$
3. For each $I_j = [\frac{j}{n+1}, \frac{j+1}{n+1})$ $(j = 0, \ldots, n)$, pick up a point $(u_j, f(u_j))$ from σ_0 such that $u_j \in I_j$ if possible and construct a sequence $\sigma_1 = \{(u_j, f(u_j))\}$ of size at most $n+1$.
4. If $|\sigma_1| = n+1$, then output the Pseudo Bernstein polynomial
$$PB_n(f; \{u_0, \ldots, u_n\})(x) = \sum_{k=0}^{n} f(u_k) \binom{n}{k} x^k (1-x)^{n-k}$$
and halt.
5. Otherwise output $\emptyset$ and halt.

Proof of Theorem 2. First, we prove that *Algorithm PBA* outputs an ε-approximation to f with probability at least $1-\delta$. By Theorem 1, if $\{u_0, \ldots, u_n\}$ is a sequence of base points, then $\sum_{k=0}^{n} f(u_k)\binom{n}{k} x^k (1-x)^{n-k}$ is ε-approximate to f. $\{t_0, \ldots, t_{m-1}\}$ is a set of points generated by independent random selection according to the uniform distribution over [0, 1]. Assume $n \geq 1$ and $0 < \delta < 1$. By Corollary 1, it is easy to see that the third step of *Algorithm PBA* produces base points $\{u_0, \ldots, u_n\}$ with probability at least $1-\delta$ when it is given a set σ_0 of size $\lceil (n+1) \ln(\frac{n+1}{\delta}) \rceil$.

Secondly, we verify that *Algorithm PBA* is a polynomial time algorithm. We assume that any of arithmetic operations is executed in constant time. It needs time of $O(m)$ to receive the sample σ_0 of size m. In Step 3, for each $t_i \in \sigma_0$ we can easily decide the interval $I_{index(t_i)}$ in which t_i is included by calculating $index(t_i) = \lfloor (n+1)t_i \rfloor$. Thus It costs time of $O(m)$ to select a sequence of base points from the set σ_0. And it is needed time of $O(n)$ to output a Pseudo Bernstein polynomial of degree n. Because n is $\lceil (\frac{3L}{\varepsilon})^2 \rceil$ and m is $\lceil (n+1) \ln(\frac{n+1}{\delta}) \rceil$, so the running time of *Algorithm PBA* is polynomial in $\frac{1}{\varepsilon}$, $\frac{1}{\delta}$, and L.

The evaluation problem in polynomial. When we intend to obtain an approximate value of f for any x in [0, 1], we may consider how many steps are required to calculate the value. We call it the evaluation problem. Let n be degree of the Bernstein polynomial. If it costs steps in size polynomial of n to calculate an approximation value for any x, then it is said that the evaluation problem of the (pseudo) Bernstein polynomial is in polynomial.

Lemma 3. *The evaluation problem is in polynomial of n for any (pseudo) Bernstein polynomial of degree n.*

Proof. We calculate the number of evaluating step of a Pseudo Bernstein polynomial.

$$\mathrm{PB}_n(f;S)(x) = \sum_{k=0}^{n} f(t_k)\binom{n}{k}x^k(1-x)^{n-k}$$

We can evaluate the value in step of order $O(n)$ for each $\binom{n}{k}$ $(k = 0, \ldots, n)$. We store the results $\binom{n}{k}$ $(k = 0, \ldots, n)$, as the table $\{d_k\}_{k=0,\ldots,n}$. We have $\{f(t_k)\}_{k=0,\ldots,n}$ as the table σ_1. We deform the polynomial above as $p(x) = \sum_{k=0}^{n} c_k(x)x^k$, where $c_k = f(t_k)d_k(1-x)^{n-k}$. It costs steps of order $O(n \log n)$ each time when any x is given. After calculating c_k $(k = 0, \ldots, n)$, we evaluate $p(x)$ by Horner's rule [1], $(\cdots(c_n x + c_{n-1})x + \cdots + c_1)x + c_0$, which costs steps of $2n$. Hence, The evaluation problem is in time of $O(n \log n)$ for any (Pseudo) Bernstein polynomial of degree n.

4 Examples of Algorithm PBA

We demonstrate an advantage of Pseudo Bernstein polynomials in approximating to bounded countinuous function as the following examples.

Example 1. A continuous function $f(t)$ has at most K points on $\mathrm{I} = [0, 1]$ and their any neighborhood I_γ where the function is not content itself with the Lipschitz condition. We denote these points by $Q = \{q_1, \ldots, q_K\}$. Let I_γ be $\bigcup_{i=1}^{\mathrm{K}}(q_i - \gamma, q_i + \gamma)$. We approximate such a function $f(t)$ with a Lipschitz continuous function as the following.

$$f(t; Q, \gamma) = \begin{cases} f(t), & \text{if } t \in \mathrm{I} - \mathrm{I}_\gamma; \\ a(q_i, t)t + b(q_i, t), & \text{if } t \in \mathrm{I}_\gamma. \end{cases}$$

where

$$a(q_i, t) = \begin{cases} a_i^-, & \text{if } t \in (q_i - \gamma, q_i); \\ a_i^+, & \text{if } t \in (q_i, q_i + \gamma), \end{cases} \qquad b(p_i, t) = \begin{cases} b_i^-, & \text{if } t \in (q_i - \gamma, q_i); \\ b_i^+, & \text{if } t \in (q_i, q_i + \gamma). \end{cases}$$

For any $\varepsilon > 0$, there are $\gamma > 0$, $a(q_i, t)$ and $b(q_i, t)$ $(i = 1, \ldots, \mathrm{K})$ such that

$$\sum_{i=1}^{\mathrm{K}} \int_{q_i-\gamma}^{q_i+\gamma} |f(t) - (a(q_i, t)t + b(q_i, t))| dt \leq \frac{\varepsilon}{2}.$$

We can bound $\|f - f(Q, \gamma)\|_1$ with γ, K and $\mathrm{M} = sup\{|f(t)| : t \in [0, 1]\}$. If $\gamma \leq \frac{\varepsilon}{8\mathrm{KM}}$, then $\|f - f(Q, \gamma)\|_1 \leq \frac{\varepsilon}{2}$. We bound the Lipschitz bound of $f(t; Q, \gamma)$ by $\mathrm{L}_\gamma = max\{\mathrm{L}_{\mathrm{I}-\mathrm{I}_\gamma}, \mathrm{L}_{\mathrm{I}_\gamma}\}$, where $\mathrm{L}_{\mathrm{I}-\mathrm{I}_\gamma}$ is the Lipschitz bound of $f(t; Q, \gamma)$ on $\mathrm{I} - \mathrm{I}_\gamma$ and $\mathrm{L}_{\mathrm{I}_\gamma} = max\{|a(q_i, t)| : 1 \leq i \leq \mathrm{K}\} \leq \frac{16\mathrm{KM}^2}{\varepsilon}$. For any $\varepsilon > 0$, it can be taken γ enough small to be $\mathrm{L}_\gamma = \mathrm{L}_{\mathrm{I}_\gamma}$, there is a Lipschitz bounded function $f(t; Q, \gamma)$ such that $\|f - f(Q, \gamma)\|_1 \leq \frac{\varepsilon}{2}$. By Theorem 1 if

$n = \lceil \left(\frac{6L_\gamma}{\varepsilon}\right)^2 \rceil$, then $\|f(Q,\gamma) - B_n(f(Q,\gamma))\|_1 \le \frac{\varepsilon}{4}$ and $\|B_n(f(Q,\gamma)) - PB_n(f(Q,\gamma))\|_1 \le \frac{\varepsilon}{4}$. By using the triangle inequality, we get $\|f - PB_n(f(Q,\gamma))\|_1 \le \varepsilon$.

Example 2. Let f be a Lipschitz continuous function on $[0,1]$ and its Lipschitz bound $L > 0$. Let P be a probability distribution on $[0,1]$ with the condition: for any c_1, $c_2 \in [0,1]$, if $c_1 < c_2$, then $\int_{c_1}^{c_2} dP(x) \neq 0$. Let a sample be denoted by σ which is of size m and according to the distribution P^m. For any $\varepsilon > 0$, degree of the Bernstein polynomial is calculated such that $n = \lceil \left(\frac{3L}{\varepsilon}\right)^2 \rceil$. It is supposed that we obtain the sequence of positive numbers $\{\eta_n\}$ as the following.

$$\eta_n = \min_{0 \le k \le n} \int_{I_k} dP(x),$$

where $I_k = [\frac{k}{n+1}, \frac{k+1}{n+1})$. If $m = \lceil \frac{1}{\eta_n} \ln(\frac{n+1}{\delta}) \rceil$, then for any k $(= 0,1,\ldots,n)$ there exits $t_k \in \sigma$ such that $t_k \in I_k$ with probalility at least $1-\delta$. *Algorithm PBA* is capable of getting a sequence of base points from the sample σ. This means that $\sup_{x \in [0,1]} |B_n(f)(x) - PB_n(f;S)(x)| < \frac{\varepsilon}{2}$. Thus we have

$$\int_0^1 |f(x) - PB_n(f;S)(x)|\, dP(x) \le \varepsilon \int_0^1 dP(x) = \varepsilon.$$

Therefore *Algorithm PBA* produces an ε-approximation to f with probability $1-\delta$ from a sample σ of size $m = \lceil \frac{1}{\eta_n} \ln(\frac{n+1}{\delta}) \rceil$ under a fixed, but unknown and non-uniformly distribution P over [0, 1].

If we shall approximate a function f on a machine with finite resources, we must consider the accumulation of rounding errors on the machine, which may be over the expected accuracy. Therefore, we consider to establish robustness in Bernstein polynomials.

5 (Pseudo) Bernstein polynomials have some robustness with rounding error on machine epsilon

Let f be a continuous function defined on X and its range is Y. Here $X = [0,1]$ and $Y = [-M, M]$, M is constant. When we calculate the value of a Bernstein polynomial for any $x \in X$, a rounding error of x and the rounding errors of $f(k/n) \in Y$ $(k = 0,1,\ldots,n)$, they may cause the difference between an approximation value of f and its true value, which is beyond the accuracy ε that we expect. While evaluating the value of a Bernstein polynomial for a given x, we intend to bound the accumulation of the rounding errors within the accuracy ε. First, we state the meaning of machine epsilons, ε_x and ε_y, in this section. They express ability of resolution which the machine equips for X and Y. If a difference between u and v or $f(u)$ and $f(v)$ is less than ε_x or ε_y , i.e., that $|u - v| < \varepsilon_x$ $(u, v \in X)$, $|f(u) - f(v)| < \varepsilon_y$ $(f(u), f(v) \in Y)$, then they are seem to be exactly same of value on the machine. Assume that : they are binary-coded in finite length, exactly expressed themselves, and have no error for a given x and the values of $f(k/n)$ $(k = 0,1,\ldots,n)$. If we evaluate the value of the polynomial as we prevent rounding errors which are caused by lack of figures in the process of calculation, in appearance, then there is no accumulation of rounding errors in the result of calculation. Secondly, we decide an upper bound of the accumulation of errors, and we provide robustness by estimating degree of the Bernstein polynomial by the next Lemma 4. It will become larger degree of Bernstein polynomial to achieve the accuracy of approximation.

Lemma 4. *Let f be a continuous, Lipschitz bounded function with* L. *For any $\varepsilon > 0$, which is accuracy of approximation to f, if machine epsilon ε_x of domain X is less than* $\log\left(\frac{\varepsilon}{4\mathrm{L}}\right)$, *$\varepsilon_y$ that of range Y is less than* $\log\left(\frac{\varepsilon}{4}\right)$, *and degree of a Bernstein polynomial is greater than* $\left(\frac{12\mathrm{L}}{\varepsilon}\right)^2$, *then we can establish the bound of total error in a Pseudo Bernstein polynomial within ε.*

Proof. Let f be a Lipschitz bounded function with L. For any $u, v \in [0,1]$ and all Bernstein polynomials of degree n, the difference between values of the Bernstein polynomial is bounded that

$$\begin{aligned}
|\mathrm{B}_n(f)(u) - \mathrm{B}_n(f)(v)| &\le |f(u) - \mathrm{B}_n(f)(u)| + |f(v) - \mathrm{B}_n(f)(v)| + |f(u) - f(v)| \\
&\le 2\alpha + \mathrm{L}|u - v| \\
&\le 2\alpha + \mathrm{L}\varepsilon_x
\end{aligned}$$

where $\alpha \ge sup\{|f(u) - \mathrm{B}_n(f)(u)| : u \in [0,1]\}$ and $\varepsilon_x \ge |u - v|$. If $|f(t_k) - g(t_k)| \le \varepsilon_y (k = 0, 1, \ldots, n)$, then the difference between $\mathrm{PB}_n(f; S)$ and $\mathrm{PB}_n(g; S)$ for any $u \in [0, 1]$ is that

$$\begin{aligned}
|\mathrm{PB}_n(f;S)(u) - \mathrm{PB}_n(g;S)(u)| &= \left|\sum_{k=0}^{n}\{f(t_k) - g(t_k)\}\binom{n}{k}u^k(1-u)^{n-k}\right| \\
&\le \left|\sum_{k=0}^{n}\varepsilon_y\binom{n}{k}u^k(1-u)^{n-k}\right| = \varepsilon_y.
\end{aligned}$$

For any $\alpha > 0$ and any continuous f of Lipschitz bounded, there exists a positive integer n such that $|f(u) - \mathrm{B}_n(f)(u)| \le \alpha$ $(u \in [0,1])$. And there is a sequence of base points S such that $|\mathrm{B}_n(f)(u) - \mathrm{PB}_n(f;S)(u)| \le \alpha$ $(u \in [0,1])$. So, by using above four inequarities, we can bound that

$$\begin{aligned}
|f(u) - \mathrm{PB}_n(g)(v)| &\le |f(u) - \mathrm{B}_n(f)(u)| + |\mathrm{B}_n(f)(u) - \mathrm{B}_n(f)(v)| \\
&\quad + |\mathrm{B}_n(f)(v) - \mathrm{PB}_n(f;S)(v)| + |\mathrm{PB}_n(f;S)(v) - \mathrm{PB}_n(g;S)(v)| \\
&\le 4\alpha + \mathrm{L}\varepsilon_x + \varepsilon_y.
\end{aligned}$$

We regard ε_x, ε_y as machine epsilon of domain X and range Y. We also have the bound $\alpha \le \frac{3\mathrm{L}}{2\sqrt{n}}$ from Popoviciu's estimation to the modulus of continuity of f. In order to provide robustness for Pseudo Bernstein polynomials, we show that $\frac{6L}{\sqrt{n}} + L\varepsilon_x + \varepsilon_y \leq \varepsilon$ for $\varepsilon > 0$ that is the accurancy of approxitmation to f. Therefore, $\frac{\varepsilon}{4} \geq L\varepsilon_x$, $\frac{\varepsilon}{4} \geq \varepsilon_y$, and degree of Bernstein polynomials is set to $n \geq \left(\frac{12L}{\varepsilon}\right)^2$, then Pseudo Bernstein polynomials can be provided its robustness for the rounding errors on machine epsilons. Hence, for any $\varepsilon > 0$, we confirm the robustness to let machine epsilons of domain ε_x and range ε_y to 2^{m_x} and 2^{m_y} , where $m_x = \lfloor \log(\frac{\varepsilon}{4L}) \rfloor$ and $m_y = \lfloor \log(\frac{\varepsilon}{4}) \rfloor$ respectively.

6 Conclusion

The *Algorithm PBA* views the domain of functions by the lattice points $\{0, \frac{1}{n}, \ldots, \frac{n-1}{n}, 1\}$ and look around the whole to obtain the information of the target function as the sequence $\{f(0), f(\frac{1}{n}), \ldots, f(\frac{n-1}{n}), f(1)\}$. For $\varepsilon > 0$ and the Lipschitz bound $L > 0$ on the functions, it is that $n = \lceil (\frac{3L}{\varepsilon})^2 \rceil$. The *Algorithm PBA* requires a sample of size $\lceil (n+1) \ln \left(\frac{n+1}{\delta}\right) \rceil$, and produces a Pseudo Bernstein polynomial which is ε-approximate to f with probability at least $1 - \delta$. The size of the polynomial is $O((\frac{1}{\varepsilon})^2)$. This says that the speed of convergence of Bernstein polynomials is not very fast. But Bernstein polynomials have the robustness for errors stemed from machine epsilons. If an essence of approximate learning is interpolating for a sample, then our choice have some sort of validity in the context of approximate learning.

As a further subject we will examine an approximation algorithm with linear approximation, such as the interpolating spline methods [6], under the condition that a sample is according to non-uniformly distribusion which has continuous density function.

References

[1] Aho, A. V., Hopcroft, J. E. and Ullman, J. D., *The Desigin and Analysis of Computer Algorithms* , Addison-Wesley, 1975.

[2] Blumer, A., Ehrenfeucht, A., Haussler, D. and Warmuth, M. K., Learnability and the Vapnik-Chervonenkis dimension, *Journal of Association for Computing Machinery*, Vol. 36, 1989, pp. 929-965.

[3] Benedek, G. M. and Itai, A., Learnability with respect to fixed distributions, *Theoretical Computer Science*, Vol. 86, 1991, pp. 377-389.

[4] Feinerman, R. P. and Newman, D. J., *Polynomial Approximation*, The Williams & Wilkins, 1974.

[5] Takenouchi, Y. and Nishishiraho, T., *Theory of Approximation*, Bai-fu-kan, 1985 (In Japanese).

[6] Wahba, G., Interpolating spline methods for density estimation I. equi-spaced knots. *The Annals of Statistics*,Vol. 3, No. 1, 1975, pp. 30-48.

ON PAC LEARNABILITY OF FUNCTIONAL DEPENDENCIES

Tatsuya Akutsu
Mechanical Engineering Laboratory
1-2 Namiki, Tsukuba, Ibaraki 305, Japan

Atsuhiro Takasu
National Center for Science Information Systems
3-29-1 Otsuka, Bunkyo, Tokyo 112, Japan

Abstract

This paper presents a PAC learning framework for inferring a set of functional dependencies from example tuples. A simple algorithm which outputs a set of all the functional dependencies consistent with example tuples is considered. Let T be a set of the whole tuples. The error is defined as the minimum $\sum_{t \in V} p(t)$ where V is a set such that a set of functional dependencies is consistent with $T - V$ and $p(t)$ denotes the probability of a tuple t. The number of examples required to infer a set of functional dependencies whose error does not exceed ε with probability at least $1 - \delta$ under an arbitrary probability distribution is shown to be $O(\frac{\sqrt{\ln \frac{1}{\delta}}}{\varepsilon}\sqrt{|T|})$.

1 Introduction

The concept of the functional dependency is very important and very familiar in the field of relational databases [3, 9]. In the relational database, data is represented with a set of relations each of which is a N-ary table. Each row of the table is called a tuple that consists of N components and each column is referred to as an attribute. The relation scheme is defined as a list of attributes and represent a structure of the table like the record format in a file. Various constraints are used to design and maintain the relational database. Among them, the functional dependency is very important constraint. It means that for any relation with the same relation scheme, values of a set of attributes functionally determines ones of another set of attributes. Functional dependencies are useful for reducing the disk space and checking the illegal manipulation of databases.

Functional dependencies are discussed mainly in the deductive way so far, that is, when dependencies are known, how to utilize them. However, from a practical viewpoint, there are many cases where dependencies are not known. Moreover, only a part of the data is usually stored in the database system. In such a case, a method to extract dependencies embedded in the data is very useful. So, this paper discusses the problem of inferring all the functional dependencies consistent with the whole tuples from a part of tuples. Of course, it is impossible to find all the functional dependencies correctly from a part of tuples. However, there is a possibility to find functional dependencies probably approximately correctly. Approximately correct functional dependencies would be useful if a subset of tuples which violates the functional dependencies is handled as an exception.

A lot of studies have been done for learning rules in database systems [4, 6, 7, 8]. Honeymann studied a method for testing the consistency of functional dependencies [4]. However, in his framework, all the data are assumed to be known. Quinlan and Rivest studied a method for inferring decision trees from tuples [7]. Piatetsky-Shapiro studied a method for deriving rules from a part of data [6]. The forms of rules are limited to $Cond(t) \rightarrow (t[A_i] = b)$ and $Cond(t) \rightarrow (b_1 \leq t[A_i] \leq b_2)$, where $t[A_i]$ denotes the ith attribute value of a tuple t. Sonoo et al. studied another method for deriving rules from a part of data [8]. The form of rules is limited to $(t[A_i] = a) \rightarrow (t[A_j] = b_1 \vee b_2 \vee \cdots \vee b_k)$. Functional dependencies can not be described in either form. As far as we know, there is no work for learning a set of functional dependencies from example tuples.

By the way, the framework of PAC (Probably Approximately Correct) learning was proposed by Valiant [10] and a lot of studies have been done [2, 5]. The framework seems to be applicable for functional dependencies. However, the framework can not be applied directly since there is a following problem:

- Whether a functional dependency is consistent or not is determined not for one tuple, but for a set of tuples. That is, positive or negative is not defined for one example.

Therefore, we developed a PAC learning framework for functional dependencies. In this paper, we do not focus on the time complexity or the polynomiality, but focus on the number of examples. We consider a simple algorithm which outputs a set of all the functional dependencies consistent with example tuples. We show that the number of examples required to infer a set of functional dependencies whose error does not exceed ε with probability at least $1 - \delta$ under an arbitrary and unknown probability distribution is $O(\frac{\sqrt{\ln \frac{1}{\delta}}}{\varepsilon}\sqrt{n})$ where n is the number of the whole tuples.

2 An Algorithm to Test a Functional Dependency

To learn a set of functional dependencies, we use a simple algorithm as most PAC learning algorithms do [2, 5, 10]. That is, we derive all the functional dependencies consistent with example tuples. In section 2 and 3, we consider the case of testing one functional dependency. In section 4, we consider the case of deriving a set of functional dependencies.

2.1 Functional Dependency

To begin with, we overview the concept of the functional dependency [9]. In this paper, we fix the domain of the relation to $D_1 \times D_2 \times \cdots \times D_N$ where every D_i is a (sufficiently large) finite set. An element of $D_1 \times D_2 \times \cdots \times D_N$ is called as a tuple. A_i denotes the

ith attribute and $t[A_i]$ denotes the ith attribute value of a tuple t. $t[A_{i_1}A_{i_2}\cdots A_{i_k}]$ denotes a subtuple $(t[A_{i_1}], t[A_{i_2}], \cdots, t[A_{i_k}])$. A relation T is defined as a subset of $D_1 \times D_2 \times \cdots \times D_N$.

[**Definition 1**] A functional dependency $A_{i_1}A_{i_2}\cdots A_{i_p} \rightarrow A_q$ is consistent with a set of tuples S if, for every $t \in S$ and $s \in S$, $(1 \leq \forall h \leq p)(t[A_{i_h}] = s[A_{i_h}]) \implies t[A_q] = s[A_q]$ holds.

Of course, a functional dependency can be defined as $A_{i_1}A_{i_2}\cdots A_{i_p} \rightarrow A_{j_1}A_{j_2}\cdots A_{j_q}$. However, we consider only the form of Definition 1 since

$$(\forall j_k)(A_{i_1}A_{i_2}\cdots A_{i_p} \rightarrow A_{j_k}) \iff A_{i_1}A_{i_2}\cdots A_{i_p} \rightarrow A_{j_1}A_{j_2}\cdots A_{j_q}$$

holds. In section 2 and 3, we fix a functional dependency to $A_{i_1}A_{i_2}\cdots A_{i_p} \rightarrow A_q$ without loss of generality. F denotes $A_{i_1}A_{i_2}\cdots A_{i_p} \rightarrow A_q$. Moreover, for a tuple t, $L(t)$ and $R(t)$ denote $t[A_{i_1}A_{i_2}\cdots A_{i_p}]$ and $t[A_q]$, respectively.

Note that the consistency of a functional dependency is defined not for a tuple, but for a set of tuples. For example, consider a set of tuples $\{(a,1,p),(a,2,q),(b,1,p)\} \subset D_1 \times D_2 \times D_3$ and a functional dependency $A_1 \rightarrow A_2$. In this case, it is meaningless to discuss about whether a tuple $(a,1,p)$ (or $(a,2,q)$ or $(b,1,p)$) is consistent with $A_1 \rightarrow A_2$ or not. However, it is meaningful to discuss about whether a subset of tuples is consistent with $A_1 \rightarrow A_2$ or not. For example, $\{(a,1,p),(b,1,p)\}$ is consistent and $\{(a,1,p),(a,2,q)\}$ is inconsistent. Since the conventional PAC learning framework is based on the property that the consistency of a concept is determined for each example, it can not be applied for our case. Moreover, the error can not be defined in a usual way since positive or negative is not defined for each tuple. Thus, the conventional PAC learning framework can not be applied directly for functional dependencies.

2.2 An Algorithm

The following algorithm (Algorithm A1) tests the consistency of a functional dependency F with S ($|S| = m$). In this paper, a set of example tuples S is a multi set, that is, S may contains more than one identical tuples.

Procedure $TestFD(S, F)$
  begin
    for all tuples $t \in S$ **do**
      if there is a tuple $s \in S$ such that $(L(t) = L(s) \wedge R(t) \neq R(s))$ **then**
        return FALSE;
    return TRUE
  end

There is an efficient implementation of Algorithm A1. If tuples are sorted according to $A_{i_1}A_{i_2}\cdots A_{i_p}A_q$ and N can be considered as a constant, the test can be done in $O(m \log m)$ time including the time for sorting. Of course, any algorithm to test the consistency of a functional dependency can be used in place of Algorithm A1.

3 PAC Learnability of a Functional Dependency

In this section, we consider the number of example tuples for testing F whose error is at least ε with the probability at least $1-\delta$. For that purpose, we consider the probability that F is consistent with example tuples although F is inconsistent with the whole tuples (i.e. all the tuples in the relation). That is, we consider the probability that Algorithm A1 returns TRUE although F is inconsistent with the whole tuples. In such a case, we say that Algorithm A1 makes error. Note that error is one-sided since Algorithm A1 never return FALSE when F is consistent with the whole tuples.

3.1 Preliminaries

[Definition 2] For a set of tuples U, $vs(U,F)$ denotes a set $\{U_1, U_2, \cdots, U_r\}$ which satisfies the following conditions:

- $(\forall U_i)(\forall s \in U_i)(s \in U)$,
- $(\forall U_i)(\forall t \in U_i)(\forall s \in U_i)(L(t) = L(s))$,
- $(\forall i)(\forall j)(i \neq j \Longrightarrow U_i \cap U_j = \emptyset)$,
- $(\forall U_i)(\exists t \in U_i)(\exists s \in U_i)(R(t) \neq R(s))$,
- Every U_i is a maximal set.

[Definition 3] For a set of tuples U, a set of violation pairs $pairs(U,F)$ is defined as

$$pairs(U,F) \equiv \{\ U_i \mid |U_i| = 2 \wedge U_i \in vs(U,F)\ \}\ .$$

Hereafter, T ($|T| = n$) denotes a set of all the tuples in the relation and $S \subset T$ ($|S| = m$) denotes a multi set of example tuples in T. For each tuple $t \in T$, $p(t)$ denotes the probability that t appears when we pick an element of T as an example.

[Definition 4] For a set of tuples T, a probability distribution D and a functional dependency F, the error $\varepsilon(T,D,F)$ is defined as $\sum_{t\in V} p(t)$ where V satisfies the following conditions:

(1) F is consistent with $T - V$,

(2) $\sum_{t\in V} p(t)$ is the minimum among such sets that satisfy the condition (1).

We will explain the reason why we define the error as above. The error in Valiant's PAC model is defined as a symmetric difference between a target concept C and a hypothesis H. In other word, the error is defined as $\sum p(t)$ where t is an element such that $(t \in C \wedge t \notin H) \vee (t \notin C \wedge t \in H)$ holds. If such elements are removed, H coincides with C. In the case of this paper, a set of tuples which should be removed is not determined uniquely. Any set V such that F is consistent with $T - V$ seems to be appropriate. However, for example, we consider a functional dependency $A_1 \rightarrow A_2$, a relation $T = \{(a,1,p),(a,2,q),(b,1,p)\}$ and the corresponding probabilities $(0.8, 0.1, 0.1)$. In this case, both $\{(a,1,p)\}$ and $\{(a,2,q)\}$ satisfy the condition. However, it is reasonable to consider $\{(a,2,q)\}$ as V since $p((a,1,p)) = 0.8$ and $p((a,2,q)) = 0.1$. Therefore, the error defined in Definition 4 seems reasonable.

[Definition 5] $P(m,T,D,F)$ denotes the probability that F is inconsistent with S where a (multi) set of example tuples S ($|S| = m$) is given according to D. Moreover, $Q(m,T,D,F)$ denotes $1 - P(m,T,D,F)$.

Note that the probability that Algorithm A1 makes error is $Q(m,T,D,F)$.

[Example] In this example, we consider a domain of the relation $D_1 \times D_2 \times D_3$ and a functional dependency $A_1 \rightarrow A_2$. Let T be

$$\{(a,1,p),(a,1,q),(a,2,p),(a,2,r),(a,3,p),(b,1,p),(b,2,q),(c,2,p),(c,3,p),(d,1,p),(d,1,q)\}$$

and a distribution D be $(0.2, 0.1, 0.1, 0.1, 0.1, 0.1, 0.1, 0.04, 0.06, 0.05, 0.05)$. Then, $vs(T,F) = \{\ \{(a,1,p),(a,1,q),(a,2,p),(a,2,r),(a,3,p)\}, \{(b,1,p),(b,2,q)\}, \{(c,2,p),(c,3,p)\}\ \}$ and $pairs(T,F) = \{\ \{(b,1,p),(b,2,q)\},\ \{(c,2,p),(c,3,p)\}\ \}$. Moreover, $\varepsilon(T,D,F) = 0.44$ for $V = \{(a,2,p),(a,2,r),(a,3,p),(b,1,p),(c,2,p)\}$ or $V = \{(a,2,p),(a,2,r),(a,3,p),(b,2,q),(c,2,p)\}$.

3.2 Upper Bound of the Number of Examples

In this subsection, we show an upper bound of the number of example tuples for testing F such that $\varepsilon(T,D,F) \geq \varepsilon$ ($\varepsilon > 0$) with the probability at least $1 - \delta$. For that purpose, we derive an upper bound of $Q(m,T,D,F)$. First, we show that we may consider only the simple case by using several propositions and lemmas.

[Proposition 1] For a fixed T, $P(m,T,D,F)$ depends on only the probabilities of the tuples in $vs(T,F)$. □

[Lemma 1] For any (T,D), there exists a (T',D') which satisfies the following conditions:

- $P(m,T,D,F) \geq P(m,T',D',F)$,
- $|T'| \leq 2|T| - 1$,
- $\frac{1}{2}\varepsilon(T,D,F) \leq \varepsilon(T',D',F)$,
- $vs(T',F) = pairs(T',F)$,
- $(\forall U_i \in vs(T',F))(\forall s,t \in U_i)(p(s) = p(t))$.

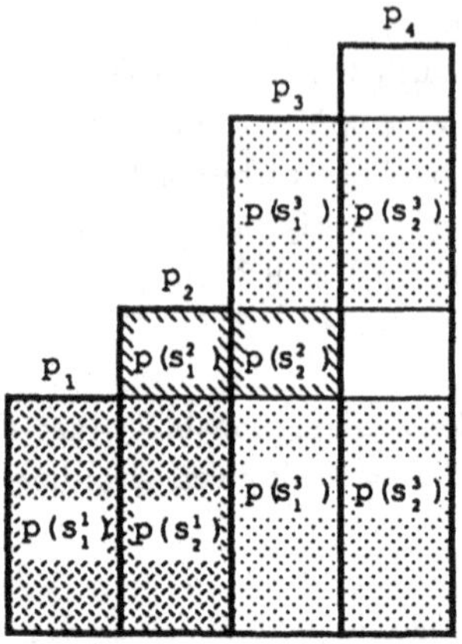

Figure 1: Construction of s^i_j in Lemma 1.

(Proof) Let $\{t^1_1, \cdots, t^1_{n_1}, t^2_1, \cdots, t^2_{n_2}, \cdots, t^k_1, \cdots, t^k_{n_k}\}$ be an arbitrary element of $vs(T, F)$. We assume w.l.o.g. (without loss of generality) that $(\forall p, q)(R(t^i_p) = R(t^i_q))$ and $(\forall i, j, p, q)(i \neq j \rightarrow R(t^i_p) \neq R(t^j_q))$ hold.

Let $p_i = \sum_{j=1}^{n_i} p(t^i_j)$. We assume w.l.o.g. $p_1 \leq p_2 \leq \cdots \leq p_k$. We construct (T', D') such that T' includes a set of tuples $\{s^1_1, s^1_2, s^2_1, s^2_2, \cdots, s^{k-1}_1, s^{k-1}_2\}$ and the following conditions are satisfied (see also Fig.1):

- $(\forall i)(L(s^i_1) = L(s^i_2) \wedge R(s^i_1) \neq R(s^i_2))$,
- $(\forall s^i_1)(\forall t \in T')((t \neq s^i_1 \vee t \neq s^i_2) \rightarrow L(t) \neq L(s^i_1))$,
- $p(s^1_1) = p(s^1_2) = p_1$,
- $(\forall i \geq 1)(p(s^{i+1}_1) = p(s^{i+1}_2) = p_{i+1} - p(s^i_1))$.

It is easy to see that such (T', D') satisfies the conditions of the lemma. □

By Lemma 1, we consider only the case where $vs(T, F) = \{\{t_1, t'_1\}, \cdots, \{t_k, t'_k\}\}$ and $(\forall t_i)(p(t_i) = p(t'_i))$ hold until Theorem 1. We assume w.l.o.g. that $(\forall i)(p(t_i) \leq p(t_{i+1}))$ holds. Note that $\varepsilon(T, D, F) = \sum_{i=1}^{k} p(t_i)$ holds. We assume that $n = |T|$ is an odd number until Theorem 1. The similar results holds in the case where n is even since this case can be handled by adding a dummy tuple of the probability 0.

[Proposition 2] Let D be a distribution such that $p(t_1) \leq p(t_2) \leq \cdots \leq p(t_k)$ and $\sum_{i=1}^{k} p(t_i) = \varepsilon$ hold. Let D' be a distribution such that $p(t_1) = p(t_2) = \cdots = p(t_k)$ and

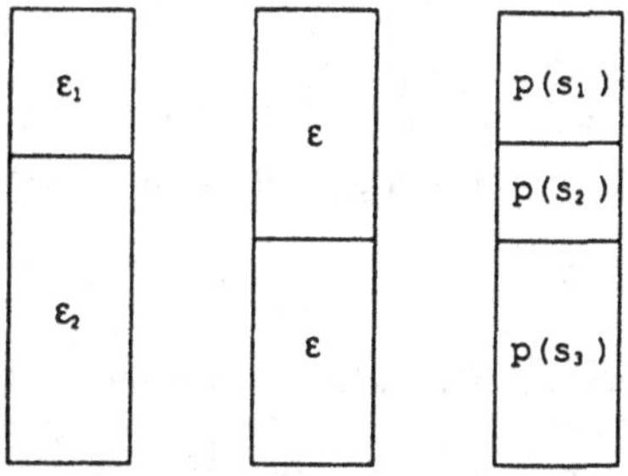

Figure 2: Construction of s_i in Proposition 2.

$\sum_{i=1}^{k} p(t_i) = \varepsilon$ hold. Then, $P(m,T,D,F) \geq P(m,T,D',F)$ holds. Moreover, the equality holds only if $p(t_1) = p(t_2) = \cdots = p(t_k)$ holds under D.

(Proof) First, we consider the case for $k = 2$ (see Fig.2). Let $\varepsilon_1 = p(t_1)$ and $\varepsilon_2 = p(t_2)$ under D. Moreover, we assume $\varepsilon_1 < \varepsilon_2$. Let $\varepsilon = p(t_1) = p(t_2)$ under D'. Note that $\varepsilon_1 + \varepsilon_2 = 2\varepsilon$ holds. In place of $\{\{t_1, t_1'\}, \{t_2, t_2'\}\}$, we consider a set of tuples $\{\{s_1, s_1'\}, \{s_2, s_2'\}, \{s_3, s_3'\}\}$ such that $p(s_1) = p(s_1') = \varepsilon_1$, $p(s_2) = p(s_2') = \varepsilon - \varepsilon_1$ and $p(s_3) = p(s_3') = \varepsilon$. A set of pairs $\{\{s_2, s_3'\}, \{s_2', s_3\}, \{s_1, s_1'\}, \{s_2, s_2'\}, \{s_3, s_3'\}\}$ corresponds to a set of violation pairs for (T, D). A set of pairs $\{\{s_2, s_1'\}, \{s_2', s_1\}, \{s_1, s_1'\}, \{s_2, s_2'\}, \{s_3, s_3'\}\}$ corresponds to a set of violation pairs for (T, D'). Since $p(s_1) < p(s_3)$ and $p(s_1') < p(s_3')$ holds, $P(m,T,D,F) > P(m,T,D',F)$ holds.

Next, we consider the general case ($k > 2$). If $p(t_i) < p(t_{i+1})$ holds for some i under D, $P(m,T,D,F)$ can be reduced by picking two pairs $\{t_i, t_i'\}$, $\{t_{i+1}, t_{i+1}'\}$ and applying the proposition for $k = 2$. Therefore, the proposition is proved. □

By Proposition 2, $P(m,T,D,F)$ with $\varepsilon(T,D,F) = \varepsilon$ for a fixed ε is the minimum when there are $\frac{n-1}{2}$ violation pairs and the probability of each tuple in violation pairs is $\frac{2\varepsilon}{n-1}$. $P(m,k,\hat{\varepsilon},F)$ denotes $P(m,T,D,F)$ such that there are k violation pairs where $\hat{\varepsilon} = p(t_i) = p(t_i') = \frac{\varepsilon(T,D,F)}{k}$. Note that $P(m,k,\hat{\varepsilon},F)$ does not depend on n (from Proposition 1). $Q(m,k,\hat{\varepsilon},F)$ is defined as $1 - P(m,k,\hat{\varepsilon},F)$.

[Proposition 3] The following inequalities hold:

$$k_1 \geq k_2 \Longrightarrow Q(m,k_1,\hat{\varepsilon},F) \leq Q(m,k_2,\hat{\varepsilon},F),$$
$$\hat{\varepsilon}_1 \geq \hat{\varepsilon}_2 \Longrightarrow Q(m,k,\hat{\varepsilon}_1,F) \leq Q(m,k,\hat{\varepsilon}_2,F).$$

□

[Proposition 4] For $m > 0$ and $0 \leq x \leq 1$, $(1-x)^m \leq 1 - mx + \dfrac{m^2}{2}x^2$.

(Proof) Let $f(x) = (1-x)^m$ and $g(x) = 1 - mx + \frac{m^2}{2}x^2$. Since $f(x) \leq g(x)$ holds trivially for $m = 1$ and $m = 2$, we consider the case where $m > 2$. Obviously, $f(0) = g(0) = 1$ holds.

Since

$$\begin{aligned} g'(x) - f'(x) &= -m + m^2x + m(1-x)^{m-1} \\ &\geq -m + m^2x + m(1-(m-1)x) = mx \geq 0 \end{aligned}$$

holds, the proposition is proved. □

[Lemma 2] For any even number m such that $m > 3$ and $2m < k$ hold,

$$Q(m,k,\hat{\varepsilon},F) \leq (1 - \frac{km}{4}\hat{\varepsilon}^2)^{\frac{m}{2}} .$$

(Proof) Consider the first element t of example tuples. If t is not in the violation pairs, the probability that no violation pair appears in the rest $m-1$ tuples is $Q(m-1,k,\hat{\varepsilon},F)$. If t is in the violation pairs (i.e. $t = t_i$), the probability that t'_i does not appear and no other violation pair appears in the rest $m-1$ tuples is $(1-\hat{\varepsilon})^{m-1}Q(m-1,k-1,\frac{\hat{\varepsilon}}{1-\hat{\varepsilon}},F)$. Therefore, the following equality holds:

$$Q(m,k,\hat{\varepsilon},F) = (1-2k\hat{\varepsilon})Q(m-1,k,\hat{\varepsilon},F) + 2k\hat{\varepsilon}(1-\hat{\varepsilon})^{m-1}Q(m-1,k-1,\frac{\hat{\varepsilon}}{1-\hat{\varepsilon}},F) .$$

By applying Proposition 3,

$$\begin{aligned} Q(m,k,\hat{\varepsilon},F) &\leq (1-2k\hat{\varepsilon})Q(m-1,k-1,\hat{\varepsilon},F) + 2k\hat{\varepsilon}(1-\hat{\varepsilon})^{m-1}Q(m-1,k-1,\hat{\varepsilon},F) \\ &= (1-2k\hat{\varepsilon}+2k\hat{\varepsilon}(1-\hat{\varepsilon})^{m-1})Q(m-1,k-1,\hat{\varepsilon},F) . \end{aligned}$$

By using Proposition 4,

$$\begin{aligned} Q(m,k,\hat{\varepsilon},F) &\leq (1-2k(m-1)\hat{\varepsilon}^2 + k(m-1)^2\hat{\varepsilon}^3)Q(m-1,k-1,\hat{\varepsilon},F) \\ &\leq (1-k(m-1)\hat{\varepsilon}^2)Q(m-1,k-1,\hat{\varepsilon},F) \end{aligned}$$

holds. The last inequality comes from

$$-k(m-1)\hat{\varepsilon}^2 + k(m-1)^2\hat{\varepsilon}^3 = k(m-1)\hat{\varepsilon}^2((m-1)\hat{\varepsilon}-1) \leq 0 , \qquad (1)$$

since we assume $2m < k$ and $\hat{\varepsilon} = \frac{\varepsilon}{k}$.

By applying this inequality repeatedly (it is easy to see that an inequality similar to (1) also holds in each step), we get:

$$\begin{aligned} Q(m,k,\hat{\varepsilon},F) &\leq (1-k(m-1)\hat{\varepsilon}^2)(1-(k-1)(m-2)\hat{\varepsilon}^2)\cdots \\ &\quad (1-(k-\frac{m}{2}+1)(\frac{m}{2})\hat{\varepsilon}^2)Q(\frac{m}{2},k-\frac{m}{2},\hat{\varepsilon},F) \\ &\leq (1-(k-\frac{m}{2})(\frac{m}{2})\hat{\varepsilon}^2)^{\frac{m}{2}} \\ &\leq (1-\frac{km}{4}\hat{\varepsilon}^2)^{\frac{m}{2}} . \end{aligned}$$

□

[Theorem 1] For any (T,D,F) such that $\varepsilon(T,D,F) \geq \varepsilon$, the number of examples required to test F correctly with the probability at least $1-\delta$ is $O(\frac{\sqrt{\ln\frac{1}{\delta}}}{\varepsilon}\sqrt{n})$.

(Proof) From Proposition 1, Lemma 1 and Proposition 2, $Q(m,T,D,F) \leq Q(m,k,\hat{\varepsilon},F)$ holds for any (T,D) such that $|T| = n$ and $\varepsilon(T,D,F) = \varepsilon$ where $\hat{\varepsilon} = \frac{\varepsilon}{2k}$ and $k = n-1$. Note that we consider T' such that $|T'| = 2n-1$. By means of the following inequalities, we get the upper bound of m.

$$\begin{aligned}
Q(m,n-1,\hat{\varepsilon},F) &< \delta, \\
(1 - \frac{(n-1)m}{4}\hat{\varepsilon}^2)^{\frac{m}{2}} &< \delta, \text{ (from Lemma 2)} \\
\frac{m}{2}\ln(1 - \frac{(n-1)m}{4}\hat{\varepsilon}^2) &< -\ln\frac{1}{\delta}, \\
\frac{m}{2}(-\frac{(n-1)m}{4}\hat{\varepsilon}^2) &< -\ln\frac{1}{\delta}, \text{ (from } \ln(1-x) \leq -x \text{ for } 0 < x < 1) \\
m &> \sqrt{\frac{8\ln\frac{1}{\delta}}{(n-1)\hat{\varepsilon}^2}} = \frac{\sqrt{32\ln\frac{1}{\delta}}}{\varepsilon}\sqrt{n-1}\ .
\end{aligned}$$

Since m is assumed to be even in Lemma 2, $\frac{\sqrt{32\ln\frac{1}{\delta}}}{\varepsilon}\sqrt{n-1} + 2$ tuples are sufficient. □

3.3 Lower Bound of the Number of Examples

In this section, we show a lower bound of the number of example tuples for Algorithm A1.

[Theorem 2] The number of examples, which is required to test F correctly with the probability at least $1-\delta$ by Algorithm A1 for every (T,D) such that $\varepsilon(T,D,F) \geq \varepsilon$ and $|T| = n$ hold, is $\Omega(\frac{\sqrt{1-\delta}}{\varepsilon}\sqrt{n})$.

(Proof) To show the lower bound, we consider only one (T,D) such that $vs(T,F) = \{\{t_1,t_2\},\{t_3,t_4\},\cdots,\{t_{n-3},t_{n-2}\}\}$ and $(\forall 1 \leq i \leq n-2)(p(t_i) = \frac{2\varepsilon}{n-2})$ hold. Let $\hat{\varepsilon}$ denote $\frac{2\varepsilon}{n-2}$. For each $j \leq \frac{n-2}{2}$, the probability that $(t_{2j-1} \in S \wedge t_{2j} \in S)$ holds is at most $(m\hat{\varepsilon})^2$ since $Prob(t_{2j} \in S | t_{2j-1} \in S) \leq m\hat{\varepsilon}$. Thus, $P(m,T,D,F) \leq \frac{n-2}{2}(m\hat{\varepsilon})^2$ holds.

If $1 - \frac{n-2}{2}(m\hat{\varepsilon})^2 > \delta$, $Q(m,T,D,F)$ becomes larger than δ. Therefore, by solving $1 - \frac{n-2}{2}(m\hat{\varepsilon})^2 < \delta$, we get $m > \frac{\sqrt{1-\delta}}{\sqrt{2}\varepsilon}\sqrt{n-2}$. □

4 PAC Learnability of Functional Dependencies

In this section, we consider the original problem, that is, to infer a set of all the functional dependencies consistent with the whole data from example tuples probably approximately correctly. For the problem, we consider a simple algorithm. It enumerates all the functional dependencies and then tests the consistency of each dependency by means of Algorithm A1. We call it as Algorithm A2. Note that the error of Algorithm A2 is one-sided as in the case of Algorithm A1. Although Algorithm A2 is not a polynomial time algorithm, it works in $O(m\log(m))$ time if we fix the domain of the relation. Of course, any algorithm which generates a set of all the functional dependencies consistent with example tuples can be used in place of Algorithm A2. For example, efficient algorithms for orthogonal range queries can be applied [11] and we can get more efficient algorithms. However, it seems very difficult to develop an algorithm whose time complexity is a polynomial of N since the number of the functional dependencies is exponential.

[Definition 6] For a set of tuples T, a distribution of tuples D and a set of functional dependencies FS, the error $\varepsilon(T, D, FS)$ is defined as $\sum_{t \in V} p(t)$ where V satisfies the following conditions:

(1) each functional dependency in FS is consistent with $T - V$,

(2) $\sum_{t \in V} p(t)$ is the minimum among such sets that satisfy the condition (1).

Next, we count the number of all the functional dependencies. It is easy to see that the number is at most $\sum_{i=1}^{N} N \binom{N}{i} < N2^N$. Then, we get the following result.

[Theorem 3] The number of examples required to infer a set of functional dependencies FS such that $\varepsilon(T, D, FS) \leq \varepsilon$ holds and all the functional dependencies consistent with the whole tuples are included by FS with the probability at least $1 - \delta$ is

$$O((\frac{N2^N}{\varepsilon})\sqrt{\ln \frac{N2^N}{\delta}\sqrt{n}}) .$$

(Proof) Note that Algorithm A2 outputs at most $N2^N$ functional dependencies. If the error of each functional dependency does not exceed $\frac{\varepsilon}{N2^N}$, the total error does not exceed ε. If the probability that each functional dependency whose error exceeds $\frac{\varepsilon}{N2^N}$ is consistent with example tuples is at most $\frac{\delta}{N2^N}$, the probability that at least one functional dependency whose error exceeds $\frac{\varepsilon}{N2^N}$ is derived is at most δ. Thus, by replacing ε and δ with $\frac{\varepsilon}{N2^N}$ and $\frac{\delta}{N2^N}$ respectively in Theorem 1, we get the theorem. □

5 Discussions

In this paper, a PAC learning framework for functional dependencies is presented. The number of examples required to infer a set of functional dependencies whose error does not exceed ε with probability at least $1 - \delta$ is studied.

The bound shown in Theorem 3 is not practical since a constant is too large even for $\varepsilon = 0.1$ and $\delta = 0.1$. However, if we consider the case where the distribution of tuples is uniform and the same tuple does not appear more than once in examples, the upper bound is much smaller [1]. Therefore, the sample complexity should be analyzed in more realistic situations.

A simple method of learning functional dependencies is considered in this paper. However, other methods may work better. Therefore, other methods for finding a set of functional dependencies probably approximately correctly should be developed and examined.

Although only the functional dependencies are considered in this paper, our framework may be applied to other types of constraints such as multi-valued dependencies and more general logical formulae. Although PAC learning for formulae in propositional logic are studied very well, few results about formulae in predicate logic are known. To study extensions of our framework would be valuable in the field of the computational learning theory.

Acknowledgment

We would like to thank Satoshi Kobayashi and Akira Utsumi for helpful discussions.

References

[1] T. Akutsu and A. Takasu. "On PAC learnability of functional dependencies in relational databases". In *Proceedings of the 6th Conference of Japanese Society of Artificial Intelligence*, pp. 327–330, (in Japanese) 1992.

[2] A. Blumer, A. Ehrenfeucht, D. Haussler, and M. Warmuth. "Learnability and the Vapnik-Chervonenkis dimension". *Journal of the ACM*, Vol. 36, pp. 929–965, 1990.

[3] E. F. Codd. "A relational model for large shared data banks". *Communications of the ACM*, Vol. 13, pp. 377–387, 1970.

[4] P. Honeyman. "Testing satisfaction of functional dependencies". *Journal of the ACM*, Vol. 29, pp. 668–677, 1982.

[5] B. K. Natarajan. *"Machine Learning - A Theoretical Approach"*. Morgan Kaufmann, CA, 1991.

[6] G. Piatetsky-Shapiro. "Discovery and analysis of strong rules in databases". In *Proceedings of Advanced Database System Symposium '89*, pp. 135–142, 1989.

[7] J. R. Quinlan and R. L. Rivest. "Inferring decision trees using the minimum description length principle". *Information and Computation*, Vol. 80, pp. 227–248, 1989.

[8] K. Sonoo, H. Kawano, S. Nishio, and T. Hasegawa. "Accuracy evaluation of rules derived from sample data in VLKD". In *Proceedings of the 5th Conference of Japanese Society of Artificial Intelligence*, pp. 181–184, (in Japanese) 1991.

[9] J. D. Ullman. *"Principles of Database and Knowledge-Base Systems - Volume 1"*. Computer Science Press, 1988.

[10] L. G. Valiant. "A theory of the learnable". *Communications of the ACM*, Vol. 27, pp. 1134–1142, 1984.

[11] D. E. Willard. "New data structures for orthogonal range queries". *SIAM Journal on Computing*, Vol. 14, pp. 232–253, 1985.

Protein Secondary Structure Prediction Based on Stochastic-Rule Learning

Hiroshi Mamitsuka† Kenji Yamanishi†
†C&C Information Technology Research Labs., NEC Corporation
1-1, Miyazaki 4-chome, Miyamaeku, Kawasaki, Kanagawa 216, Japan
E-mail:mami@ibl.cl.nec.co.jp, yamanisi@ibl.cl.nec.co.jp

ABSTRACT

This paper proposes a new strategy for predicting α-helix regions for any given protein sequence, on the basis of the theory of learning stochastic rules. We confine our study to the problem of predicting where α-helix regions are located in a given protein sequence, rather than the conventional three-state prediction problem, i.e., that of predicting to which among the three-states (α-helix, β-sheet, or coil) each of the amino acids in the sequence corresponds.

Our strategy consists of three steps: generation of training examples, learning, and prediction.

In the learning phase, we construct a rule for secondary-structure prediction from training examples. Here a rule is represented not as a deterministic rule but as a *stochastic rule*, i.e., a probability distribution which assigns, to each region in a sequence, a probability that it corresponds to α-helix. Each stochastic rule used here is further represented as the product of a number of *stochastic rules with finite partitioning* developed by Yamanishi. Optimal stochastic rules with finite partitioning are obtained from training examples by Laplace estimation of real-valued parameters and by model selection based on the minimum description length (MDL) principle. We allow our stochastic rules to make use of not only the characters themselves of amino acids but also their physico-chemical properties (i.e., numerical attributes, e.g. hydrophobicity, molecular weight, etc).

In the prediction phase, when given a test sequence, the likelihood that any given region (i.e., any subsequence of amino acids) in the test sequence corresponds to α-helix is calculated with the stochastic rules constructed in the learning phase.

We evaluate the predictive performance of our strategy from experimental viewpoints. In generating training examples, examples of α-helix regions are drawn from hemoglobin sequences alone. Experimental results show that the prediction accuracy rate of our prediction strategy was 94.8% for hemoglobin α-chain (1HBSα), 68.5% for parvalbumin β (1CDP), and 73.6% for lysozyme c (1LYM), a significant rate over the rate achieved with the Garnier-Osguthorpe-Robson's (GOR) method.

1 Introduction

We consider the problem of predicting a secondary structure for a given protein sequence. A protein sequence is a sequence of amino acids, each of which corresponds to one of three states: α-helix, β-sheet, or coil. We refer to such a correspondence as a *secondary structure* of a protein sequence. Let a protein sequence whose secondary structure is unknown be given and referred to as a *test sequence*. The problem of predicting secondary structure for the test sequence is an issue of how to predict accurately which state corresponds to each of amino acids in the sequence.

One of most successful approaches to the secondary structure prediction is the construction (from training examples) of a classification rule which represents the general relationship between the states and the amino acids. In this case, the prediction of secondary structures may be essentially reduced to the problem of *learning from examples*.

There exist a number of well-known methods for predicting secondary structures for protein sequences: e.g. Chou-Fasman's method (the CF method [1],[2]); Garnier-Osguthorpe-Robson's method (the GOR method [3]); Gibrat-Garnier-Robson's method (the GGR method [4]); Qian-Sejnowski's method (the QS method [5]); Bohr et al's method (the BO method [6]), King & Sternberg method (the KS method [7]), etc.

The CF method uses heuristics based on the statistical observations of secondary structures. The GOR method and the GGR method use learning phases based on information-theoretic techniques, e.g., directional information, pair information, etc. The QS method and the BO method use techniques for neural network learning. The KS method uses machine learning techniques, e.g. generation of production rules, etc.

In this paper, we develop a new strategy for predicting the secondary structure of a given protein sequence, based on the theory of learning stochastic rules. As with all of the above methods, our strategy also consists of generation of training examples, a learning phase, and a prediction phase. However, unlike them, our strategy is unique in the following ways:

1) Rather than predicting to which state individual amino acids correspond, our strategy predicts the probability of correspondence to α-helix for regions (i.e., subsequences of amino acids) in a given test sequence. When all of amino acids in a region correspond to α-helix, we say that the state of the region is α-helix.

2) A rule for secondary structure prediction is represented not as a deterministic rule but as a *stochastic rule* [8]. Here a stochastic rule is a probability distribution which assigns, to each region in a sequence, a probability that its state is α-helix. Each stochastic rule used here is represented as the product of a number of *stochastic rules with finite partitioning*, which is developed in [8]. Here a stochastic rule with finite partitioning is defined as a rule specified by a set of disjoint cells and by a real-valued probability parameter vector whose elements are in $[0,1]$. The method for learning them is based on Yamanishi's theory for learning stochastic rules [8]. More precisely, in our learning phase, probability parameters are estimated using *Laplace estimation* of probability parameters, and the number of disjoint cells is optimized using the *minimum description length (MDL) principle*, which was developed by Rissanen [9],[10].

3) In the learning phase, we allow our stochastic rules to make use of not only the characters themselves of amino acids but also their physico-chemical properties (i.e. numerical attributes, e.g. hydrophobicity, molecular weight, etc). Such rules may reasonably be expected to predict secondary structures for an unknown future test sequence more accurately than those constructed based on the characters themselves of the amino acids.

4) In the prediction phase, when given a test sequence, the likelihood that the state of any given region in the test sequence is α-helix is calculated with the stochastic rules constructed in the learning phase.

We evaluate the predictive performance of our strategy from experimental viewpoints. We will show that our prediction phase has better predictive performance for hemoglobin α-chain (1HBSα), parvalbumin β (1CDP) and lysozyme c (1LYM) than the GOR method, which is considered to be the most populous and powerful strategy among the conventional methods.

2 Outline of Predicting Protein Secondary Structure

The goal of our strategy is to predict locations of α-helix regions for any given protein sequence whose secondary structure is unknown. Here a *region* refers to a subsequence of amino acids of which a test sequence consists. Our strategy is decomposed into *generation of training examples*, a *learning phase* and a *prediction phase*.

In the prediction phase, when given a protein sequence (which we call a *test sequence*), by moving a *test window* of a fixed size from the left to the right through the test sequence, we calculate a likelihood of α-helix for each region specified by the test window. Here the likelihood of α-helix for a given region in the test sequence refers to a probability that the state of the region is α-helix. From the observations on the likelihoods, the state of any region such that the likelihood of α-helix for it is not less than some given threshold is regarded as α-helix.

In order to calculate likelihoods in the prediction phase, we have to construct a *stochastic rule*, i.e., a probabilistic model which assigns each region in a protein sequence a probability that its state is α-helix. Hence in the learning phase, we must learn such a stochastic rule from a number of protein sequences drawn from a given protein database. In our strategy, a stochastic rule is represented as the product of a number of stochastic rules, each of which is specified by a real-valued probability parameter vector and a finite number of disjoint cells. The learning phase includes the estimation of real-valued parameters and the optimization of the number of disjoint cells and of their assignment.

Prior to the learning phase, we must generate training examples from which a stochastic rule is learned. Assume that there exist a number of α-helix regions in a protein sequence (Figure 1), where an α-helix region is the region whose state is α-helix. For each of α-helix regions, we

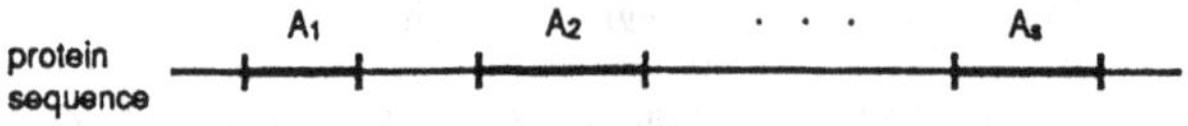

Figure 1: α-helix regions in a protein sequence

prepare positive and negative examples of it. Here a positive example refers to a region whose state is α-helix of interest, and a negative example refers to a region whose state is not α-helix region.

As mentioned above, in our strategy, first we generate training examples, next the learning phase is implemented, and then the prediction phase is implemented (Figure 2).

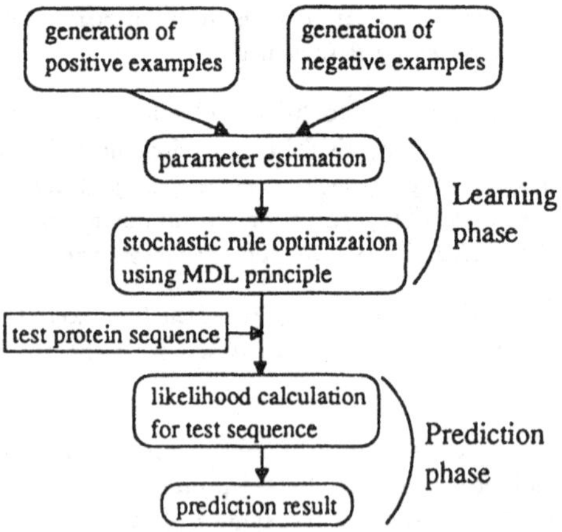

Figure 2: Outline of the prediction

3 Generation of Training Examples

We prepare a number of training positive and negative examples of α-helix regions in order to construct a stochastic rule predictor. In this section, let us describe a method for generating training examples.

Let a protein sequence which belongs to a specific family (e.g. a hemoglobin family etc.) be fixed. In it, in general, there exist a number of α-helix regions (see again Figure 1). Let such

α-helix regions be $A_1, \cdots, A_s$, where s is the total number of α-helix regions and the lengths of the regions may be different one another. Below, for a given region A, $|A|$ denotes the number of amino acids of which A consists, and we refer to it as a *size* of the region A.

For example, let us consider a hemoglobin sequence of length 287 which consists of 1 α-chain and 1 β-chain. This sequence is a subsequence of an ordinary hemoglobin sequence which consists of 2 α-chains and 2 β-chains. In this sequence, there exist 15 α-helix regions, and the sizes of the regions are as follows: 16, 16, 7, 20, 9, 19, 21, 15, 16, 7, 7, 20, 9, 19, 21. The former 7 regions are included in an α-chain and the latter 8 regions are included in a β-chain.

For each of α-helix regions, we must generate positive and negative examples of it. Here a *positive example* of any given α-helix region A ($|A| = \ell$) is a region whose state is A and whose size is ℓ. On the other hand, the *negative example* of any given α-helix region A ($|A| = \ell$) is a region whose state is not α-helix and whose size is ℓ.

3.1 Generation of Positive Examples

Below, let us describe how to generate positive examples of a given α-helix region.

First let a protein sequence be given where the locations of all α-helix regions in it are known. We call this sequence an *original sequence*. We denote α-helix regions in it as $A_1, \cdots, A_s$.

Next we align $N-1$ protein sequences (belonging to the same family as the original sequence) relative to the original sequence. Here for any sequence to be aligned, the locations of α-helix regions are assumed to be unknown. After alignment, from each of aligned sequences, we obtain a *positive example of A_i* as a region which corresponds to A_i in the original sequence (Figure 3). Therefore, by alignment of $N-1$ examples, for each A_i, we obtain $N-1$ positive examples

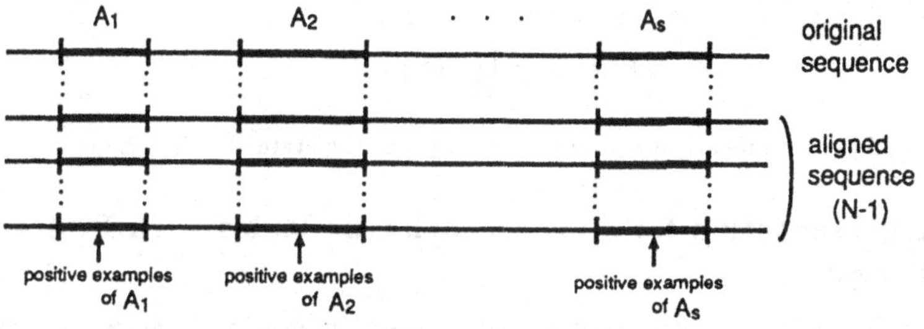

Figure 3: Generation of positive examples

of A_i ($i = 1, \cdots, s$). Letting A_i itself in the original sequence be also regarded as a positive example of A_i, we, after all, obtain N positive examples of A_i in total ($i = 1, \cdots, s$).

3.2 Generation of Negative Examples

Below, let us describe how to generate negative examples of a given α-helix region. For each

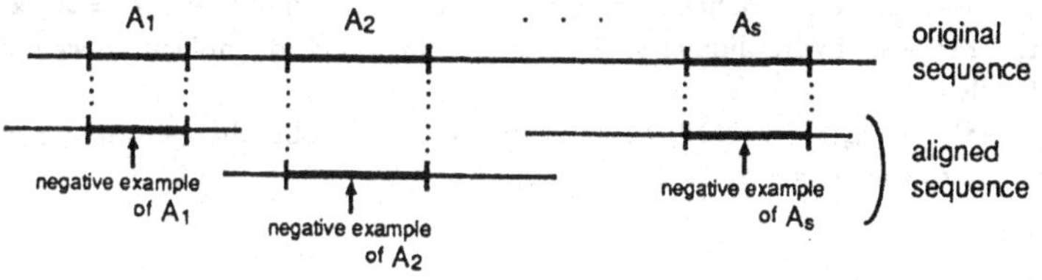

Figure 4: Generation of negative examples

α-helix region A_i in the original sequence, we generate N regions of the same size as A_i from the database, e.g. the Protein Data Bank(PDB), each of which is not included in any α-helix region. We regard each of regions obtained in this way as a *negative example* of A_i $(i = 1, \cdots, s)$ (Figure 4). For each A_i, we generate N negative examples of A_i $(i = 1, \cdots, s)$.

4 Learning Phase

In this section, let us describe the learning phase in detail.

4.1 Basic Probabilistic Structure

First let us describe a basic probabilistic structure which we use in the training. $\mathcal{X} \stackrel{\text{def}}{=} \mathcal{X}_1 \times \mathcal{X}_2 \times \cdots \times \mathcal{X}_n$. For example, $\mathcal{X}$ =the set of all amino acids such that $|\mathcal{X}| = 20$. For another example, $\mathcal{X} = \mathcal{X}_1 \times \mathcal{X}_2$ where $\mathcal{X}_1$ denotes the range of hydrophobicity and $\mathcal{X}_2$ denotes the range of molecular weight. In the latter example, the protein sequence can be represented as a sequence of 2-dimensional numerical vectors. We call $\mathcal{X}$ a *domain for a single residue.*

Let $\mathcal{S} \stackrel{\text{def}}{=} \mathcal{X}^W = \mathcal{X} \times \cdots \times \mathcal{X}$ (W-tuple of $\mathcal{X}$) be the set of regions of size W. We define $P(\alpha \mid *)$ by a mapping from a region $S \in \mathcal{S}$ to a probability value in $[0,1]$:

$$\begin{array}{cccc} P(\alpha \mid *): & \mathcal{S} & \longrightarrow & [0,1] \\ & \Psi & & \Psi \\ & S & \longrightarrow & P(\alpha \mid S). \end{array}$$

Let $P(\alpha \mid X_i)$ be the probability that state of S is α-helix when the i-th residue in S is X_i. We assume that $P(\alpha \mid S)$ is always represented in a form of the product of $P(\alpha \mid X_i)$s $(i = 1, \cdots, W)$ as follows:

$$P(\alpha \mid S) = \prod_{i=1}^{W} P(\alpha \mid X_i), \tag{1}$$

where $P(\alpha \mid S)$ can be regarded as the probability that the state of S is α-helix.

4.2 Representation for Rules: Stochastic Rules with Finite Partitioning

Each $P(\alpha \mid X_i)$ is further constrained so that it takes a form of a *stochastic rule with finite partitioning* [8], which is expressed as follows: Let the size W of region be given. For a fixed $i(\in \{1, \cdots, W\})$, let $\{C_j(i)\}_j$ be a finite set of disjoint cells of $\mathcal{S}$ such that $\mathcal{S} = \cup_{j=1}^{m} C_j(i)$, $C_j(i) \cap C_k(i) = \phi\,(j \neq k)$ where m is the number of the disjoint cells and let $p_j(i) \in [0,1]$ $(j = 1, \cdots, m)$. For each i, $P(\alpha \mid X_i)$ is defined as

$$\begin{array}{llll} \textbf{For} & j := 1 & \textbf{to} & m \\ \textbf{if} & X_i \in C_j(i) & \textbf{then} & P(\alpha \mid X_i) = p_j(i). \end{array} \tag{2}$$

Figure 5 shows an example of stochastic rules with finite partitioning over $\mathcal{X} = \mathcal{X}_1 \times \mathcal{X}_2$ where $\mathcal{X}_1$ denotes the range of hydrophobicity, $\mathcal{X}_2$ denotes the range of molecular weight and the number m of cells is 9.

Let $\theta_i \stackrel{\text{def}}{=} (p_1(i), \cdots, p_m(i)) \in [0,1]^m$ be an m-dimensional probability parameter vector. We write a stochastic rule specified by θ_i as $P(\alpha \mid X_i : \theta_i)$.

4.3 Parameter Estimation

Next, letting m (the number of cells) be fixed, we consider the problem of estimating the probability parameter vector θ from given examples. We use a *Laplace estimator* [11], which we denote as $\hat{\theta}_i = (\hat{p}_1(i), \cdots, \hat{p}_m(i))$ $(i = 1, \cdots, W)$.

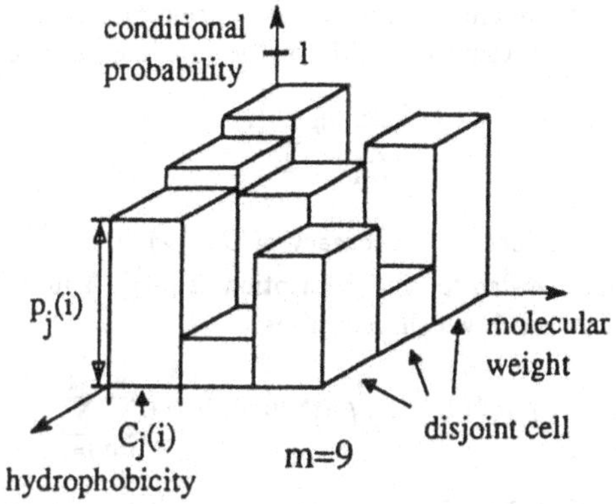

Figure 5: An example of stochastic rules with finite partitioning

Let $2N$ training examples of an α-helix region be given where for the i-th residue in the region, the number of examples that fell into the j-th cell is $N_j(i)$ and the number of α-helix examples that fell into the j-th cell is $N_j^+(i)$. For each $i \in \{1, \cdots, W\}$, a Laplace estimator is calculated as:

$$\hat{p}_j(i) = \frac{N_j^+(i) + 1}{N_j(i) + 2} \qquad (j = 1, \cdots, m). \tag{3}$$

Notice here that $2N = \sum_{j=1}^{m} N_j(i)$ for each $i \in \{1, \cdots, W\}$ and that the total number of amino acids included in an α-helix region whose size is W is $2NW$.

4.4 Optimization of Cell Structure Using the MDL Principle

Next. let us consider the problem of selecting an optimal number of cells and their assignment by which a stochastic rule with finite partitioning is specified. In general, the larger the number m of cells is, the better a stochastic rule fits the given training examples. However, a stochastic rule whose number of cells is too large, may be affected by statistical irregularities of training examples, and such a rule may not be able to predict the secondary structure for future test examples more accurately than smaller-sized-cell rules. Therefore it is important to address an issue of how to select an optimal number of cells and their assignment on the basis of the given examples.

We use the MDL (Minimum Description Length) principle developed by Rissanen [9], [10] as a criterion for selecting an optimal number of cells.

Applying the MDL principle to our optimization problem, an optimal number $\hat{m}$ of cells is obtained as an m which minimizes

(description length for examples relative to a given rule with m cells)
+(description length for the rule itself).

Let W be the size of an α-helix region of interest. Letting m_i $(i = 1, \cdots, W)$ be the number of disjoint cells of the stochastic rule at the i-th residue in the window, hereafter, assume that $m_1 = \cdots = m_W$. The description length for $2N$ training examples of size W (in which $2NW$ amino acids are included) relative to a given rule with m cells is calculated as the minus logarithm of the likelihood of the rule for examples as follows:

$$-\sum_{i=1}^{W}\sum_{j=1}^{m} \log\{\hat{p}_j(i)^{N_j^+(i)}(1-\hat{p}_j(i))^{N_j(i)-N_j^+(i)}\},$$

which is the codelength of a Shannon code for $2NW$ training amino acids.

On the other hand, the description length for the rule itself is calculated as

$$\sum_{i=1}^{W}\sum_{j=1}^{m}\frac{\log N_j(i)}{2},$$

since each $\hat{p}_j(i)$ can be described with accuracy of $O(\frac{1}{\sqrt{N_i}})$ $(i = 1,\cdots,W,\ j = 1,\cdots,m)$, and thus $(\log N_j(i))/2$ bits are needed for the description of $\hat{p}_j(i)$. Thus, an optimal number $\hat{m}$ of disjoint cells is obtained as an $\hat{m}$ which minimizes

$$-\sum_{i=1}^{W}\sum_{j=1}^{m}\log\{\hat{p}_j(i)^{N_j^+(i)}(1-\hat{p}_j(i))^{N_j(i)-N_j^+(i)}\}+\sum_{i=1}^{W}\sum_{j=1}^{m}\frac{\log N_j(i)}{2}. \tag{4}$$

For example, assume that $\mathcal{X} = \mathcal{X}_1 \times \mathcal{X}_2$ where $\mathcal{X}_1$ denotes the range of hydrophobicity and $\mathcal{X}_2$ denotes the range of molecular weight. If $\hat{m}$ is less than 20, then it implies that 20 kinds of amino acids are categorized into $\hat{m}$ groups based on their similarity of hydrophobicity and molecular weight.

5 Prediction Phase

In this section, let us describe a prediction phase in detail.

5.1 Calculation of Likelihoods

Let a test sequence, for which we predict α-helix regions, be given. We prepare a ***test window*** of size t where t is not more than the size W of an α-helix region. By moving the test window from the left of the test sequence to the right, we can examine all of regions of size t specified by the test window for the given test sequence (Figure 6). We call a region specified by the test window a ***test region***.

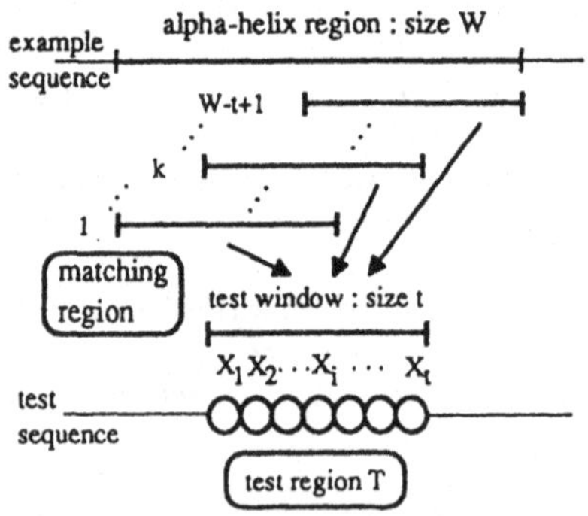

Figure 6: Test window

For any given test region T in the test sequence, we calculate *the likelihood* of α-helix for T, which refers to the probability that the state of T is α-helix. Below let us describe a method for calculating the likelihoods in details.

Let W be the size of a given α-helix region. Let $\hat{\theta}_i = (\hat{p}_1(i),\cdots,\hat{p}_m(i))$ $(i = 1,\cdots,W)$ be a probability parameter vector of the stochastic rule at the i-th residue of the region (Figure 6). Notice here that $\hat{\theta}_i$ $(i = 1,\cdots,W)$ has been estimated in the learning phase.

By moving the test window of size t from the left of the given α-helix region of size W to the right, we can examine $W-t+1$ regions of size t to compare these regions with test regions on the test sequence (Figure 6). We call the regions *matching regions*. Notice that a product of t stochastic rules, each of which an m-dimensional probability parameter is associated with $(i = 1, \cdots, W)$, corresponds to a single α-helix region.

For a given test region T of the test sequence, which is specified by a test window, we calculate the likelihood of α-helix for T with regard to each matching region of the α-helix region. We define an $(m \times t)$-dimensional parameter vector ω_k $(k = 1, \cdots, W - t + 1)$ associated with the k-th matching region by:

$$\omega_k = (\hat{\theta}_k, \cdots, \hat{\theta}_{k+t-1}) \quad (k = 1, \cdots, W - t + 1).$$

For a given test region T of the test sequence, we define the *k-th likelihood* $P(\alpha \mid T : \omega_k)$ as:

$$P(\alpha \mid T : \omega_k) = \prod_{i=1}^{t} P(\alpha \mid X_i : \hat{\theta}_{k+i-1}), \tag{5}$$

where X_i denotes the i-th residue of T, and $P(\alpha \mid X_i : \hat{\theta}_{k+i-1})$ is calculated as:

$$\begin{array}{llll} \textbf{For} & j := 1 \ \textbf{to} \ m & & \\ \textbf{if} & X_i \in C_j(i) & \textbf{then} & P(\alpha \mid X_i : \hat{\theta}_{k+i-1}) = \hat{p}_j(k+i-1), \end{array} \tag{6}$$

where $\{C_j(i)\}_j$ is a set of disjoint cells specifying the stochastic rule at the i-th residue.

We define the likelihood associated with the k-th matching region for T (which we call *likelihood of α-helix for T*), $P(\alpha \mid T)$, as

$$P(\alpha \mid T) \stackrel{\text{def}}{=} \max\{P(\alpha \mid T : \omega_1), \cdots, P(\alpha \mid T : \omega_{W-t+1})\}. \tag{7}$$

For a given test region T of the test sequence, we regard the likelihood of α-helix for T as the probability that the state of T is α-helix.

In the above statement, we have assumed that there exists a single α-helix region. If there exist s α-helix regions $A_1, \cdots, A_s$ $(s > 1)$, by giving them indices, we have s likelihoods of α-helix for T; $P^{(1)}(\alpha \mid T), \cdots, P^{(s)}(\alpha \mid T)$, where $P^{(j)}(\alpha \mid T)$ denotes the likelihood of α-helix for T corresponding to the j-th α-helix region. In this case, for example, we define the *likelihood of α-helix for T*, which we denote as $\bar{P}(\alpha \mid T)$, as

$$\bar{P}(\alpha \mid T) \stackrel{\text{def}}{=} \max\{P^{(1)}(\alpha \mid T), \cdots, P^{(s)}(\alpha \mid T)\}. \tag{8}$$

By the calculation of (8), if $\bar{P}(\alpha \mid T) = P^{(i)}(\alpha \mid T)$ and the value of $\bar{P}(\alpha \mid T)$ is high, then we see that the state of T is with high probability α-helix, and the property of α-helix corresponding to T is the same as that of A_i. Of course, we can use various kinds of analogues of (8).

5.2 Prediction Curve and Prediction Error Rate

In the above subsection, we have introduced a method for calculating a likelihood of α-helix for a given test "region" of size t in the test sequence. In this section, let us derive a method for calculating a likelihood for a "single amino acid" and introduce a concept of a prediction curve.

Let a test sequence of amino acids be given and let its total number of amino acids included in it be L. Let a size of the test window be t. First notice that for any amino acid x in the test sequence, there exist t test windows that include x (Figure 7). We denote such test windows including x as $T_1(x), \cdots, T_t(x)$. We may define a likelihood (of α-helix) for x, which we denote as $P(\alpha \mid x)$, as follows:

$$P(\alpha \mid x) \stackrel{\text{def}}{=} \max\{P(\alpha \mid T_1(x)), \cdots, P(\alpha \mid T_t(x))\}. \tag{9}$$

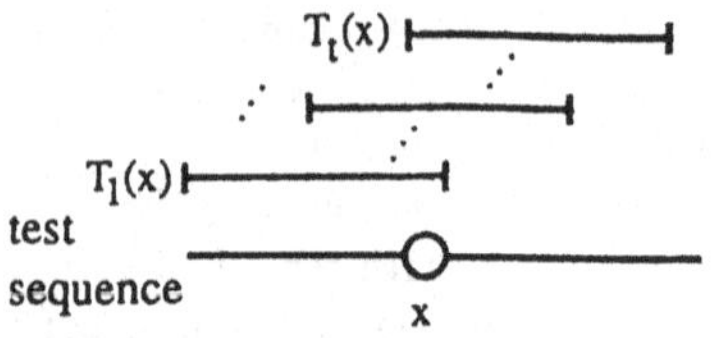

Figure 7: Prediction phase

Further let us define a function $\hat{f}(x): \mathcal{X} \longrightarrow [0,1]$ by

$$\hat{f}(x) = P(\alpha \mid x) \quad \text{for any } x, \tag{10}$$

and we call $\hat{f}(x)$ a ***prediction curve*** (of α-helix regions in the test sequence). Figure 8 shows an example of a prediction curve, which assigns each amino acid x a likelihood that x corresponds to α-helix.

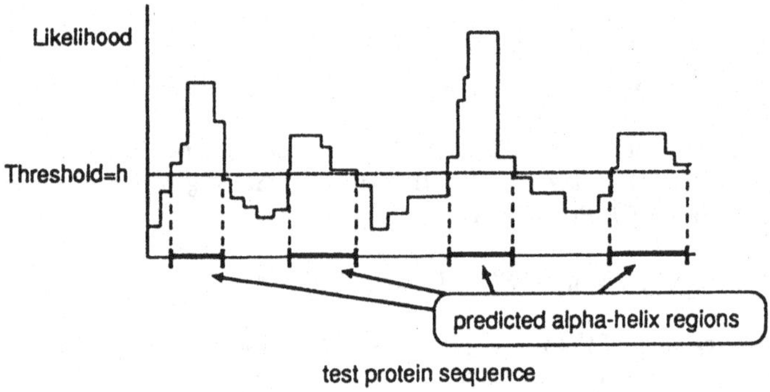

Figure 8: An example of prediction curves

For a given prediction curve $\hat{f}(x)$, we recognize α-helix regions in the following way. Let h be a given ***threshold***. For example, we can let $h = (1/2)^t$ where t is the size of the test window. For any amino acid x in the test sequence, if $\hat{f}(x) \geq h$, we consider that x corresponds to α-helix; otherwise we consider that x doesn't correspond to α-helix. Then we obtain a set of α-helix amino acids, which we denote as $\mathcal{D}^+$, in a form of $\mathcal{D}^+ = \{x : \hat{f}(x) \geq h\}$. The final output of our prediction phase is a partitioning of the whole set $\mathcal{D}$ of amino acids in the test sequence into $\mathcal{D}^+$ and $\mathcal{D}^- = \mathcal{D} - \mathcal{D}^+$.

Next, let us define a prediction error rate for any given prediction phase. For any given prediction curve $\hat{f}(x)$, we define a *quantized function* $\bar{f}(x)$ *associated with* $\hat{f}(x)$ by: For any x, if $\hat{f}(x) \geq h$, $\bar{f}(x) = 1$; otherwise $\bar{f}(x) = 0$. For a given test sequence, we call a curve plotting ***true*** locations of α-helix regions as a ***true curve***. A true curve $f^*(x)$ is formally defined as follows: For any x, if x corresponds to α-helix, $f^*(x) = 1$; otherwise $f^*(x) = 0$.

From a prediction curve $\hat{f}(x)$ and a true curve $f^*(x)$, we define the deviation $d(\hat{f}, f^*)$ between $\hat{f}$ and f^* as:

$$d(\hat{f}, f^*) \stackrel{\text{def}}{=} \frac{1}{L}\sum_{i=1}^{L} \mid \bar{f}(x) - f^*(x) \mid . \tag{11}$$

where $\bar{f}$ is a quantized function $\bar{f}(x)$ associated with $\hat{f}(x)$. We call $d(\hat{f}, f^*)$ a ***prediction accuracy rate*** of $\hat{f}$.

6 Experimental Results

In this section, let us show experimental results obtained by applying our strategy to predicting α-helix regions for various kinds of protein sequence.

We used hemoglobin sequences alone each of which consists of 1 α-chain and 1 β-chain to get positive examples. Recall that there exist 15 α-helix regions in such a hemoglobin sequence and that the total length of a hemoglobin sequence is 287 (see Section 3).

In generating training examples, we prepared 128 positive and 128 negative examples for each of 15 α-helix regions. 128 hemoglobin sequences were used to generate positive examples of α-helix regions. They were drawn from the PIR-1 (Protein Identification Resource-1) database, and were more than 70% homologous one another both in α-chain and β-chain. Sequences including negative examples of α-helix are drawn from the PDB (Protein Data Bank) database and are not always hemoglobin sequences. We selected negative examples so that each example is 20%-40% homologous to the α-helix region corresponding to it in the original sequence.

In the learning phase, let us a domain $\mathcal{X}$ for a single residue be $\mathcal{X}_A$: the whole set of characters of amino acids such that $|\mathcal{X}_A| = 20$ or $\mathcal{X}_B = \mathcal{X}_1 \times \mathcal{X}_2$ where $\mathcal{X}_1$ denotes the range of hydrophobicity and $\mathcal{X}_2$ denotes the range of molecular weight. Here we used the hydrophobicity values of amino acids determined by Fauschere et al.[12]. Below, E_1 denotes the prediction strategy with $\mathcal{X} = \mathcal{X}_A$. E_2 denotes the prediction strategy with $\mathcal{X} = \mathcal{X}_B$ and the number m of cells is set to be 14 where m is not optimized by the MDL principle. E_3 denotes the prediction strategy in which $\mathcal{X} = \mathcal{X}_B$ and their number m of cells is optimized using the MDL principle over the range from 9 to 18. Using the three strategies E_1, E_2 and E_3, we predicted locations of α-helix regions for three types of proteins: hemoglobin α-chain (1HRBα), parvalbumin β (1CDP), and lysozyme c (1LYM).

In the prediction phase, we let the size t of the test window be 7 and we let a threshold be $h = (1/2)^t$ where t is a size of a test window.

In the case where a test sequence is 1HRBα which is randomly drawn from the pool of the hemoglobin α-chain sequences used to generate positive examples, the prediction accuracy rates of E_1, E_2, and E_3 are all at least 90%, and the maximum rate is about 94.8%.

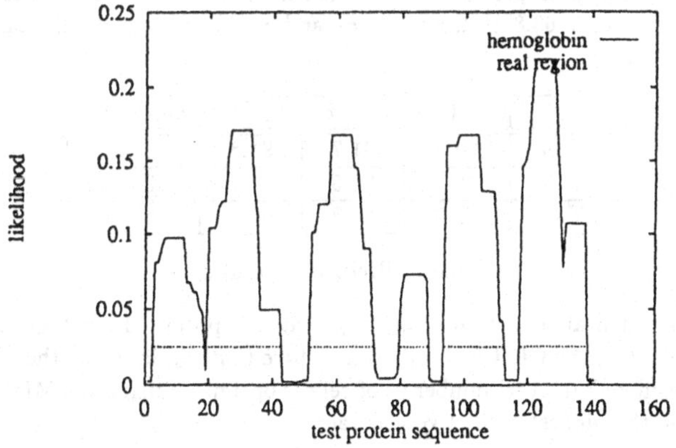

Figure 9: α-helix prediction curve of 1HRBα

In Figure 9, the real line shows a prediction curve of E_3 for 1HRBα, and a dotted line indicates true α-helix regions in 1HRBα. From the results, we may say that we could construct a rule which recognizes α-helix regions for hemoglobin sequence themselves with accuracy of more than 90%.

In the case where a test sequence is 1CDP which is randomly drawn from PDB, the prediction accuracy rates of E_1, E_2, and E_3 are all higher than that of the GOR method, whose rate is 52.8%. The highest prediction accuracy rate is about 68.5%, which is attained by E_2.

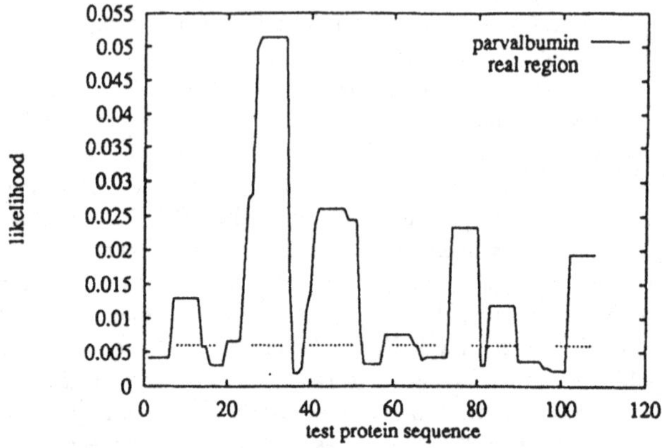

Figure 10: α-helix prediction curve of 1LYM

In Figure 10, a real line shows a prediction curve of E_2 for 1CDP, and a dotted line indicates true α-helix regions in 1CDP. There exist 6 peaks which exceed the threshold in the prediction curve; each of 4 peaks corresponds to a single true α-helix region, and other 2 peaks correspond to a single true α-helix region.

It should be noticed that 1CDP is less than 20% homologous to the hemoglobin sequence used as an original sequence when positive examples were generated, but that our prediction phase recognized 5 α-helix regions in 1CDP.

In the case where a test sequence is 1LYM which is randomly drawn from PDB, the highest prediction accuracy rate of our strategy is 73.6%, which is achieved by E_1. This predictive performance is a significant improvement over the rate achieved with the GOR method, whose prediction accuracy rate is 69.8%. However, E_2 and E_3 have lower prediction accuracy rates than the GOR method.

	E_1 (%)	E_2 (%)	E_3 (%)	GOR (%)
1HRBα	94.4	91.7	94.8	36.9
1CDP	60.2	68.5	65.7	52.8
1LYM	73.6	58.9	65.1	69.8

Table 1: Prediction results

From our experimental results, we may say that our proposed strategy has significantly better performance than the GOR method. Notice here that E_3, in which the physico-chemical properties are considered and the number m of cells is optimized using the MDL principle, does not always achieve the highest predictive accuracy.

7 Conclusion

We have developed a new strategy for predicting α-helix regions for any given protein sequence. Our strategy is characterized most specifically by likelihood-prediction per each test region and by the learning phase in which the theory of learning stochastic rules is used. Further, in the learning phase, we have utilized not only the characters of amino acids themselves but also their numerical attributes: hydrophobicity and molecular weight, and a method for optimizing the number of cells specifying the rule based on the MDL principle has also been proposed.

Experimental results have shown that even when a rule is trained from only hemoglobin sequences, our strategy has the following prediction accuracy rates: 94.8% for 1HBSα, 68.5% for 1CDP, and 73.6% for 1LYM. It has turned out that for these proteins, our strategy has higher predictive accuracy than the GOR method, which is the most populous and powerful method among conventional secondary-structure prediction methods.

The following issues remain open: .

1)When we make use of any physico-chemical properties of amino acids other than hydrophobicity and molecular weight, can our strategy achieve higher predictive performance?

2)How accurately can our strategy predict β-sheet regions or any regions other than α-helix?

3)When we construct from examples a stochastic rule with "variable-length" cells per at each residue, can our strategy achieve higher predictive performance?

These problems will be dealt in future papers.

Acknowledgment The authors wish to express their sincere gratitude to Mr.Nakamura, Mr.Kaneko, and Dr.Abe of C & C Information Technology Research Laboratories, NEC Corporation, Mr.Konagaya of C & C System Research Laboratories, NEC Corporation, for their support and helpful advice.

References

[1] P.Y. Chou and G.D. Fasman. Conformational parameters for amino acids in helical, β-sheet, and random coil regions calculated from proteins. *Biochemistry*, 13(2):211–221, 1974.

[2] P.Y. Chou and G.D. Fasman. Prediction of protein conformation. *Biochemistry*, 13(2):222–245, 1974.

[3] J. Garnier, D.J. Osguthorpe, and B. Robson. Analysis of the accuracy and implication of simple methods for predicting the secondary structure of globular proteins. *J.Mol.Biol.*, 120:97–120, 1978.

[4] J.F. Gibrat, J. Garnier, and B. Robson. Further developments of protein secondary structure prediction using information theory. *J.Mol.Biol.*, 198:425–443, 1987.

[5] N.Qian and T.J.Sejnowski. Predicting the secondary structure of globular proteins using neural network models. *J.Mol.Biol.*, 202:865–884, 1988.

[6] H. Bohr, J. Bohr, S. Brunek, M.J.R. Cotterill, B. Lautrup, L. Norskov, H.O.Olsen, and B.S.Petersen. Protein secondary structure and homology by neural networks. *FEBS Letters*, 241(1,2):223–228, 1988.

[7] R.D. King and M.J.E. Sternberg. Machine learning approach for the prediction of protein secondary structure. *J.Mol.Biol.*, 216:441–457, 1990.

[8] K. Yamanishi. A learning criterion for stochastic rules. In *Proceedings of the Third Annual Workshop on Computational Learning Theory*, pages 67–81. Morgan Kaufmann, 1990. To appear in *Machine Learning*.

[9] J. Rissanen. Modeling by shortest data description. *Automatica*, 14:465–471, 1978.

[10] J. Rissanen. *Stochastic complexity in statistical inquiry*, volume 15 of *Comp. Sci.* 1989. World Scientific.

[11] F. Schreiber. The bayes laplace statistics of the multinomial distributions. *AEU*, 39(5):293–298, 1985.

[12] J.L. Fauchere and V. Pliska. Hydrophobic parameters of amino acid side chains from the partitioning of N-acetyl-amino acid amides. *Eur.J.Med.Chem.Chim.Ther.*, 18:369–375, 1983.

Notes on the PAC Learning of Geometric Concepts with Additional Information

Ken-ichiro Kakihara[1] and Hiroshi Imai[2]
Department of Information Science
University of Tokyo
Tokyo 113, Japan

Abstract

In the PAC learning model, the learner must output good approximate results with only the information that tells the learner whether the example is "positive" or "negative." This restriction results in narrowing the area of the learnable concepts or in increasing the size of examples required for successful learning. However, if the learner receives some additional information about the example besides being positive or negative, e.g., real values corresponding to the degree of the positiveness (or negativeness), a larger class may become learnable or the number of necessary examples may be reduced.

In the case of learning geometric concepts, such additional information may be given, for instance, how close a given positive or negative example is to the boundary of a target geometric concept. Using this type of additional information for geometric concepts consisting of complex boundaries, the learner may identify the geometric concept more rigorously or with sampling the smaller number of examples for the required accuracy. In the case of neural networks of threshold functions, some of values of weighted sum of inputs at nodes, say a value at the output node, may be output instead of simply producing outputs of either 0 or 1 after comparing the values with the threshold values at nodes. Such values represent the closeness of an input to the threshold boundary.

This note investigates the effect of such additional information in computational learning theory. Two types of networks of threshold-like functions are considered, and it is demonstrated how some additional information is profitable to learn these networks more efficiently.

1 Additional Information

In learning by examples, examples are mostly labeled as positive or negative according as they are contained in a target concept or not. Using these labeled examples, learning algorithms try to identify the unknown target concept in a reasonable time. Positive and/or negative labels are minimally necessary for learning. However, in many learning processes, besides just being positive or negative, some additional information is almost always provided. If the additional information is utilized in learning, more efficient learning algorithms may be developed.

---

[1]Currently at the Management and Coordination Agency, Japan
[2]E-mail: imai@is.s.u-tokyo.ac.jp

In this note, we consider the use of additional information in the framework of the PAC learning model due to Valiant [9]. For this purpose, we have to first analyze what type of additional information is available in practice. There seem to be many candidates for that. Here, we first consider learning a linear threshold function, which is a very useful function in logic and learning theory.

As an example, let us consider an entrance examination of a university. Some applicants are permitted to enter the university and others are not based on their scores. Assume three applicants A, B, C had scores on the four subjects as follows:

- A(85,70,55,45,F)
- B(70,80,60,95,S)
- C(50,55,75,90,S)

The last letter means the result of the examination (S is success (positive) and F is failure (negative)). In this case the criterion determining which applicants pass the examination is the following inequality (x_i is the ith score of the examination):

$$2x_1 + 3x_2 + x_3 + 5x_4 \geq 700 .$$

Here to guess which applicants are likely to be permitted is a kind of learning process. The object to be learned is the criterion inequality which decides the successful applicants. This type of learning problem is formalized as follows.

Let $\boldsymbol{x} \in \mathbf{R}^n$ be the input value vector, $\tau_{\boldsymbol{w},\theta}(\boldsymbol{x})$ be the threshold function defined as follows:

$$\tau_{\boldsymbol{w},\theta}(\boldsymbol{x}) = \begin{cases} 1 & \text{if } \boldsymbol{w}^{\mathrm{T}}\boldsymbol{x} - \theta \geq 0 \\ 0 & \text{otherwise} . \end{cases}$$

where $\boldsymbol{w} \in \mathbf{R}^n$ is the weight vector and θ is the threshold value. In many cases, the input vector to the threshold function is restricted to a 0-1 vector, but here we do not impose such restriction.

The value of the above linear threshold function is determined by the value of $\boldsymbol{w}^{\mathrm{T}}\boldsymbol{x} - \theta$, which will be called a linear value of the function. Here $\tau_{\boldsymbol{w},\theta}(\boldsymbol{x})$ defines the label of $\boldsymbol{x}$ as positive or negative (the function outputs 1 and 0 accordingly), however, the linear value itself may be output as some other information obtained in the selection stage. Such other information may be called *additional* information. In this case, apparently, n linear values as additional information are sufficient to learn the linear threshold function exactly by solving a linear equation.

A linear threshold function is one of the simplest forms, and may be too trivial to learn. In the note, we will consider two types of networks of threshold-like functions, adopt appropriate additional information as above, and then investigate the effect of having additional information.

This note proceeds as follows. In section 2, two types of functions f and g treated in the note are described. f has connection with learning convex polytopes, and g is a simple three-layered network of threshold functions. Section 3 considers the function f. We first investigate the sample complexity for learning a polytope with the value of f as additional information. By virtue of the additional information, a better bound for the sample complexity is obtained by easy arguments. We also relate this learning problem to probing convex polytopes, and mention that the exact learning is possible in this case. Section 4 handles the function g, and shows that, given some appropriate additional information, more rigorous learning becomes possible, and, in some regularized case, the complexity of a learning algorithm may be reduced. Section 5 concludes this note.

2 Two Networks of Functions

Although a single threshold function can be trivially learned with the above-mentioned additional information, a network of linear threshold-like functions such as a neural network with n inputs and one output, is not so easy to learn. That is, the problem of identifying the network from examples with the linear value of the single output as additional information is never trivial. In this case, the network may be regarded as a complicated black box with a single output, and the additional information given at the output node just provides partial information of the complicated network. This additional information at the output node may be regarded as the degree of positiveness. Having additional information of this type is rather natural, although the output value may be an approximate one in practical cases.

Investigating whether this kind of additional information is useful in the learning process enriches the framework of PAC learning in two ways: First, using the additional information, the sample complexity may be reduced by developing an algorithm making use of it. Second, in some cases such as learning a linear threshold function, a more rigorous or even exact solution may be obtained with the use of the additional information.

One might expect that a class of learnable concepts in the PAC model may be extended larger by virtue of the additional information, but this might be possible only when very strong additional information is considered. Therefore, we are here interested in only improving some complexity, such as the sample complexity and the time complexity of a learning algorithm, in the PAC model.

2.1 Function f: learning convex polytopes

We will consider these issues by taking the following two types of simply-structured networks as target functions. First, consider a function $f: \mathbf{R}^n \to \mathbf{R}$ given by

$$f(\boldsymbol{x}) = \min_{j=1,\dots,k} \{ \boldsymbol{w}_j^{\mathrm{T}} \boldsymbol{x} - \theta_j \}$$

where $\boldsymbol{x} \in \mathbf{R}^n$ is an input vector, $\boldsymbol{w}_j \in \mathbf{R}^n$ and θ_j $(j = 1, \dots, k)$ are weights and threshold-like values, respectively, for the j-th hidden unit among k hidden units. Figure 1(a) depicts this network as a three-layered network with $n = 2$ and $k = 3$. Note that the units in the hidden layer are not real threshold functions but they output the linear values. Also, note that in this network weights from the hidden layer to the output unit can be regularized to one by transferring them to the weights for the input of the hidden units.

This function can be regarded as describing a convex n-dimensional polytope $P = \{ \boldsymbol{x} \mid \boldsymbol{w}_i^{\mathrm{T}} \boldsymbol{x} - \theta_i \leq 0 \}$ since $\boldsymbol{x} \in P$ iff $f(\boldsymbol{x}) \leq 0$. That is,

$$f_0(\boldsymbol{x}) = \begin{cases} 1 & \text{if } f(\boldsymbol{x}) \leq 0 \\ 0 & \text{otherwise ,} \end{cases}$$

is a characteristic function for the polytope P. Learning a convex polytope itself is a standard problem, and is much easier than learning a neural network of threshold functions, but through investigating the use of additional information for this function would clarify how useful the additional information may be.

2.2 Function g: a simple network of threshold functions

Consider a function $g : \mathbf{R}^n \to \{0, 1, \dots, k\}$ defined by

$$g(\boldsymbol{x}) = \sum_{j=1}^{k} \tau_{\boldsymbol{w}_j, \theta_j}(\boldsymbol{x})$$

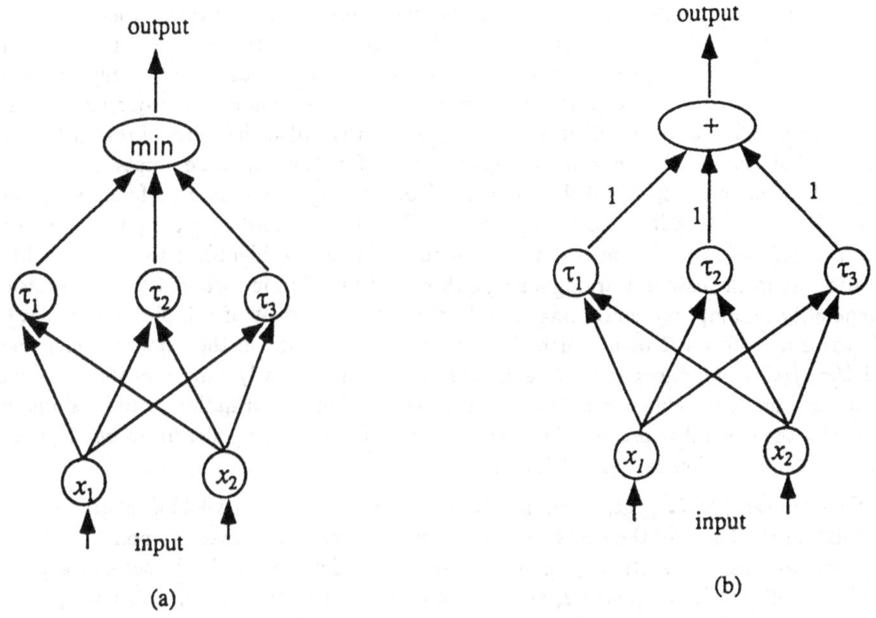

Figure 1: Functions f and g

where k is the number of threshold functions in the hidden layer, w_j and θ_j are the weight vector and threshold value, respectively, of the j-th threshold function ($j = 1, \ldots, k$). See Figure 1(b). A function $g_\theta(x)$ defined as

$$g_\theta(\boldsymbol{x}) = \begin{cases} 1 & \text{if } g(\boldsymbol{x}) - \theta \geq 0 \\ 0 & \text{otherwise .} \end{cases}$$

is a network of linear threshold functions, a very special case of neural networks. In g, the output threshold function is restricted to a simple one, i.e., non-weighted summation of outputs of the hidden units. This, of course, makes this function g less interesting from the standpoint of neural networks, but the function g still reflects the structure of the hidden layer, that is, the arrangement of k hyperplanes in the n-dimensional space corresponding to the threshold functions (Edelsbrunner [4]). We will show that, when the value of g is given as additional information in learning g_θ, more rigorous solution is obtained and this also benefits developing a more efficient learning algorithm.

3 Learning Convex Polytopes

3.1 Producing a good approximate solution using the additional information

For the k threshold functions in the hidden layer, we may consider the intersection $\tilde{P}$ of half-spaces in the $(n+1)$-dimensional space naturally determined by them:

$$\tilde{P} = \bigcap_{j=1}^{k} \{ (\boldsymbol{x}, y) \in \mathbf{R}^{n+1} \mid \boldsymbol{x} \in \mathbf{R}^n,\ y \leq \boldsymbol{w}_j^{\mathrm{T}} \boldsymbol{x} - \theta_j \}$$

The n-dimensional polytope P is the intersection of this polytope $\tilde{P}$ with hyperplane $y = 0$. The boundary $\partial\tilde{P}$ of $\tilde{P}$ has at most k facets. Projecting down the boundary to the original space $\mathbf{R}^n$ vertical to the y-axis yields a partition of $\mathbf{R}^n$ into at most k convex regions, which will be called the Voronoi partition. Each region in the partition is the projection of a facet of $\tilde{P}$ corresponding to a threshold function τ_{w_j,θ_j}, and is called the Voronoi region V_j of the j-th threshold function. The Voronoi region of some function might be empty.

By these definitions, if $x \in \mathbf{R}^n$ is in the Voronoi region of the j-th function, $f(x)$ is $w_j^T x - \theta_j$, the linear value of the j-th function. If $n+1$ or more example points are in the j-th Voronoi region, then the j-th threshold function can be identified by solving a linear equation, as in the case of learning a single threshold function in section 1. We here assume a generality assumption concerning sampling points: for a set of sampled points $\{x_i\}$ in $\mathbf{R}^n$, there are no $n+2$ points such that these points are not in the same Voronoi region and $\{(x_i, f(x_i))\}$ for them are collinear. This assumption greatly simplifies the subsequent discussion, and is suitable to clarify the role of the additional information in an ideal manner.

Under the generality assumption, the following simple learning algorithm employing the additional information can be considered.

1. Given m samples $(x_i, f(x_i))$, compute their convex hull in the $(n+1)$-dimensional space;
2. Find all the facets of the convex hull containing at least $n+2$ sample points;
3. Compute the intersection $\tilde{Q}$ of lower half-spaces determined by hyperplanes passing through facets found in step 2, and compute the intersection Q of $\tilde{Q}$ and $y = 0$;
4. $\tilde{P} \subseteq \tilde{Q}$, $P \subseteq Q$, and $\tilde{Q}$ and Q are good guesses of $\tilde{P}$ and P, respectively.

Due to the generality assumption, all the hyperplanes in step 3 correspond to facets of $\tilde{P}$, and also since the lower half-spaces corresponding to these hyperplanes completely include the polytope $\tilde{P}$, $\tilde{P} \subseteq \tilde{Q}$ and $P \subseteq Q$ hold. This algorithm may run in $O(m^{\lfloor (n+1)/2 \rfloor})$ time for $n > 2$.

It should be noted that this algorithm does not necessary produce a guess consistent with the given examples in a sense that samples points in Voronoi regions containing at most $n+1$ samples are ignored. Nevertheless, about the guess obtained by this algorithm may be good as stated in the following.

Theorem 1 *(i) For any* ϵ, δ *with* $0 < \epsilon, \delta < 1$, *if*

$$m \geq \frac{2}{\epsilon}\left(\log\frac{1}{\delta} + k(n+2)\right)$$

independent draws x_i $(i = 1, \ldots, m)$ *are given under any distribution* D *in* $\mathbf{R}^n$ *and* $(x_i, f(x_i))$ $(i = 1, \ldots, m)$ *are given as examples for the polytope* P, *the above algorithm finds a solution polytope* Q *such that* $P \subseteq Q$ *and*

$$\Pr(x \in \mathbf{R}^n,\ x \in Q - P) < \epsilon$$

with probability $1 - \delta$, *where* Pr *is under* D.

(ii) Furthermore, all the threshold functions whose Voronoi region has probability at least ϵ *under* D *are exactly found with probability* $1 - \delta$.

Outline of Proof: When a learner receives at least $n+2$ examples in V_j, the j-th threshold function is correctly found.

Suppose that, without loss of generality, for some k', each of V_j $(j = 1, \ldots, k')$ has at most $n+1$ sample points. When $\epsilon' \equiv \sum_{j=1}^{k'} \Pr(V_j) < \epsilon$, no problem happens. Otherwise, for fixed V_j ($j = 1, \ldots, k'$), the probability of an event that each V_j contains at most $n+1$ samples is less than or equal to the probability of an event that the union of V_j contains at most $k'(n+1)$ samples, which is bounded, using the Chernoff bound, by

$$\exp\left(-\frac{(m\epsilon' - k'(n+1))^2}{2m\epsilon'}\right).$$

The number of combinations of choosing k' regions from k Voronoi regions is at most 2^k. By solving an inequality

$$2^k \cdot \exp\left(-\frac{(m\epsilon' - k'(n+1))^2}{2m\epsilon'}\right) \leq \delta$$

and using $\epsilon' \geq \epsilon$ and $k' \leq k$, the bound in (i) is obtained.

(ii) is obvious from the above arguments. □

Everything is very simple by utilizing the Voronoi partition, the additional information, explicitly. Furthermore, the bound in (i) is better than that obtained by the general PAC method by Blumer, Ehrenfeucht, Haussler and Warmuth [2] (also consult Hasegawa and Kakihara [5] for a little improved bound). Moreover, we can guarantee that easily searchable functions are exactly computed with high probability by the additional information.

3.2 Probing Convex Polytopes

In the preceding subsection, examples are given based on some probability distribution and errors are estimated under the same distribution, which is the basis of the PAC learning model. However, instead of the PAC model, we may apply the Exact learning by queries (Angluin [1]) to this case. That is, by an oracle or black box, the value of $f(x)$ is returned for any query point $x \in \mathbf{R}^n$, and by providing a finite number of points as queries the polytope may be found exactly. At a first glance, just having the value $f(x)$ for x may seem too crude to identify the polytope, but this can be accomplished by asking a sequence of queries whose number is proportional to the combinatorial complexity of the polytope.

In fact, this kind of operation for identifying convex polytopes is considered as probing convex polytopes in computational geometry. There, a so-called finger probe is considered. By using the result of Dobkin, Edelsbrunner and Yap [3], we can show the following, where the proof is omitted in this version.

Theorem 2 *With $O(n^{\lfloor k+1 \rfloor})$ queries, the function f can be learned exactly.*

4 Learning a Simple Network of Threshold Functions

Learning or synthesizing a network of threshold functions is a traditional problem (Muroga [7]). In general, learning a complex network of threshold functions is very hard from the viewpoint of computational complexity (e.g., Lin and Vitter [8]). However, with the additional information, more accurate solution may be found in the PAC model, and some simple cases may be solved efficiently.

The k threshold functions of n inputs in the hidden layer naturally induces the arrangement $\mathcal{A}$ of k hyperplanes in the n-dimensional space. Any point x in a cell of the arrangement has the same value of $g(x)$. In the PAC framework, from examples $(x, g(x))$ drawn by some probability distribution, another arrangement of k hyperplanes which is "consistent" with these examples is shown to be a good approximate solution.

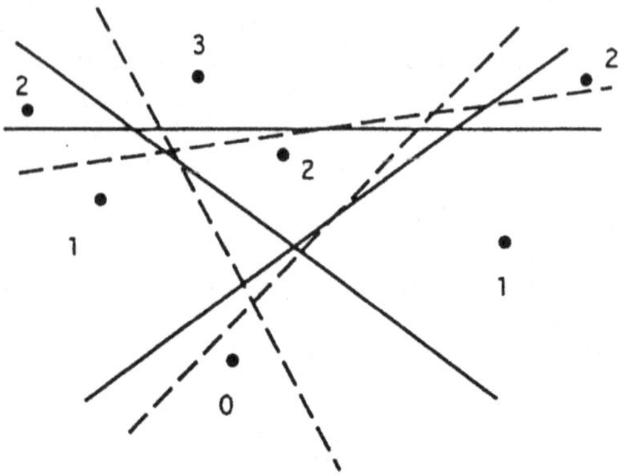

Figure 2: Consistent arrangements

For the arrangement $\mathcal{A}$, consider a set S of points such that each cell of $\mathcal{A}$ has exactly one point in the set S. S is said to cover all the cells of the arrangement. Then, another arrangement $\mathcal{A}'$ of k hyperplanes is said to be consistent with respect to S if there is a one-to-one correspondence between two sets of hyperplanes such that the partition of S induced by a hyperplane is the same with that by the corresponding one. See Figure 2. Denote the function g corresponding to the arrangements $\mathcal{A}$ and $\mathcal{A}'$ by $g_{\mathcal{A}}$ and $g_{\mathcal{A}'}$, respectively. Then, the following theorem holds.

Theorem 3 *For any* $x \in \mathbf{R}^n$, $|g_{\mathcal{A}}(x) - g_{\mathcal{A}'}(x)| \leq 2$.

The proof is omitted in this version. In the PAC model, if examples cover all the cells of the arrangement and examples in each cell are correctly clustered to produce one representative point to form a cover S of points, by constructing an arrangement $\mathcal{A}'$ consistent with respect to S, a very good approximate function $g_{\mathcal{A}'}$ may be obtained. About the probability that examples cover all the cells, similar arguments in section 3 may be applied. Concerning learning the function g_θ, the PAC framework provides a bound about an error of the approximate solution. With the additional information, we can obtain much more well-structured solution due to Theorem 3 in this case.

We now turn to an algorithmic issue. Finding an arrangement $\mathcal{A}'$ consistent with $\mathcal{A}$ with respect to S is a purely geometrical problem, and computational geometric algorithms may be used. Consider the two-dimensional case in the (x, y)-plane. We say that the function g is canonical if each threshold function in the hidden layer becomes 1 when $w_x x + w_y y - \theta \geq 0$ and $w_y > 0$. We then have the following.

Theorem 4 *Suppose $n = 2$, the function g is canonical and a cover S of points is given. Then, finding the consistent arrangement $\mathcal{A}'$ can be done in $O(n^2 \log n)$ time by the plane-sweep method.*

Outline of Proof: In this case, the κ-belt for $\kappa = 0, \ldots, n$ is naturally considered as the collection of points x with $g(x) = \kappa$ (cf. [4]). In the two-dimensional case, each belt has the monotonicity in respect to the x-coordinate. Therefore, performing the plane sweep by a vertical sweep line, points in S in the same belt are hit in that order. Using this fact, a simple plane sweep can work correctly to form all the belts simultaneously. □

5 Concluding Remarks

In this note, some roles of additional information are investigated. When considering additional information, one of the most important points would be what kind of information is considered as additional one from the practical point of view. In this respect, it may be said that two types of additional information treated here are not yet so practical ones, but it is hoped that, even so, the additional information and its use to improve learning processes demonstrated so far have shown some usefulness of considering it. In Kakihara [6], many other types of additional information in learning planar concepts are considered with some further discussion on the issue of additional information.

Acknowledgment

We would like to thank other members of our laboratory, especially Mr. Susumu Hasegawa, for helpful discussion. The authors are deeply indebted to anonymous referees who suggested many useful things concerning the original version of the paper, which has made it possible to make this more compact but hopefully nicer version compared with the original one. This research was supported in part by the Grant-in-Aid for Scientific Research on Priority Areas, "Knowledge Science," 03245201, 04229201, of the Ministry of Education, Science and Culture of Japan.

References

[1] Angluin, D., "Queries and Concept Learning," *Machine Learning*, Vol. 2, 1988, pp. 319–342.

[2] Blumer, A., A. Ehrenfeucht, D. Haussler, and M. K. Warmuth, "Learnability and the Vapnik-Chervonenkis Dimension," *Journal of the ACM*, Vol. 36, No. 4, 1989, pp. 929–965.

[3] Dobkin, D., H. Edelsbrunner, and C. K. Yap, "Probing Convex Polytopes," *Proceedings of the 18th Annual ACM Symposium on Theory of Computing*, 1986, p.424–432.

[4] Edelsbrunner, H., "*Algorithms in Combinatorial Geometry*," Springer-Verlag, Berlin, 1987.

[5] Hasegawa, S., and K. Kakihara, "A Generalization of ϵ-Approximations for k-Label Space and Its Application," Manuscript, 1992.

[6] Kakihara, K., "*Effect of Additional Information in Computational Learning Theory*," Graduation Thesis, Department of Information Science, University of Tokyo, March 1992.

[7] Muroga, S., "*Threshold Logic and Its Applications*," Wiley-Interscience, New York, 1971.

[8] Lin, J.-H., and Vitter, J. S., "Complexity Issues in Learning by Neural Nets," *Proceedings of the 2nd Annual Workshop on Computational Learning Theory*, 1989, pp.118–133.

[9] Valiant, L. G., "A Theory of the Learnable," *Communications of the ACM*, Vol. 27, No. 11, 1984, pp. 1134–1142.

Index of Authors

Lecture Notes in Artificial Intelligence (LNAI)

Vol. 568: H.-J. Bürckert, A Resolution Principle for a Logic with Restricted Quantifiers. X, 116 pages. 1991.

Vol. 587: R. Dale, E. Hovy, D. Rösner, O. Stock (Eds.), Aspects of Automated Natural Language Generation. Proceedings, 1992. VIII, 311 pages. 1992.

Vol. 590: B. Fronhöfer, G. Wrightson (Eds.), Parallelization in Inference Systems. Proceedings, 1990. VIII, 372 pages. 1992.

Vol. 592: A. Voronkov (Ed.), Logic Programming. Proceedings, 1991. IX, 514 pages. 1992.

Vol. 596: L.-H. Eriksson, L. Hallnäs, P. Schroeder-Heister (Eds.), Extensions of Logic Programming. Proceedings, 1991. VII, 369 pages. 1992.

Vol. 597: H. W. Guesgen, J. Hertzberg, A Perspective of Constraint-Based Reasoning. VIII, 123 pages. 1992.

Vol. 599: Th. Wetter, K.-D. Althoff, J. Boose, B. R. Gaines, M. Linster, F. Schmalhofer (Eds.), Current Developments in Knowledge Acquisition - EKAW '92. Proceedings. XIII, 444 pages. 1992.

Vol. 604: F. Belli, F. J. Radermacher (Eds.), Industrial and Engineering Applications of Artificial Intelligence and Expert Systems. Proceedings, 1992. XV, 702 pages. 1992.

Vol. 607: D. Kapur (Ed.), Automated Deduction – CADE-11. Proceedings, 1992. XV, 793 pages. 1992.

Vol. 610: F. von Martial, Coordinating Plans of Autonomous Agents. XII, 246 pages. 1992.

Vol. 611: M. P. Papazoglou, J. Zeleznikow (Eds.), The Next Generation of Information Systems: From Data to Knowledge. VIII, 310 pages. 1992.

Vol. 617: V. Mařík, O. Štěpánková, R. Trappl (Eds.), Advanced Topics in Artificial Intelligence. Proceedings, 1992. IX, 484 pages. 1992.

Vol. 619: D. Pearce, H. Wansing (Eds.), Nonclassical Logics and Information Processing. Proceedings, 1990. VII, 171 pages. 1992.

Vol. 622: F. Schmalhofer, G. Strube, Th. Wetter (Eds.), Contemporary Knowledge Engineering and Cognition. Proceedings, 1991. XII, 258 pages. 1992.

Vol. 624: A. Voronkov (Ed.), Logic Programming and Automated Reasoning. Proceedings, 1992. XIV, 509 pages. 1992.

Vol. 627: J. Pustejovsky, S. Bergler (Eds.), Lexical Semantics and Knowledge Representation. Proceedings, 1991. XII, 381 pages. 1992.

Vol. 633: D. Pearce, G. Wagner (Eds.), Logics in AI. Proceedings. VIII, 410 pages. 1992.

Vol. 636: G. Comyn, N. E. Fuchs, M. J. Ratcliffe (Eds.), Logic Programming in Action. Proceedings, 1992. X, 324 pages. 1992.

Vol. 638: A. F. Rocha, Neural Nets. A Theory for Brains and Machines. XV, 393 pages. 1992.

Vol. 642: K. P. Jantke (Ed.), Analogical and Inductive Inference. Proceedings, 1992. VIII, 319 pages. 1992.

Vol. 659: G. Brewka, K. P. Jantke, P. H. Schmitt (Eds.), Nonmonotonic and Inductive Logic. Proceedings, 1991. VIII, 332 pages. 1993.

Vol. 660: E. Lamma, P. Mello (Eds.), Extensions of Logic Programming. Proceedings, 1992. VIII, 417 pages. 1993.

Vol. 667: P. B. Brazdil (Ed.), Machine Learning: ECML – 93. Proceedings, 1993. XII, 471 pages. 1993.

Vol. 671: H. J. Ohlbach (Ed.), GWAI-92: Advances in Artificial Intelligence. Proceedings, 1992. XI, 397 pages. 1993.

Vol. 679: C. Fermüller, A. Leitsch, T. Tammet, N. Zamov, Resolution Methods for the Decision Problem. VIII, 205 pages. 1993.

Vol. 681: H. Wansing, The Logic of Information Structures. IX, 163 pages. 1993.

Vol. 689: J. Komorowski, Z. W. Raś (Eds.), Methodologies for Intelligent Systems. Proceedings, 1993. XI, 653 pages. 1993.

Vol. 695: E. P. Klement, W. Slany (Eds.), Fuzzy Logic in Artificial Intelligence. Proceedings, 1993. VIII, 192 pages. 1993.

Vol. 698: A. Voronkov (Ed.), Logic Programming and Automated Reasoning. Proceedings, 1993. XIII, 386 pages. 1993.

Vol. 699: G.W. Mineau, B. Moulin, J.F. Sowa (Eds.), Conceptual Graphs for Knowledge Representation. Proceedings, 1993. IX, 451 pages. 1993.

Vol. 723: N. Aussenac, G. Boy, B. Gaines, M. Linster, J.-G. Ganascia, Y. Kodratoff (Eds.), Knowledge Acquisition for Knowledge-Based Systems. Proceedings, 1993. XIII, 446 pages. 1993.

Vol. 727: M. Filgueiras, L. Damas (Eds.), Progress in Artificial Intelligence. Proceedings, 1993. X, 362 pages. 1993.

Vol. 728: P. Torasso (Ed.), Advances in Artificial Intelligence. Proceedings, 1993. XI, 336 pages. 1993.

Vol. 743: S. Doshita, K. Furukawa, K. P. Jantke, T. Nishida (Eds.), Algorithmic Learning Theory. Proceedings, 1992. X, 260 pages. 1993.

Vol. 744: K. P. Jantke, T. Yokomori, S. Kobayashi, E. Tomita (Eds.), Algorithmic Learning Theory. Proceedings, 1993. XI, 423 pages. 1993.

Vol. 745: V. Roberto (Ed.), Intelligent Perceptual Systems. VIII, 378 pages. 1993.

Vol. 746: A. S. Tanguiane, Artificial Perception and Music Recognition. XV, 210 pages. 1993.

Lecture Notes in Computer Science

Vol. 711: A. M. Borzyszkowski, S. Sokołowski (Eds.), Mathematical Foundations of Computer Science 1993. Proceedings, 1993. XIII, 782 pages. 1993.

Vol. 712: P. V. Rangan (Ed.), Network and Operating System Support for Digital Audio and Video. Proceedings, 1992. X, 416 pages. 1993.

Vol. 713: G. Gottlob, A. Leitsch, D. Mundici (Eds.), Computational Logic and Proof Theory. Proceedings, 1993. XI, 348 pages. 1993.

Vol. 714: M. Bruynooghe, J. Penjam (Eds.), Programming Language Implementation and Logic Programming. Proceedings, 1993. XI, 421 pages. 1993.

Vol. 715: E. Best (Ed.), CONCUR'93. Proceedings, 1993. IX, 541 pages. 1993.

Vol. 716: A. U. Frank, I. Campari (Eds.), Spatial Information Theory. Proceedings, 1993. XI, 478 pages. 1993.

Vol. 717: I. Sommerville, M. Paul (Eds.), Software Engineering - ESEC '93. Proceedings, 1993. XII, 516 pages. 1993.

Vol. 718: J. Seberry, Y. Zheng (Eds.), Advances in Cryptology - AUSCRYPT '92. Proceedings, 1992. XIII, 543 pages. 1993.

Vol. 719: D. Chetverikov, W.G. Kropatsch (Eds.), Computer Analysis of Images and Patterns. Proceedings, 1993. XVI, 857 pages. 1993.

Vol. 720: V.Mařík, J. Lažanský, R.R. Wagner (Eds.), Database and Expert Systems Applications. Proceedings, 1993. XV, 768 pages. 1993.

Vol. 721: J. Fitch (Ed.), Design and Implementation of Symbolic Computation Systems. Proceedings, 1992. VIII, 215 pages. 1993.

Vol. 722: A. Miola (Ed.), Design and Implementation of Symbolic Computation Systems. Proceedings, 1993. XII, 384 pages. 1993.

Vol. 723: N. Aussenac, G. Boy, B. Gaines, M. Linster, J.-G. Ganascia, Y. Kodratoff (Eds.), Knowledge Acquisition for Knowledge-Based Systems. Proceedings, 1993. XIII, 446 pages. 1993. (Subseries LNAI)

Vol. 724: P. Cousot, M. Falaschi, G. Filè, A. Rauzy (Eds.), Static Analysis. Proceedings, 1993. IX, 283 pages. 1993.

Vol. 725: A. Schiper (Ed.), Distributed Algorithms. Proceedings, 1993. VIII, 325 pages. 1993.

Vol. 726: T. Lengauer (Ed.), Algorithms — ESA '93. Proceedings, 1993. IX, 419 pages. 1993.

Vol. 727: M. Filgueiras, L. Damas (Eds.), Progress in Artificial Intelligence. Proceedings, 1993. X, 362 pages. 1993. (Subseries LNAI).

Vol. 728: P. Torasso (Ed.), Advances in Artificial Intelligence. Proceedings, 1993. XI, 336 pages. 1993. (Subseries LNAI).

Vol. 729: L. Donatiello, R. Nelson (Eds.), Performance Evaluation of Computer and Communication Systems. Proceedings, 1993. VIII, 675 pages. 1993.

Vol. 730: D. B. Lomet (Ed.), Foundations of Data Organization and Algorithms. Proceedings, 1993. XII, 412 pages. 1993.

Vol. 731: A. Schill (Ed.), DCE - The OSF Distributed Computing Environment. Proceedings, 1993. VIII, 285 pages. 1993.

Vol. 732: A. Bode, M. Dal Cin (Eds.), Parallel Computer Architectures. IX, 311 pages. 1993.

Vol. 733: Th. Grechenig, M. Tscheligi (Eds.), Human Computer Interaction. Proceedings, 1993. XIV, 450 pages. 1993.

Vol. 734: J. Volkert (Ed.), Parallel Computation. Proceedings, 1993. VIII, 248 pages. 1993.

Vol. 735: D. Bjørner, M. Broy, I. V. Pottosin (Eds.), Formal Methods in Programming and Their Applications. Proceedings, 1993. IX, 434 pages. 1993.

Vol. 736: R. L. Grossman, A. Nerode, A. P. Ravn, H. Rischel (Eds.), Hybrid Systems. VIII, 474 pages. 1993.

Vol. 737: J. Calmet, J. A. Campbell (Eds.), Artificial Intelligence and Symbolic Mathematical Computing. Proceedings, 1992. VIII, 305 pages. 1993.

Vol. 738: M. Weber, M. Simons, Ch. Lafontaine, The Generic Development Language Deva. XI, 246 pages. 1993.

Vol. 739: H. Imai, R. L. Rivest, T. Matsumoto (Eds.), Advances in Cryptology - ASIACRYPT '91. X, 499 pages. 1993.

Vol. 740: E. F. Brickell (Ed.), Advances in Cryptology - CRYPTO '92. Proceedings, 1992. X, 565 pages. 1993.

Vol. 741: B. Preneel, R. Govaerts, J. Vandewalle (Eds.), Computer Security and Industrial Cryptography. ESAT, 1991. VIII, 275 pages. 1993.

Vol. 742: S. Nishio, A. Yonezawa (Eds.), Object Technologies for Advanced Software. Proceedings, 1993. X, 543 pages. 1993.

Vol. 743: S. Doshita, K. Furukawa, K. P. Jantke, T. Nishida (Eds.), Algorithmic Learning Theory. Proceedings, 1992. X, 260 pages. 1993. (Subseries LNAI)

Vol. 744: K. P. Jantke, T. Yokomori, S. Kobayashi, E. Tomita (Eds.), Algorithmic Learning Theory. Proceedings, 1993. XI, 423 pages. 1993. (Subseries LNAI)

Vol. 745: V. Roberto (Ed.), Intelligent Perceptual Systems. VIII, 378 pages. 1993. (Subseries LNAI)

Vol. 746: A. S. Tanguiane, Artificial Perception and Music Recognition. XV, 210 pages. 1993. (Subseries LNAI)